Advances in Intelligent Systems and Computing

Volume 1050

The series "Advances in Intelligent Systems and Computing" contains publications on theory, applications, and design methods of Intelligent Systems and Intelligent Computing. Virtually all disciplines such as engineering, natural sciences, computer and information science, ICT, economics, business, e-commerce, environment, healthcare, life science are covered. The list of topics spans all the areas of modern intelligent systems and computing such as: computational intelligence, soft computing including neural networks, fuzzy systems, evolutionary computing and the fusion of these paradigms, social intelligence, ambient intelligence, computational neuroscience, artificial life, virtual worlds and society, cognitive science and systems, Perception and Vision, DNA and immune based systems, self-organizing and adaptive systems, e-Learning and teaching, human-centered and human-centric computing, recommender systems, intelligent control, robotics and mechatronics including human-machine teaming, knowledge-based paradigms, learning paradigms, machine ethics, intelligent data analysis, knowledge management, intelligent agents, intelligent decision making and support, intelligent network security, trust management, interactive entertainment, Web intelligence and multimedia.

The publications within "Advances in Intelligent Systems and Computing" are primarily proceedings of important conferences, symposia and congresses. They cover significant recent developments in the field, both of a foundational and applicable character. An important characteristic feature of the series is the short publication time and world-wide distribution. This permits a rapid and broad dissemination of research results.

**** Indexing: The books of this series are submitted to ISI Proceedings, EI-Compendex, DBLP, SCOPUS, Google Scholar and Springerlink ****

More information about this series at http://www.springer.com/series/11156

Leszek Borzemski · Jerzy Świątek ·
Zofia Wilimowska
Editors

Information Systems Architecture and Technology: Proceedings of 40th Anniversary International Conference on Information Systems Architecture and Technology – ISAT 2019

Part I

Editors
Leszek Borzemski
Faculty of Computer Science
and Management
Wrocław University of Science
and Technology
Wrocław, Poland

Jerzy Świątek
Faculty of Computer Science
and Management
Wrocław University of Science
and Technology
Wrocław, Poland

Zofia Wilimowska
University of Applied Sciences in Nysa
Nysa, Poland

ISSN 2194-5357 ISSN 2194-5365 (electronic)
Advances in Intelligent Systems and Computing
ISBN 978-3-030-30439-3 ISBN 978-3-030-30440-9 (eBook)
https://doi.org/10.1007/978-3-030-30440-9

This Springer imprint is published by the registered company Springer Nature Switzerland AG
The registered company address is: Gewerbestrasse 11, 6330 Cham, Switzerland

Preface

We are pleased to present before you the proceedings of the 2019 40th Anniversary International Conference Information Systems Architecture and Technology (ISAT), or ISAT 2019 for short, held on September 15–17, 2019 in Wrocław, Poland. The conference was organized by the Department of Computer Science, Faculty of Computer Science and Management, Wrocław University of Science and Technology, Poland, and the University of Applied Sciences in Nysa, Poland.

The International Conference on Information Systems Architecture and Technology has been organized by the Wrocław University of Science and Technology from the eighties of the last century. Most of the events took place in Szklarska Poręba and Karpacz—charming small towns in the Karkonosze Mountains, Lower Silesia in the southwestern part of Poland. This year 2019, we celebrate the 40th anniversary of the conference in Wrocław—the capital of Lower Silesia, a city with a thousand-year history. A beautiful and modern city that is developing dynamically and is a meeting point for people from all over the world. It is worth noting that Wrocław is currently one of the most important centers for the development of modern software and information systems in Poland.

The past four decades have also been a period of dynamic development of computer science, which we can recall when reviewing conference materials from these years—their shape and content were always created with current achievements of national and international IT.

The purpose of the ISAT is to discuss a state-of-art of information systems concepts and applications as well as architectures and technologies supporting contemporary information systems. The aim is also to consider an impact of knowledge, information, computing and communication technologies on managing of the organization scope of functionality as well as on enterprise information systems design, implementation, and maintenance processes taking into account various methodological, technological, and technical aspects. It is also devoted to information systems concepts and applications supporting the exchange of goods and services by using different business models and exploiting opportunities offered by Internet-based electronic business and commerce solutions.

ISAT is a forum for specific disciplinary research, as well as on multi-disciplinary studies to present original contributions and to discuss different subjects of today's information systems planning, designing, development, and implementation.

The event is addressed to the scientific community, people involved in a variety of topics related to information, management, computer and communication systems, and people involved in the development of business information systems and business computer applications. ISAT is also devoted as a forum for the presentation of scientific contributions prepared by MSc. and Ph.D. students. Business, Commercial, and Industry participants are welcome.

This year, we received 141 papers from 20 countries. The papers included in the three proceedings volumes have been subject to a thoroughgoing review process by highly qualified peer reviewers. The final acceptance rate was 60%. Program Chairs selected 85 best papers for oral presentation and publication in the 40th International Conference Information Systems Architecture and Technology 2019 proceedings.

The papers have been clustered into three volumes:

Part I—discoursing about essential topics of information technology including, but not limited to, Computer Systems Security, Computer Network Architectures, Distributed Computer Systems, Quality of Service, Cloud Computing and High-Performance Computing, Human–Computer Interface, Multimedia Systems, Big Data, Knowledge Discovery and Data Mining, Software Engineering, E-Business Systems, Web Design, Optimization and Performance, Internet of Things, Mobile Systems, and Applications.

Part II—addressing topics including, but not limited to, Pattern Recognition and Image Processing Algorithms, Production Planning and Management Systems, Big Data Analysis, Knowledge Discovery, and Knowledge-Based Decision Support and Artificial Intelligence Methods and Algorithms.

Part III—is gain to address very hot topics in the field of today's various computer-based applications—is devoted to information systems concepts and applications supporting the managerial decisions by using different business models and exploiting opportunities offered by IT systems. It is dealing with topics including, but not limited to, Knowledge-Based Management, Modeling of Financial and Investment Decisions, Modeling of Managerial Decisions, Production and Organization Management, Project Management, Risk Management, Small Business Management, Software Tools for Production, Theories, and Models of Innovation.

We would like to thank the Program Committee Members and Reviewers, essential for reviewing the papers to ensure a high standard of the ISAT 2019 conference, and the proceedings. We thank the authors, presenters, and participants of ISAT 2019 without them the conference could not have taken place. Finally, we thank the organizing team for the efforts this and previous years in bringing the conference to a successful conclusion.

We hope that ISAT conference is a good scientific contribution to the development of information technology not only in the region but also internationally. It happens, among others, thanks to cooperation with Springer Publishing House, where the AISC series is issued from 2015. We want to thank Springer's people who deal directly with the publishing process, from publishing contracts to the delivering of printed books. Thank you for your cooperation.

September 2019

Leszek Borzemski
Jerzy Świątek
Zofia Wilimowska

Organization

ISAT 2019 Conference Organization

General Chair

Leszek Borzemski, Poland

Program Co-chairs

Leszek Borzemski, Poland
Jerzy Świątek, Poland
Zofia Wilimowska, Poland

Local Organizing Committee

Leszek Borzemski (Chair)
Zofia Wilimowska (Co-chair)
Jerzy Świątek (Co-chair)
Mariusz Fraś (Conference Secretary, Website Support)
Arkadiusz Górski (Technical Editor)
Anna Kamińska (Technical Secretary)
Ziemowit Nowak (Technical Support)
Kamil Nowak (Website Coordinator)
Danuta Seretna-Sałamaj (Technical Secretary)

International Program Committee

Leszek Borzemski (Chair), Poland
Jerzy Świątek (Co-chair), Poland
Zofia Wilimowska (Co-chair), Poland
Witold Abramowicz, Poland
Dhiya Al-Jumeily, UK
Iosif Androulidakis, Greece
Patricia Anthony, New Zealand
Zbigniew Banaszak, Poland
Elena N. Benderskaya, Russia
Janos Botzheim, Japan
Djallel E. Boubiche, Algeria
Patrice Boursier, France
Anna Burduk, Poland
Andrii Buriachenko, Ukraine

Udo Buscher, Germany
Wojciech Cellary, Poland
Haruna Chiroma, Malaysia
Edward Chlebus, Poland
Gloria Cerasela Crisan, Romania
Marilia Curado, Portugal
Czesław Daniłowicz, Poland
Zhaohong Deng, China
Małgorzata Dolińska, Poland
Ewa Dudek-Dyduch, Poland
Milan Edl, Czech Republic
El-Sayed M. El-Alfy, Saudi Arabia
Peter Frankovsky, Slovakia
Naoki Fukuta, Japan
Bogdan Gabryś, UK
Piotr Gawkowski, Poland
Manuel Graña, Spain
Katsuhiro Honda, Japan
Marian Hopej, Poland
Zbigniew Huzar, Poland
Natthakan Iam-On, Thailand
Biju Issac, UK
Arun Iyengar, USA
Jürgen Jasperneite, Germany
Janusz Kacprzyk, Poland
Henryk Kaproń, Poland
Yury Y. Korolev, Belarus
Yannis L. Karnavas, Greece
Ryszard Knosala, Poland
Zdzisław Kowalczuk, Poland
Lumír Kulhanek, Czech Republic
Binod Kumar, India
Jan Kwiatkowski, Poland
Antonio Latorre, Spain
Radim Lenort, Czech Republic
Gang Li, Australia
José M. Merigó Lindahl, Chile
Jose M. Luna, Spain
Emilio Luque, Spain
Sofian Maabout, France
Lech Madeyski, Poland
Zygmunt Mazur, Poland
Elżbieta Mączyńska, Poland
Pedro Medeiros, Portugal
Toshiro Minami, Japan
Marian Molasy, Poland
Zbigniew Nahorski, Poland
Kazumi Nakamatsu, Japan
Peter Nielsen, Denmark
Tadashi Nomoto, Japan
Cezary Orłowski, Poland
Sandeep Pachpande, India
Michele Pagano, Italy
George A. Papakostas, Greece
Zdzisław Papir, Poland
Marek Pawlak, Poland
Jan Platoš, Czech Republic
Tomasz Popławski, Poland
Edward Radosinski, Poland
Wolfgang Renz, Germany
Dolores I. Rexachs, Spain
José S. Reyes, Spain
Leszek Rutkowski, Poland
Sebastian Saniuk, Poland
Joanna Santiago, Portugal
Habib Shah, Malaysia
J. N. Shah, India
Jeng Shyang, Taiwan
Anna Sikora, Spain
Marcin Sikorski, Poland
Małgorzata Sterna, Poland
Janusz Stokłosa, Poland
Remo Suppi, Spain
Edward Szczerbicki, Australia
Eugeniusz Toczyłowski, Poland
Elpida Tzafestas, Greece
José R. Villar, Spain
Bay Vo, Vietnam
Hongzhi Wang, China
Leon S. I. Wang, Taiwan
Junzo Watada, Japan
Eduardo A. Durazo Watanabe, India
Jan Werewka, Poland
Thomas Wielicki, USA
Bernd Wolfinger, Germany
Józef Woźniak, Poland
Roman Wyrzykowski, Poland
Yue Xiao-Guang, Hong Kong
Jaroslav Zendulka, Czech Republic
Bernard Ženko, Slovenia

ISAT 2019 Reviewers

Hamid Al-Asadi, Iraq
S. Balakrishnan, India
Zbigniew Banaszak, Poland
Agnieszka Bieńkowska, Poland
Grzegorz Bocewicz, Poland
Leszek Borzemski, Poland
Janos Botzheim, Hungary
Krzysztof Brzostowski, Poland
Anna Burduk, Poland
Wojciech Cellary, Poland
Haruna Chiroma, Malaysia
Grzegorz Chodak, Poland
Piotr Chwastyk, Poland
Gloria Cerasela Crisan, Romania
Anna Czarnecka, Poland
Mariusz Czekała, Poland
Yousef Daradkeh, Saudi Arabia
Grzegorz Debita, Poland
Anna Jolanta Dobrowolska, Poland
Jarosław Drapała, Poland
Maciej Drwal, Poland
Tadeusz Dudycz, Poland
Grzegorz Filcek, Poland
Mariusz Fraś, Poland
Piotr Gawkowski, Poland
Dariusz Gąsior, Poland
Arkadiusz Górski, Poland
Jerzy Grobelny, Poland
Krzysztof Grochla, Poland
Houda Hakim Guermazi, Tunisia
Biju Issac, UK
Jerzy Józefczyk, Poland
Ireneusz Jóźwiak, Poland
Krzysztof Juszczyszyn, Poland
Tetiana Viktorivna Kalashnikova, Ukraine
Jan Kałuski, Poland
Anna Maria Kamińska, Poland
Radosław Katarzyniak, Poland
Agata Klaus-Rosińska, Poland
Grzegorz Kołaczek, Poland
Iryna Koshkalda, Ukraine
Zdzisław Kowalczuk, Poland
Dorota Kuchta, Poland
Binod Kumar, India
Jan Kwiatkowski, Poland
Wojciech Lorkiewicz, Poland
Marek Lubicz, Poland
Zbigniew Malara, Poland
Mariusz Mazurkiewicz, Poland
Vojtěch Merunka, Czech Republic
Rafał Michalski, Poland
Bożena Mielczarek, Poland
Peter Nielsen, Denmark
Ziemowit Nowak, Poland
Donat Orski, Poland
Michele Pagano, Italy
Jonghyun Park, Korea
Agnieszka Parkitna, Poland
Marek Pawlak, Poland
Dolores Rexachs, Spain
Paweł Rola, Poland
Stefano Rovetta, Italy
Abdel-Badeeh Salem, Egypt
Joanna Santiago, Portugal
Danuta Seretna-Sałamaj, Poland
Anna Sikora, Spain
Marcin Sikorski, Poland
Jan Skonieczny, Poland
Malgorzata Sterna, Poland
Janusz Stokłosa, Poland
Grażyna Suchacka, Poland
Joanna Szczepańska, Poland
Edward Szczerbicki, Australia
Jerzy Świątek, Poland
Kamila Urbańska, Poland
Jan Werewka, Poland
Zofia Wilimowska, Poland
Marek Wilimowski, Poland
Bernd Wolfinger, Germany
Józef Woźniak, Poland
Krzysztof Zatwarnicki, Poland
Jaroslav Zendulka, Czech Republic
Chunbiao Zhu, People's Republic of China

ISAT 2019 Keynote Speaker

Professor Cecilia Zanni-Merk, Normandie Université, INSA Rouen, LITIS, Saint-Etienne-du-Rouvray, France

Topic: **On the Need of an Explainable Artificial Intelligence**

Contents

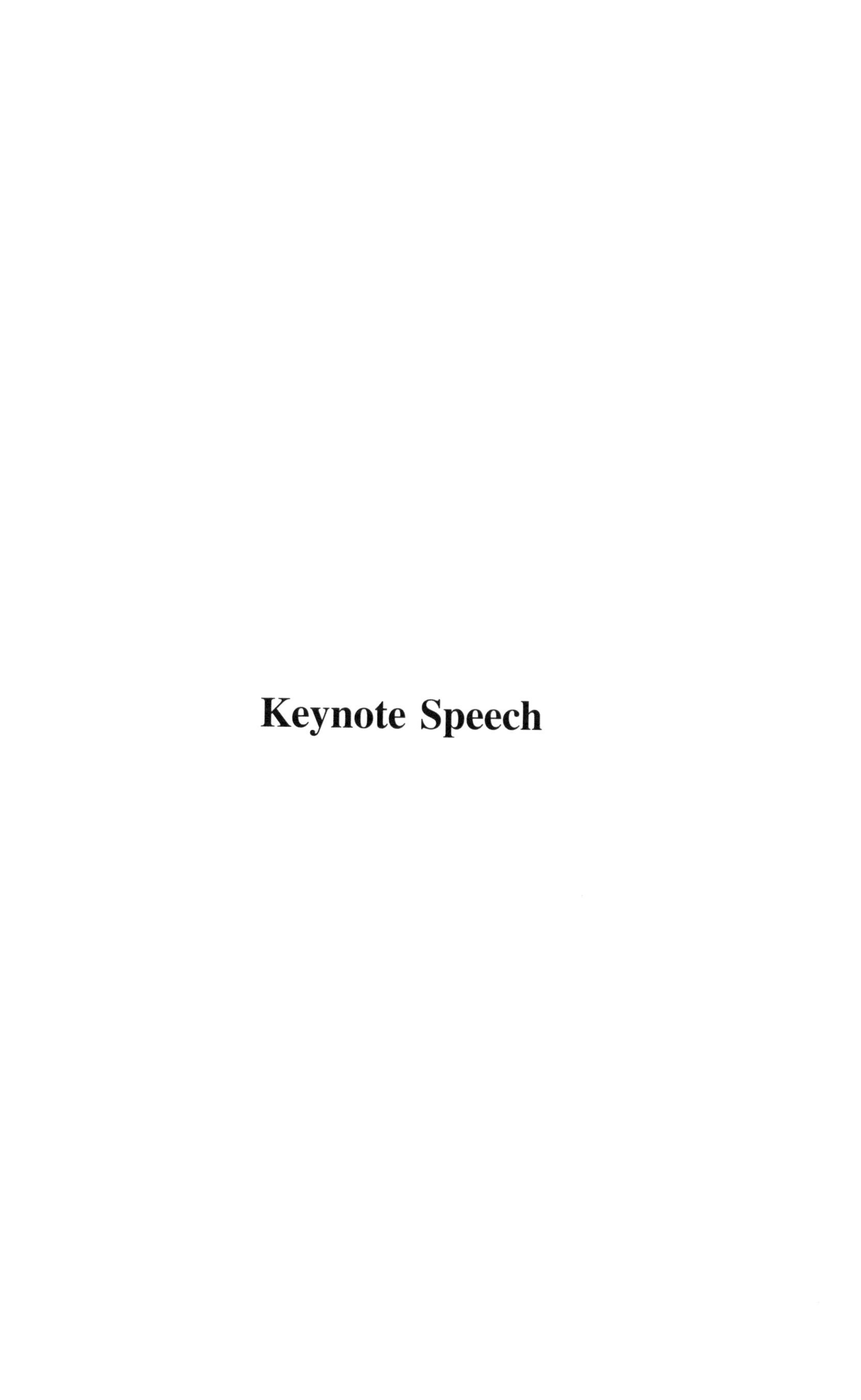

Keynote Speech

On the Need of an Explainable Artificial Intelligence

Cecilia Zanni-Merk(✉)

Normandie Université, INSA Rouen, LITIS, St-Etienne-Du-Rouvray, France
cecilia.zanni-merk@insa-rouen.fr

Abstract. This plenary talk will explore a fascinating nex research field: the Explainable Artificial Intelligence (or XAI), whose goal if the building of explanatory models, to try and overcome the shortcomings of pure statistical learning by providing justifications, understandable by a human, for decisions or predictions made by them.

Artificial Intelligence (AI) applications are increasingly present in the professional and private worlds. This is due to the success of technologies such as machine learning (and, in particular, deep learning approaches) and automatic decision-making that allow the development of increasingly robust and autonomous AI applications. Most of these applications are based on the analysis of historical data; they learn models based on the experience recorded in this data to make decisions or predictions.

However, automatic decision-making by means of Artificial Intelligence now raises new challenges in terms of human understanding of processes resulting from learning, of explanations of the decisions that are made (a crucial issue when ethical or legal considerations are involved) and, also, of human-machine communication.

To meet these needs, the field of Explainable Artificial Intelligence has recently developed.

Indeed, according to the literature, the notion of intelligence can be considered under four aspects: (a) the ability to perceive rich, complex and subtle information, (b) the ability to learn in a particular environment or context; (c) the ability to abstract, to create new meanings and (d) the ability to reason, for planning and decision-making.

These four skills are implemented by what is now called the Explainable Artificial Intelligence (or XAI) with the goal of building explanatory models, to try and overcome shortcomings of pure statistical learning by providing justifications, understandable by a human, for decisions or predictions made.

During this talk we will explore this fascinating new research field.

L. Borzemski et al. (Eds.): ISAT 2019, AISC 1050, pp. 3–3, 2020.
https://doi.org/10.1007/978-3-030-30440-9_1

Architectures and Models of IT Systems

Context-Aware Indexing and Retrieval for Cognitive Systems Using SOEKS and DDNA

Caterine Silva de Oliveira[1(✉)], Cesar Sanin[2], and Edward Szczerbicki[1,2]

[1] The University of Newcastle, University Drive, Callaghan, NSW 2208, Australia
Caterine.SilvadeOliveira@uon.edu.au
[2] Gdansk University of Technology, Gdansk, Poland
Cesar.Sanin@newcastle.edu.au

Abstract. Visual content searching, browsing and retrieval tools have been a focus area of interest as they are required by systems from many different domains. Context-based, Content-Based, and Semantic-based are different approaches utilized for indexing/retrieving, but have their drawbacks when applied to systems that aim to mimic the human capabilities. Such systems, also known as Cognitive Systems, are still limited in terms of processing different sources of information (especially when structured in different ways) for decision making purposes. This issue becomes significantly greater when past information is retrieved and taken in account. We address this issue by proposing a Structuralized Context-Aware Indexing and Retrieval using Set of Experience Knowledge Structure (SOEKS) and Decisional DNA (DDNA). SOEKS and DDNA allow the creation of a multi-modal space composed of information from different sources, such as contextual, visual, auditory etc., in a form of a structure and explicit experiential knowledge. SOKES is composed by fields that allow this experiences to participate in the processes of similarity, uncertainty, impreciseness, or incompleteness measures and facilitate the indexing and retrieval of knowledge in Cognitive Systems.

Keywords: Set of Experience Knowledge Structure (SOEKS) · Decisional DNA (DDNA) · Cognitive Systems · Context-Aware · Image indexing/Retrieval

1 Introduction

Cognitive Systems have gained substantial interest from academia and industry during the past few decade [1]. One of the main reasons for that is the potential of such technologies to revolutionize human life since they intend to work robustly under complex scenes, which environmental conditions may vary, adapting to a comprehensive range of unforeseen changes, and exhibiting prospective behavior like predicting possible visual events. The combination of these properties aims to mimic the human capabilities and create more intelligent and efficient environments [2]. However,

L. Borzemski et al. (Eds.): ISAT 2019, AISC 1050, pp. 7–16, 2020.
https://doi.org/10.1007/978-3-030-30440-9_2

perceiving the environment involves understanding the context and gathering visual and other sensorial information available and translating it into knowledge to be useful. In addition, past experiences also plays an important role when it comes to perception [3] and must also be considered as an important element in this process. Perceiving the environment such as humans do still remains a challenge, especially for real time cognitive vision applications, due to the complexity of such process that has to deal with indexing and retrieval of information in a much reduced amount of time [2].

In this paper we aim to address this issue by proposing a Structuralized Context-Aware Indexing and Retrieval [4, 5] using Set of Experience Knowledge Structure (SOEKS) [6] and Decisional DNA (DDNA) [7]. SOEKS and DDNA allow the building of a multi-modal space composed of information from different sources, such as contextual, visual, auditory etc., in a form of experiential knowledge, and is composed by fields that allow this experiences to participate in the processes of similarity, uncertainty, impreciseness, or in-completeness measures to facilitate the indexing and retrieval process.

This paper is organized as follows: In Sect. 2, some literature review on visual content indexing and retrieval is presented with special focus on content-based, context-based and semantic-based approaches. In Sect. 3 the proposed context-aware approach for Cognitive Systems is presented and SOEKS and DDNA described. In Sect. 4 the general framework and explanation of the given knowledge indexing and retrieval is given for the case of Cognitive Vision Platform for Hazard Control (CVP-HC). Finally, in Sect. 5, conclusions and future work are given.

2 Literature Review

Image and video indexing and retrieval has been an active research area over the past few decades. There are a range of researches and review papers that mention the importance, requirements and applications of visual information indexing and retrieval approaches [8–11]. In this section, the methodologies are grouped into three main categories, Context-based, Content-Based, and Semantic-based visual content indexing/retrieval. A brief overview is provided for each of those classes, pointing out its drawbacks for application in cognitive systems when applied purely.

2.1 Context-Based

Visual content searching, browsing and retrieval implementations are indispensable in various domains applications, such as remote sensing, surveillance, publishing, medical diagnoses, etc. One of the methods used for those purposes is the context-based approach [12]. Information that doesn't come directly from the visual properties of the image itself can be seen as the context of an image. The context-based approach can be tracked back to 1970s. In context-based applications, the images are manually annotated using text descriptors, key words etc., which can be used by a database management system to execute image retrieval [13]. This process is used to label both image contents and other metadata of the image, for instance, image file name, image format, image size, and image dimensions. Then, the user formulates textual or numeric queries to retrieve all images that are satisfying some of criteria based on these annotations. The similarity between images is, in this case, based on the similarity between the texts.

However, there are some drawbacks in using this approach purely. The first limitation is that the most descriptive annotations must usually be entered manually. Considerable level of human labor is required in large datasets for manual annotation [13]. The second disadvantage of this method is that the most images are very rich in its content and can have more details than those described by the user [14]. In addition, the annotator may give different descriptions to images with similar visual content and different users give different descriptions to the same image. Finally, textual annotations are language dependent [15]. It can be overcome by using a restricted vocabulary for the manual annotations, but, as mentioned previously, it is very expensive to manually index all images in a vast collection [16].

2.2 Content-Based

Content-based image retrieval (CBIR) has been introduced in the early 1980s [13]. In CBIR, images are indexed by their visual content (for instance, color, texture, shapes, etc.) instead of their context. A pioneering work was published by Chang in 1984. In his research, he presented a picture indexing and abstraction approach for pictorial database retrieval [17]. This pictorial database comprised picture objects and picture relations. To construct picture indexes, abstraction operations are formulated to perform picture object clustering and classification.

Literature on image content indexing is very large [18], and commercial products have been developed using such approach. A common approach to model image data is to extract a vector of features from each image in the database (e.g. image color pixels) and then use a distance measurement, such as Euclidean [19], between those vectors to calculate the similarity between them [13]. Nonetheless, the effectiveness of this approach is highly dependent on the quality of the feature transformation. Often it is necessary to extract many features from the database objects in order to describe them sufficiently, which results in very high-dimensional feature vectors, which demand high storage capacity and increases computational costs. In addition, there is a gap between the high-level image and the low-level image, i.e. there is a difference between what image features can distinguish and what people perceives from the image [14].

2.3 Semantic-Based

In order to overcome the limitations of Content-Based and Context Based approaches, Semantic-Based Image Retrieval (SBIR) has been proposed. SBIR can be made by extraction of low-level features of images to identify significant regions or objects based on the similar characteristics of the visual features [14]. Then, the object/region features will be used for semantic image extraction to get the semantics description of images. Semantic technologies like ontology offers promising approach to map those low level image features to high level ontology concepts [5]. Image retrieval can be performed based on the high-level concept (based on a set of textual words, which is translated to get the semantic features from the query).

For the semantic mapping process, supervised or unsupervised learning tools can be used to associate the low-level features with object concept. This procedures are combined with other techniques to close the semantic gap problem, such as using textual word through image annotation process [20]. Semantic content obtained either by textual annotation or by complex inference procedures are both based on visual content [21].

There are a number of papers that address the issue semantic mapping for images. One of the first was Gorkani and Picard [22], who used a texture orientation approach based on a multi-scale steerable pyramid to discriminate cities from landscapes. Yiu [23] applies a similar approach to classify indoor and outdoor scenes, using also color information as features. Wu and Zhu applies ontology to define high-level concepts. In their framework ontology and MPEG-7 descriptors are used to deal with problems arising from representation and semantic retrieval of images. The framework allows for the construction of incrementally multiple ontologies, and shares ontology information rather than building a single ontology for a specific domain not only between the image seekers but also between different domains [24]. For these implementations, the disadvantage is the computational complexity, which can be very high for largescale dataset [25].

3 Context-Aware Approach for Cognitive Systems

The human cognition capabilities is able to receive visual information from the environment and combine it with other sensory information to create perceptual experience [26]. The physical energy received by sense organs (such as eye, ears and nose) forms the basis of perceptual experience. In other words, the combination of sensory inputs are converted into perceptions of visual information (such as dogs, computers, flowers, cars and planes); into sights, sounds, smells, taste and touch experiences. According to Gregory [3], perceptual processes depends on perceiver's expectations and previous knowledge as well as the information available in the stimulus itself, which can come from any organ part of a sensory system. The processing all this information in a lapse of milliseconds makes the humans a very powerful "cognition machine".

In this context, Cognitive Systems has emerged attempting to meet human capabilities. Cognitive Systems have been defined as "a system that can modify its behavior on the basis of experience" [27]. However, at the present, there is no widely agreed upon definition for cognitive computing. But in general we can say that the term "cognitive system" has been used to define a new solution, software or hardware that mimics in some ways human intelligence.

However, a system capable of processing all information available such as sensor data, visual content from cameras, input signals from machines and any other contextual information available to characterize setting in analysis and at the same time retrieving past experiences for the creation of perceptions still remains a challenge. One of the main difficulties encountered in this case is processing different sources of information that comes structured in different representations at once. This issue becomes significantly greater when past information is retrieved and used in this analysis. Therefore, a knowledge representation capable of building a multi-modal space composed of information from different sources, such as contextual, visual, auditory etc., in a form of experiential knowledge would be a very useful tool to facilitate this process.

3.1 Structuralized Representation for Visual and Non-visual Content

Choosing an appropriate image representation greatly facilitates obtaining methods that efficiently learn the relevant information about the category in short time and methods that efficiently match instances of the object in large collections of images [28]. Several researchers have identified that the starting point is to establish an image/video knowledge representation for cognitive vision technologies. However, among all proposed approaches, even though they present some principles for intelligent cognitive vision, none of them provide a unique standard that could integrate image/video modularization, its virtualization, and capture its knowledge [6]. Consequently, we propose to address these issues with an experience-based technology that allows a standardization of image/video and the entities within together with any other information as a multi-source knowledge representation (required for the further development of cognitive vision) without limiting their operations to a specific domain and/or following a vendor's specification. Our representation supports mechanisms for storing and reusing experience gained during cognitive vision decision-making processes through a unique, dynamic, and single structure called Decisional DNA (DDNA) [7]. DDNA makes use of Set of Experience (SOE) in an extended version for the use of storing formal decision events related to image and video. DDNA and SOE provide a knowledge structure that has been proven to be multi-domain independent [7].

Set of Experience Knowledge Structure (SOEKS) and Decisional DNA (DDNA). The Set of Experience Knowledge Structure (SOEKS) is a knowledge representation structure which has been created to acquire and store formal decision events explicitly. SOEKS is composed by four basic elements: variables, functions, constraints, and rules. Variable are the elementary component and it is used to represent knowledge in an attribute-value manner (fundamentally, following the old-fashioned approach for knowledge representation). Functions, Constraints, and Rules are different ways of establishing relationships among these variables. Functions define relations between a dependent variable and a set of input variables to build multi-objective goals. Constraints, on the other hand, act as a way to control the performance of the system in relative to its goals by limiting the possibilities or the set of possible solutions. Lastly, rules are used to express the condition-consequence connection as "if-then-else" and are used to create inferences and associate actions with the conditions under which they should be applied [6].

The Decisional DNA is a structure that has the ability to capture decisional fingerprints companies/organization and also individuals, and has the SOEKS as its basis. Multiple Sets of Experience or multiple SOEs can be gathered, classified, ordered and then grouped into decisional chromosomes. These chromosomes accumulate decisional strategies for a specific application. The set of chromosomes comprise, what is called the Decisional DNA (DDNA) [7].

4 Knowledge Indexing and Retrieval for a CVP-HC

Cognitive Vision Platform for Hazard Control (CVP-HC) being developed to manage risky activities in industrial environments [29]. It is a scalable yet flexible system, designed to work a variety of environment setting by changing its behavior accordingly in real time (context aware). We utilize in this section the knowledge being used to feed the platform to demonstrate the framework for indexing and retrieval of knowledge based on the approach proposed in this paper. More details about the representation of visual and non-visual content in the CVP-HC can be found in [30]. In this section will be focusing in the fields of SOEKS that are relevant for the indexing and retrieval process. Figure 1 presents the overall architecture of the indexing and retrieval system.

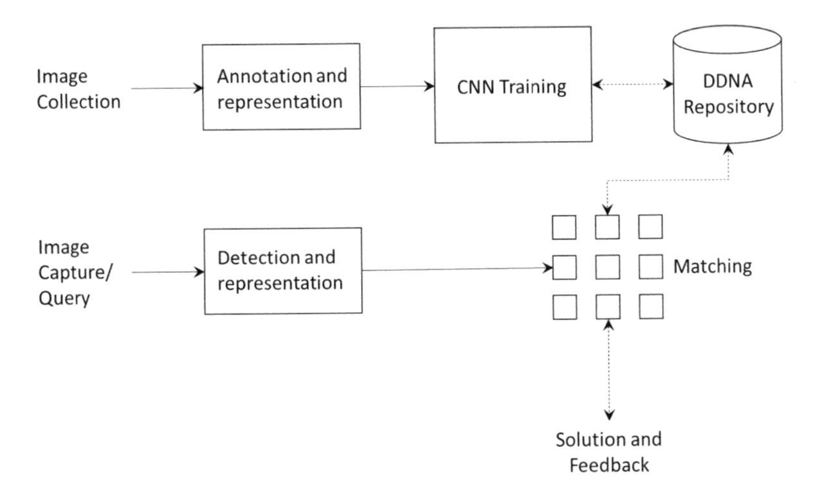

Fig. 1. Overall architecture of the indexing and retrieval system.

4.1 Weight, Range and Priority

Searching for visual similarity by simply comparing large sets of pixels (comparing a query image to images in the database, for example) is not only computationally expensive but is also very sensitive to and often adversely impacted by noise in the image or changes in views of the imaged content [31]. Variables of SOEKS includes fields that allow them to participate in the processes of similarity, uncertainty, impreciseness, or incompleteness measures to facilitate the indexing and retrieval purposes. These fields are: weight, priority, lower range and upper range [32], as shown in Fig. 2.

Weight. Each variable has a weight related. Experts may be able to decide the percentage of the weight rate. However, subjective mistakes can be made by human beings, and it will influence decision making. This is a great challenge to find a better automatic and objective way instead of the experts [33]. In the CVP-HC the weights are initialized with the same value for all variables in interaction 0 (before first training)

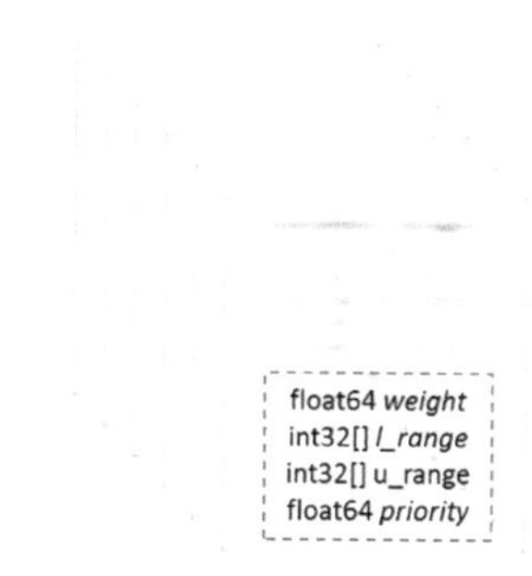

Fig. 2. Fields of SOEKS that allow them to participate in the processes of similarity, uncertainty, impreciseness, or incompleteness measures.

and automatically assigned as the contribution of each variable to the decision making process (variable importance) for next interactions. The summation of all weights for each experience is defined as 1 as shown below:

$$w_{v1} + w_{v2} + w_{v3} + \ldots + w_{vn} = 1 \tag{1}$$

Therefore, this prediction is easy to be reused and transported in different systems. Hence, knowledge will be expended and shared with different users [34]. The quantity of sets of experience has a great impact on it. If the scope of existing experience is too short for the system to learn, the weight prediction will not be precise.

After those sets of experience are loaded in memory and used to train the first classifier, the generated weights will be allocated to the related variables. A loop is, then, used to assign collected weights to the variables of each experience of the training dataset. The combination of structured knowledge and weights helps addressing the issue of calculating how similar a present and past experience is by taking in account which information is more relevant to this analysis.

This also allows to enrich the system with more knowledge as it runs without compromising the calculation of similarity between experiences of different dimensionalities.

Range and Priority. Each variable has also a priority associated p_v. For the CVP-HC system, priorities are automatically assigned as:

$$p_v = (1 - d_{c\_ulv})/w_v \tag{2}$$

Where d_{c_ulv} is degree of correspondence of each variable to a chosen upper level value and goes from 1 (completely similar) to 0 (completely dissimilar) and w_v. the weight of that variable. The priority associated with each SOE is defined by the summation of each individual priority:

$$P_{soe} = p_{v1} + p_{v2} + p_{v3} + \ldots + p_{vn} \tag{3}$$

Where p_{vn} is the priority associated with the nth element of the SOE. By having a priority associated each SOE, we can order the experiences facilitating the searching process.

In addition, in this approach new unique experiences are assigned a higher priority. By selecting experiences with higher priority during the training iterations (for learning purposes), the system can increase specificity and creating a unique DDNA for that application. Furthermore, by analyzing the curve of SOE's priorities it is possible to infer if a big change in setting has happened and the system requires a new training iteration. Finally, the reusability of a system by another company/organization can be tested over the analysis of the set of priorities of the real time collected experiences with the ones that have been used to enrich and train the system.

5 Conclusions and Future Work

The paper presents a Context-Aware approach to address the issue of indexing and retrieval of multi-modal space information systems, which processes a variety of sources such as contextual, visual, auditory, etc. at once. Such systems, also known as Cognitive Systems aim to mimic the human capabilities and therefore also make use of experiential knowledge. Our approach is an experience-based technology that allows a standardization of visual content and context together with any other information as a multi-source knowledge representation without limiting their operations to a specific domain and/or following a vendor's specification.

Our representation supports mechanisms for storing and reusing experience gained during cognitive vision decision-making processes through a unique, dynamic, and single structure called Decisional DNA (DDNA). DDNA makes use of Set of Experience (SOE) in an extended version for the use of storing formal decision events related to cognitive vision systems. The knowledge indexing and retrieval is facilitated by the use of *weight, ranges and priorities*. These fields are part of the SOEKS and allow the experiences to participate in the processes of similarity, uncertainty, impreciseness, or incompleteness measures.

The method discussed in this paper will be evaluated when the experience is enriched with other sensorial data. Suitability of reusing experiences will also be explored for different case scenarios.

References

1. Sanin, C., Haoxi, Z., Shafiq, I., Waris, M.M., de Oliveira, C.S., Szczerbicki, E.: Experience based knowledge representation for Internet of Things and Cyber Physical Systems with case studies. Future Gener. Comput. Syst. (2018)
2. Vernon, D.: The space of cognitive vision. In: Cognitive Vision Systems, pp. 7–24. Springer, Heidelberg (2006)
3. Gregory, R.L.: Eye and Brain: The Psychology of Seeing. McGraw-Hill, New York (1973)

4. Malik, S., Jain, S.: Ontology based context aware model. In: 2017 International Conference on Computational Intelligence in Data Science (ICCIDS), Chennai, pp. 1–6 (2017). https://doi.org/10.1109/ICCIDS.2017.8272632
5. Manzoor, U., Ejaz, N., Akhtar, N., Umar, M., Khan, M.S., Umar, H.: Ontology based image retrieval. In: 2012 International Conference for Internet Technology and Secured Transactions, London, pp. 288–293 (2012)
6. Sanin, C., Szczerbicki, E.: Experience-based knowledge representation SOEKS. Cybernet Sys. **40**(2), 99–122 (2009)
7. Sanin, C., Toro, C., Haoxi, Z., Sanchez, E., Szczerbicki, E., Carrasco, E., Man-cilla-Amaya, L.: Decisional DNA: a multi-technology shareable knowledge structure for decisional experience. Neurocomputing **88**, 42–53 (2012)
8. De Marsicoi, M., Cinque, L., Levialdi, S.: Indexing pictorial documents by their content: a survey of current techniques. Image Vis. Comput. **15**, 119–141 (1997)
9. Rui, Y., Huang, T., Chang, S.: Image retrieval past, present, and future. In: International Symposium on Multimedia Information Processing (1997)
10. Rui, Y., Huang, T., Chang, S.: Image retrieval: current techniques, promising directions and open issues. J. Vis. Commun. Image Represent. **10**, 39–62 (1999)
11. Muller, D.B.H., Michoux, N., Geissbuhler, A.: A review of content-based image retrieval systems in medical applications clinical benefits and future directions. Int. J. Med. Informatics **73**, 1–23 (2004)
12. Westerveld, T.: Image retrieval: content versus context. In: Content-Based Multimedia Information Access-Volume 1, pp. 276–284, April 2000
13. Raveaux, R., Burie, J.C., Ogier, J.M.: Structured representations in a content based image retrieval context. J. Vis. Commun. Image Represent. **24**(8), 1252–1268 (2013)
14. Alkhawlani, M., Elmogy, M., El Bakry, H.: Text-based, content-based, and semantic-based image retrievals: a survey. Int. J. Comput. Inf. Technol. **4**(01) (2015)
15. Tamura, H., Yokoya, N.: Image database systems: a survey. Pattern Recogn. **17**, 29–43 (1984)
16. Oard, D.W., Dorr, B.J.: A survey of multilingual text retrieval. Technical report UMIACS-TR-96-19, University of Maryland, Institute for Advanced Computer Studies (1996)
17. Liu, S.H., Chang, S.K.: Picture indexing and abstraction techniques for pictorial databases. IEEE Trans. Pattern Anal. Mach. Intell. (TPAMI) **6**(4), 475–483 (1984)
18. Datta, R., Joshi, D., Li, J., Wang, J.Z.: Image retrieval: ideas, influences, and trends of the new age. ACM Comput. Surv. **39**, 2007 (2006)
19. Danielsson, P.E.: Euclidean distance mapping. Comput. Graph. Image Process. **14**(3), 227–248 (1980)
20. Wang, H.H., Mohamad, D., Ismail, N.: Image retrieval: techniques, challenge, and trend. In: International Conference on Machine Vision, Image Processing and Pattern Analysis, Bangkok. Citeseer (2009)
21. Shanmugapriya, N., Nallusamy, R.: A new content based image retrieval system using GMM and relevance feedback. J. Comput. Sci. **10**(2), 330–340 (2013)
22. Gorkani, M.M., Picard, R.W.: Texture orientation for sorting photos "at a glance". In: International Conference on Pattern Recognition, p. 459, October 1994
23. Yiu, E.C.: Image classification using color cues and texture orientation. Doctoral dissertation, Massachusetts Institute of Technology (1996)
24. Zhu, S.C., Wu, Y., Mumford, D.: Filters, random fields and maximum entropy (FRAME): towards a unified theory for texture modeling. Int. J. Comput. Vision **27**(2), 107–126 (1998)
25. Zin, N.A.M., Yusof, R., Lashari, S.A., Mustapha, A., Senan, N., Ibrahim, R.: Content-based image retrieval in medical domain: a review. J. Phys: Conf. Ser. **1019**(1), 012044 (2018)
26. Bandura, A.: Human agency in social cognitive theory. Am. Psychol. **44**(9), 1175 (1989)

27. Hollnagel, E., Woods, D.D.: Joint Cognitive Systems: Foundations of Cognitive Systems Engineering. CRC Press (2005)
28. Amores, J., Sebe, N., Radeva, P.: Context-based object-class recognition and retrieval by generalized correlograms. IEEE Trans. Pattern Anal. Mach. Intell. **29**(10), 1818–1833 (2007)
29. de Oliveira, C.S., Sanin, C., Szczerbicki, E.: Visual content learning in a cognitive vision platform for hazard control (CVP-HC). Cybern. Syst. **50**(2), 197–207 (2019)
30. de Oliveira, C.S., Sanin, C., Szczerbicki, E.: Towards knowledge formalization and sharing in a cognitive vision platform for hazard control (CVP-HC). In: Asian Conference on Intelligent Information and Database Systems, pp. 53–61. Springer, Cham (2019)
31. Deserno, T.M., Antani, S., Long, R.: Ontology of gaps in content-based image retrieval. J. Digit. Imaging **22**(2), 202–215 (2009)
32. Sanin, C., Szczerbicki, E.: Using XML for implementing set of experience knowledge structure. In: Khosla, R., Howlett, R.J., Jain, L.C. (eds.) KES 2005. LNCS (LNAI), vol. 3681, pp. 946–952. Springer, Heidelberg (2005)
33. Sanín, C.A.M.: Smart knowledge management system. University of Newcastle (2007)
34. Wang, P., Sanin, C., Szczerbicki, E.: Enhancing set of experience knowledge structure (SOEKS) WITH A nearest neighbor algorithm RELIE-F. In: Information Systems Architecture and Technology, 13 (2012)

Designing Information Integrating Systems Based on Co-shared Ontologies

Dariusz Put(✉)

Cracow University of Economics, 27 Rakowicka Street, 30-852 Krakow, Poland
putd@uek.krakow.pl

Abstract. Data gathered by organisations have diverse architecture and are based on various data models. Cooperation between organisations often requires data exchanging which, in heterogeneous environment, is not easy task. So, various integrating systems have been proposed to overcome problems concerning data exchange in such diverse systems. The field of data integration has expanded in many directions, including schema matching, record linkage, data fusion, and many more [5, 8]. In this paper various approaches to data integration are discussed and tasks that have to be fulfilled by integrating systems are identified. Because among integrating systems are also those based on ontology, their characteristics are identified. Next, the architecture of proposed system is presented, actions performing by individual modules during the process of queries execution are characterized and the process of designing integrating systems based on the proposed model is described.

Keywords: Information integration · Ontology · Integrating systems

1 Approaches to Integration of Heterogeneous Data

The field of data integration has expanded in many directions, including tasks such as schema matching, schema mapping, record linkage, data fusion, and many more [5, 8]. Wang et al. [18] classify data integration systems by their explainability and discuss the characteristics of systems within these classes. They study and review existing approaches based on schema level integration and data level integration. Doan et al. [4] claim that in order for a data integration system to process a query over a set of data sources, the system must know, what sources are available, what data exist in each source, and how each source can be accessed. They propose a system based on modules such as query reformulator, query optimizer, execution engine and wrappers. A part of the system is schema mapping module which they define as a set of expressions that describe a relationship between a set of schemata.

Offia and Crowe [15] propose a logical data management approach using REST technology to integrate and analyse data. Data that for governance, corporate, security or

The Project has been financed by the Ministry of Science and Higher Education within "Regional Initiative of Excellence" Programme for 2019-2022. Project no.: 021/RID/2018/19. Total financing: 11 897 131,40 PLN.

L. Borzemski et al. (Eds.): ISAT 2019, AISC 1050, pp. 17–27, 2020.
https://doi.org/10.1007/978-3-030-30440-9_3

other restriction reasons cannot be copied or moved, can easily be accessed, integrated and analysed, without creating a central repository. Halevy et al. [9] propose data integration system based on schema. The schema is known as a mediated or global schema which answers queries sent by users. The system needs a mapping function that describe the semantic relationship between the mediated schema and the schema of the sources.

Asano et al. [1] describe Dejima architecture as a framework for sharing data and controlling update over multiple databases. Each peer autonomously manages its own database and collaborates with other peers through views. Fernandez and Madden [7] aim to find common representation for diverse, heterogeneous data. Specifically, they argue for an embedding in which all entities, rows, columns, and paragraphs are represented as points. They introduce Termite, a prototype to learn the best embedding from the data. Because the best representation is learned, this allows Termite to avoid much of the human effort associated with traditional data integration tasks. They have implemented a Termite-Join operator, which allows users to identify related concepts, even when these are stored in databases with different schemas.

In systems integrating heterogeneous and distributed information resources there is multidirectional variety of: data models, data and information storing systems, query languages, categories and forms of shared resources, names of instances and attributes, methods of instances modification. Several solutions have been proposed to solve this problem (see e.g. [3, 6, 12, 14]). They show that integrating model being a basic for systems linking various information resources repositories has to comprise components that: make metadata about shared resources available, enable uncomplicated search for data and information, choose demanded data and information and visualize them in friendly forms. So, integrating systems have to fulfil a considerable number of tasks, including:

- the communication with external heterogeneous data sources – a solution should enable to collect data and information stored in various repositories. The variety of resources formats has to be taken under consideration and a designed solution has to be adjusted to individual systems delivering information resources during the system operation,
- reformulation of queries formulated with the use of a query language implemented in the integrating system to query languages used in individual systems. Integrated information resources may be collected from systems based on various data models with various query languages employed. For every such a component query converter responsible for queries reformulation has to be elaborated. This problem may be solved globally, by designing a module responsible for such reformulation to any existing query languages or locally, by implementing dedicated converters in individual sub-systems by their administrators,
- linking and unifying information resources collected from heterogeneous systems – data and information received from source systems, even after processing to uniform format (e.g. JSON, XML or CSV), has to be linked before being sent to users,
- making available metadata about shared resources – a data model that will be used for metadata about shared resources description (ontology) has to be elaborated. The ontology enables uniform access to integrated resources. For the sake of the fact that for description of shared data the XML or JSON technologies are commonly used,

the choice of data model based on one of these technologies should be considered. It will facilitate cooperation with other systems, both integrating and storing data if only those systems will support one of these technologies,
- enabling users to access resources – universal solution enables to use shared resources both by previously defined and undefined users and by various applications such as information agents. System designers have to take into account this variety of actors and implement support for all of them, e.g. appropriate user interface, predefined queries, interfaces for applications,
- the fragmentation of the integration task into possibly independent sub-tasks so that the modification of one of system layers or one of its modules does not demand re-designing of other layers or modules,
- making access to various forms of information resources presentation. Information resources may have heterogeneous forms so such presentation methods enable different views to the same resources as well as fitting presentation methods to their categories (e.g. some presentation views are adjusted to present plain files, others to display resources collected from databases).

There are approaches where ontology as a basis for data integration models is proposed. Sardelich et al. [16] discuss the solution where data are represented as an ontology in order to facilitate their handling, and to allow the integration of other relevant information, such as the link between a subsidiary company and its holding or the names of senior management and their links to other companies. The ontology is used in many other proposals [2, 10, 11, 13, 16, 17].

Systems for distributed heterogeneous information management are usually complex and those based on co-shared ontology should have the following characteristics:

- they connect independent sub-systems that operate without any changes,
- they provide users with necessary, actual or historical information,
- the process of information resources searching is transparent for users,
- one part of their structure is a flexible co-shared ontology, so once defined metadata module does not have to be re-designed in the future,
- automatic modification of the ontology is possible,
- not only they enable to predefine queries during designing the system but also *ad hoc* queries may be formulated,
- the process of queries formulation is as simple as possible, so any user has an access to shared information resources and acquaintance with the query language is not necessary,
- the user query language is as simple as possible so everybody may formulate a query,
- the user query language is maximally expressive, which enables to formulate precise queries selecting only necessary information resources,
- they make possible to integrate all existing categories of information resources,
- they may be used for inter and intra-organisational integration,
- they are scalable which enables to join new sub-systems including clients and business partners repositories during the system operation,
- they do not decrease existing systems effectiveness,
- they are easy to design, implement and configure,

- they have structure consisting of layers/modules, which enables to modify any component during the system exploitation without a necessity to change other elements.

Designing systems having all these characteristics is not easy, because some of mentioned above features are contradictory (e.g. expressiveness of a query language and its simplicity, not complicated structure whereas the information integration task is complex). So, when embarking on defining framework for the most ideal integrating model it is necessary to compromise. First, tasks that have to be fulfilled during accomplishment of such a system have to be identified. Second, the architecture should be designed and tasks executed by individual components have to be established. Then, instructions concerning the modification of components of the system during its operation have to be elaborated. Another drawbacks of such systems, apart from their complexity, are:

- difficulties in maintaining actual metadata about accessible resources,
- the necessity to involve all actors of being prepared system (including business partners) into the process of its designing,
- the need for establishing standards and the necessity to compromise on this aspect (which may be difficult because individual actors manage different businesses and because of the culture and natural languages differences), which is especially important when a solution embracing cooperating organisations distributed all over the world is designed.

2 The Proposed Model for Information Integration

Diversity of IT systems and, especially, heterogeneity of data gathered by them, impede exchanging of information resources between organisations or even between departments in one company. So, organisations search for solutions which will facilitate cooperation by making exchanging of information resources self-acting. Taking under consideration such expectations, thorough analysis of integrating solution proposals discussed in scientific papers has been made. As a result, identification of tasks that have to be performed during integration of heterogeneous data and information resources have been identified. This has led to designing the proposal of solution which could be designed by cooperating organisations to make heterogeneous data and information resources available to all actors in unified form. The proposed system has to be designed and implemented by IT team but, after implementation, it does not need user participation in accessing to heterogeneous information resources gathered by individual actors of a solution for previously defined as well as undefined users.

Due to complexity of the task of heterogeneous and distributed information resources integration, especially if integrated sub-systems remain unchanged, the architecture of integrating solutions is more complex than in traditional systems. In the system integrating information resources received from a considerable number of heterogeneous sources, where, in addition, there is a possibility to formulate queries directly by final users, the necessity to fulfil tasks not existing in traditional systems has to be taken into consideration. During the process of designing such systems the structure of some of their components has to be elaborated in cooperation with representatives of all integrated sub-systems. Taking under consideration this complexity we assume that the best architecture for such integrating systems is a level/module solution, where individual tasks are fulfilled by dedicated modules. So, the proposed model consists of four layers comprising modules performing partial tasks leading to the integration and enabling users to find necessary information in the system designed on the basis of the model. The layers are (see Fig. 1):

- local information resources,
- local communication wrappers,
- global,
- user interface.

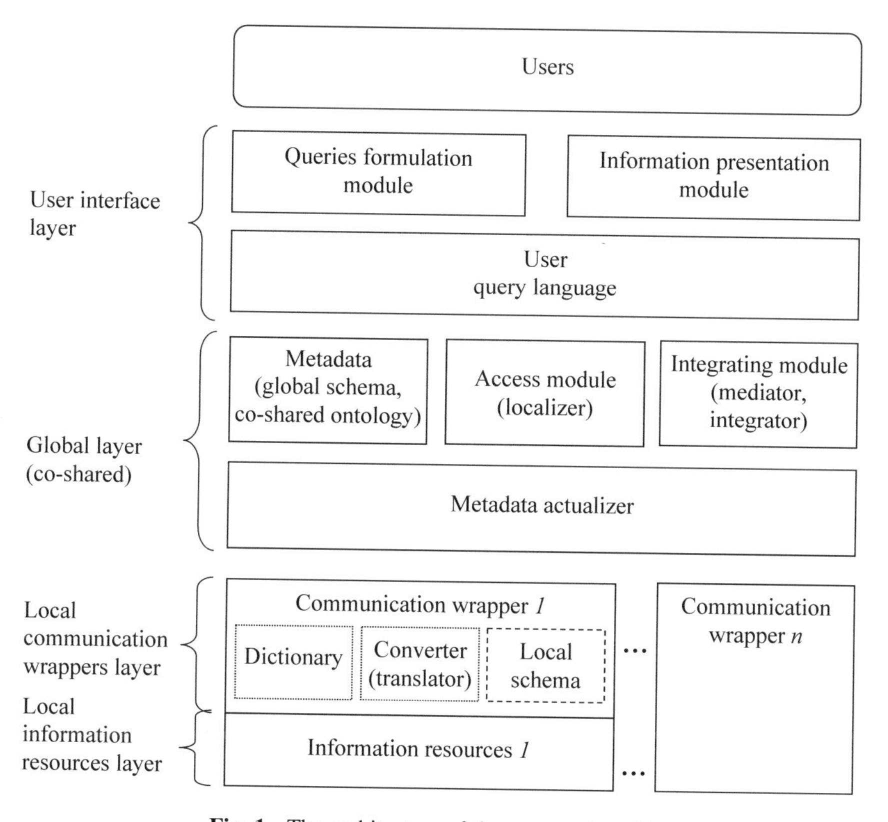

Fig. 1. The architecture of the proposed model.

3 Tasks Fulfilled by Individual Components

The process of queries execution and tasks fulfilled by components of a system based on the model during this process are illustrated in Fig. 2. Users use queries formulation module, which enables access to actual content of the ontology (also called the global schema). Next, the access module identifies sub-systems possessing information resources being representatives of concepts chosen by users during queries formulation. This module directs queries formulated by users to the chosen sub-systems. In the next stage, dictionaries in local systems change concepts and attributes names into their local equivalents and then queries translators reformulate queries to the form understandable in individual local systems. After fulfilling these tasks the queries may be executed in local sub-systems. Information resources chosen this way are sent to dictionaries which convert instances and attributes names into global ones. In such forms information resources are sent to the integrator. Receiving answers from all sub-systems the integrator merges information resources and sends them to the users, who, using application implemented in their local systems (presentation module), finally process data and information and this way receive necessary information.

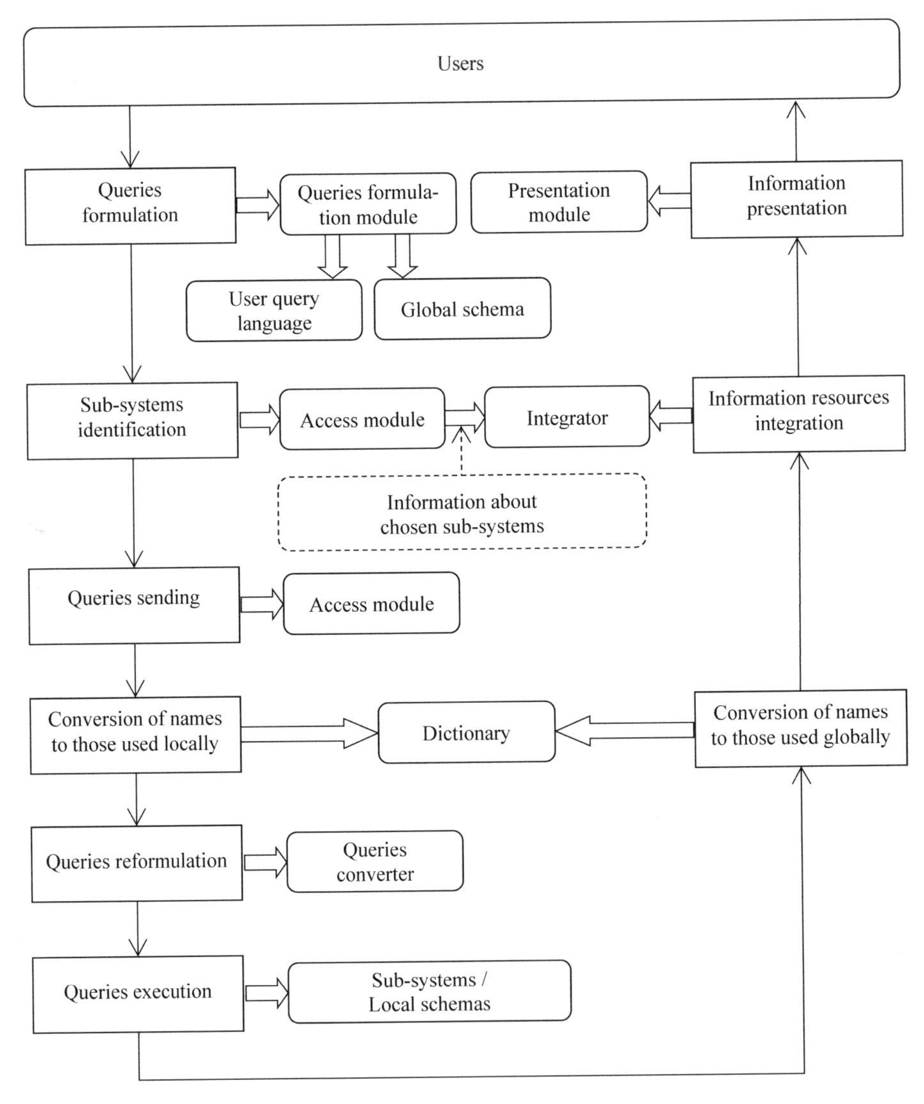

Fig. 2. The process of queries execution in the proposed model

4 The Process of Designing Integrating Systems Based on the Proposed Model

The process of designing a system based on the proposed model consists of the following steps (see Fig. 3):

1. Establishing a project team. The project team identifies tasks, watches over all actions and is responsible for implementation and subsequent maintenance.
2. Consultation with partners and a goal definition. The integrating system may involve several organisations. If it is economically justified, an attempt to convince business partners about usefulness of implementation of such solution, i.e. communication simplification, easier access to information, should be taken. It is necessary to be aware that information resources made available to business partners will be a subset of all data and information gathered in organisational systems. So, it is necessary to define and negotiate which information resources will be co-shared. On the basis of this arrangement the project goal has to be defined. The goal points out which information resources are going to be commonly accessible.
3. The identification of units gathering information resources for their own purposes. At this stage information about departments and individual employees gathering data and information that may be useful for the project should be collected.
4. Elaboration of the user query language. On the basis of characteristics of information resources user query language should be selected or elaborated. The language should have possibly uncomplicated structure, so that every user may formulate queries and queries reformulation to all languages used in integrated subsystems is possible. This decision should be connected with the elaboration of the structure of the ontology and may be done simultaneously with the identification of units gathering integrated information resources.
5. The ontology structure agreement. Framework for the structure of the global schema is defined: concepts and attributes structures as well as rules for connections between individual components are elaborated.
6. The elaboration of conceptual schemas of local information resources. Every unit gathering co-shared information analyses data and information and prepares their conceptual schemas. In the schema concepts, attributes and connections between concepts are described. The schemas include metadata about those information resources that are made available for integrating purposes.
7. The elaboration of the ontology conceptual project. The set of all local conceptual projects is the basis for elaboration of the schema of the ontology. Information resources of the same type should be identified, names of concepts and attributes have to be coordinated, the problem of their unambiguity has to be solved. The contents of the local conceptual projects will be the basis for identification of localisation of individual information resources. This module is the source for local dictionaries.
8. Inserting initial content to the ontology. On the basis of the conceptual project, logical and then physical projects have to be designed. It is necessary to take under consideration the structure of user query language, so that accessible schema elements correspond with the language elements which will be established by users during queries formulation.

9. Elaboration of the ontology modification methods. During the exploitation of the system the metadata structure will be constantly changed: new information resources may appear, the structure of existing may be changed, new local systems may be included. The elaborated modification methods have to include description of how to proceed in every of these cases, because it is essential that the ontology content is always up-to-date.
10. Elaboration of applications operating in the global layer: localizer, integrator and metadata actualizer.

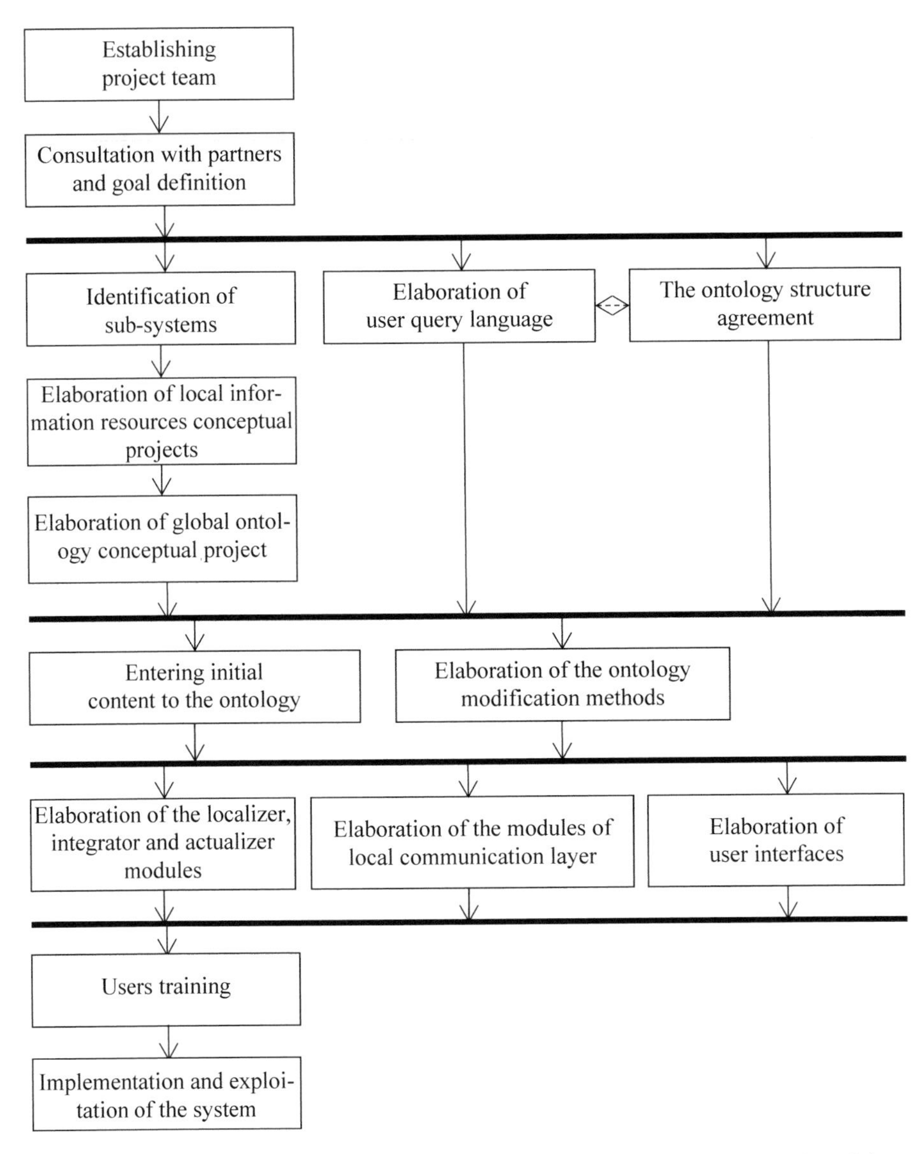

Fig. 3. The process of designing an integrating system based on the proposed model

11. Elaboration of components of the local communication wrappers layer. The administrators of local systems constituting the integrating system prepare solutions enabling an access to locally stored information resources. They have to take into consideration the structure of the ontology and the user query language and prepare modules enabling reformulation of queries formulated by users to a query language used in their systems. They also have to define procedures for concepts and attributes names conversion. There is no obligation concerning elaborated solutions, but it has to be taken under consideration that as an input to the local communication layer there are queries formulated by users (in the established query language) and as an output there are information resources chosen from local systems as a response to formulated queries. Individual solutions may be based on locally defined services, databases, local schemas, local data warehouses or data marts, etc.
12. Elaboration of solutions for users: queries formulation methods and applications enabling final processing of information resources received as a response to formulated queries. Designed methods do not have to be homogeneous but standardisation of the methods of queries formulation and information resources presentation will have an advantage – these operations will be unified which will facilitate using the system.
13. Training. The training should include queries formulation by users, information resources module and, if this possibility is defined in an individual sub-system, the way to pre-define queries and storing them in sub-system for future using.

The system implementation. The implementation is not trouble-making for individual sub-systems – they operate without any changes. The only result connected with the system implementation is a new functionality seen by users in their systems – the possibility to formulate queries to the integrating system selecting information resources from heterogeneous, distributed local systems.

5 Conclusions and Future Work

The designing of integrating systems is more complex than in case of traditional solutions. However, implementation of such systems may give a considerable number of benefits to organisations: easier access to various information resources, facilitation of data exchange with business partners, access for previously defined and undefined users, the possibility to formulate *ad hoc* queries, standardization of an access to information resources, access to data and information stored in various systems in real time, the facilitation of cooperation between business partners. The process of such a solution designing requires the elaboration of various modules that fulfil individual tasks constituting the whole process of searching information resources in heterogeneous and distributed resources, e.g.: elaboration of the user query language and the method of queries formulation, establishing the structure of the ontology and methods of its on-line modification, preparing the dictionaries, queries converters and the integrator. In the paper an attempt to describe such a model, its architecture, the process of queries formulation and execution as well as the process of its designing were discussed.

So far selected components of proposed solution have been created. The structure of ontology has been elaborated and tested with the use of some heterogeneous data. Also, the basics for the query language have been prepared. In the future, developing of other components and implementation of the system in some cooperating organisations are planned. Due to the fact that proposed integrating solution implementation does not affect existing and working systems in organisations, the whole process is transparent to them which is safe for existing systems and changing of organisation processes to adjust them to newly prepared solution is unnecessary.

References

1. Asano, Y., Herr, D., Ishihara, Y., Kato, H., Nakano, K., Onizuka, M., Sasaki, Y.: Flexible framework for data integration and update propagation: System aspect. In: 2019 IEEE International Conference on Big Data and Smart Computing (BigComp), pp. 1–5 (2019)
2. Cotfas, L.-A., Delcea, C., Roxin, I., Paun, R.: New trends in intelligent information and database systems, In: Twitter Ontology-Driven Sentiment Analysis, pp. 131–139. Springer International Publishing, Cham (2015)
3. Dittrich, J.-P., Blunschi, L., Farber, M., Girard, O.R., Shant, K.K., Salles, V., Markos, A.: From personal desktops to personal dataspaces: a report on building the iMeMex personal dataspace management system, GI-Fachtagung für Datenbanksysteme in Business, Technologie und Web (BTW), Aachen, Germany (2007)
4. Doan, A., Halevy, A.Y., Ives, Z.G.: Principles of Data Integration. Elsevier Inc. (2012)
5. Dong, X.L., Srivastava, D.: Big data integration. In: ICDE, pp. 1245–1248 (2013)
6. Fernandez, M.F., Florescu, D., Levy, A., Suciu, D.: Declarative specification of Web sites with STRUDEL. VLDB J. **9**(1), 38–55 (2000)
7. Fernandez, R.C., Madden, S.: Termite: a system for tunneling through heterogeneous data. CoRR, abs/1903.05008 (2019)
8. Golshan, B., Halevy, A., Mihaila, G., Tan, W.C.: Data integration: after the teenage years. In: SIGMOD-SIGACT-SIGAI Symposium on Principles of Database Systems, pp. 101–106 (2017)
9. Halevy, A., Rajaraman, A., Ordille, J.: Data integration: the teenage years. In: Proceedings of the 32nd International Conference on Very Large Data Bases, VLDB 2006, pp. 9–16 (2006)
10. Lee, C.-H., Wu, C.-H.: Extracting entities of emergent events from social streams based on a data-cluster slicing approach for ontology engineering. Int. J. Inf. Retrieval Res. **5**(3), 1–18 (2015)
11. Lupiani-Ruiz, E., Garcıa-Manotas, I., Valencia-Garcıa, R., Garcıa-Sanchez, F., Castellanos-Nieves, D., Fernandez-Breis, J.T., Camon-Herrero, J.B.: Financial news semantic search engine. Expert Syst. Appl. **38**(12), 15565–15572 (2011)
12. Maedche, A., Staab, S., Studer, R., Sure-Vetter, Y., Volz, R.: SEAL – tying up information integration and web site management by ontologies, Technical report, Institute AIFB, University of Karlsruhe, Germany (2002)
13. Mellouli, S., Bouslama, F., Akande, A.: An ontology for representing financial headline news. In: Web Semantics: Science, Services and Agents on the World Wide Web, vol. 8, no. 2–3, pp. 203–208 (2010)
14. Novotný, T.: A content-oriented data model for semistructured data. In: Pokorný, J., Snášel, V., Richta, K. (eds.) Proceedings of the Dateso 2007 Workshop, Amphora Research Group, Czech Republic, pp. 55–66 (2007)

15. Offia, C.E., Crowe, M.K.: A theoretical exploration of data management and integration in organisation sectors. Int. J. Database Manag. Syst. (IJDMS) **11**(1), 37–56 (2019)
16. Sardelich, M., Qomariyah, N.N., Kazakov, D.: Integrating time series with social media data in an ontology for the modelling of extreme financial events. White Rose University Consortium, Universities of Leeds, Sheffield & York (2019)
17. Shvaiko, P., Euzenat, J.: Ontology matching: state of the art and future challenges. IEEE Trans. Knowl. Data Eng. **25**, 158–176 (2013)
18. Wang, X., Haas, L.M., Meliou, A.: Explaining data integration. IEEE Data Eng. Bull. **41**, 47–58 (2018)

The Meaning of Solution Space Modelling and Knowledge-Based Product Configurators for Smart Service Systems

Paul Christoph Gembarski(✉)

Institute of Product Development, Leibniz Universität Hannover, Welfengarten 1a, 30167 Hannover, Germany
gembarski@ipeg.uni-hannover.de

Abstract. Smart service systems are understood as assembly of smart products and digital services that represent holistic solutions which are not limited to fulfill a predefined set of customer needs but to adapt to changing requirements over time. Therefore, the system recognizes operating states automatically, learns them if necessary and carries out its reconfiguration autonomously. Existing conceptualizations target at relating service consumer and service provider by operational resources where smart products, which embed communication and information processing technologies, usually take the role of the boundary object. Since a single smart product or a network of smart products must have the ability to adapt itself to the current use case, it has to be understood as instantiation of a solution space that encompasses all valid hard- and software configurations. If a reconfiguration is triggered, the smart product checks for consistency with the solution space. Only if it is approved the reconfiguration is implemented, either by reparametrization of software/control units or by the exchange of physical components. Such interrelationships are already being discussed in the field of knowledge-based product configuration in general as well as in solution space development of mass customization offerings in particular. This article aims at bridging both concepts and integrate solution space modelling and product configuration into the concept of smart service systems.

Keywords: Smart service systems · Product configuration · Solution space modelling

1 Introduction

The ongoing process of digitization leads to new business opportunities in mechanical and plant engineering, fostering new cooperative business models and intensifying the collaboration between customer and supplier [1]. A development from the recent years is the evolution from mechatronic systems to smart products which complement product components by communication and information processing technologies [2].

Going one step further, assembling smart products to a system and adding digital service components, the resulting *smart service systems* represent holistic solutions that are not limited to fulfill a predefined set of customer needs but to adapt to changing

L. Borzemski et al. (Eds.): ISAT 2019, AISC 1050, pp. 28–37, 2020.
https://doi.org/10.1007/978-3-030-30440-9_4

requirements over time [3]. Therefore, the smart service system recognizes operating states automatically, learns them if necessary and carries out its reconfiguration autonomously [4]. This strengthens the role of the underlying information systems which enable smart service systems to effectively support the "human factor" in a technical system and to increase reliability, operational safety as well as sustainability, as can be seen on the example of smart home applications [5].

From a design engineering point of view, three challenges arise with smart service systems [6]: (1) A digital degree-of-freedom has to be introduced into the single product components that is realized by (re-)configurable parameter sets, software of the corresponding control units and information systems that orchestrate the whole system. The according solution space may be restricted by other constituents of a smart service system and its use context. In order to be fail-safe, such constraints must be formalized and validated beforehand. (2) Since the digital degree-of-freedom is anyway limited through the physical characteristics of the product component itself, the exchange of a component has to be triggered if its possibilities are exceeded. The new component should log on the system automatically and offer its resources. This calls for a completely modular architecture that has to be kept stable over the lifecycle of the smart service system. (3) Since the smart service system is not only established by physical but also service components, a multi-perspective development approach is needed, going away from either goods- or service-dominant to a completely customer-centric solution perspective.

Single of these challenges are already tackled in a different context. *Mass Customization* as a business model focusses on delivering individual customer solutions with the cost-effectiveness of mass produced offerings, independently from being a physical product, a service or a product-service system [7]. The success of Mass Customization is achieved with three key capabilities [8]: (1) *Choice navigation* describes the (information system based) support of the customer in the identification and specification of his requirements as well as the guidance through the specification process itself [9–11]. (2) *Solution space development* aims at defining the options and degrees of freedom that the customer determines according to his needs, taking into account boundary conditions of manufacturing and distribution [12, 13]. (3) *Robust process design* focuses on the ability to reuse and recombine organizational units or individual resources in the value chain to enable them to efficiently produce variety [14–16]. A key tool for managing the resulting complexity is knowledge-based product configuration [17–19].

This article aims at conceptually transferring solution space modelling and knowledge-based product configuration into the field of smart service systems and in particular to extend existing conceptualizations of smart service systems. Therefore, after describing the theoretical background in the following Sect. 2, smart product solution spaces are introduced as part of a smart service system in Sect. 3. Afterwards, the concept is applied on two examples in Sect. 4 and discussed in Sect. 5 as well as referred to the three challenges mentioned above. The final Sect. 6 presents a conclusion and further research potential.

2 Theoretical Background

2.1 Smart Service Systems

The terms *smart service* and *smart service system* are originally coined by service science and information system literature. The first is considered as a special type of service that is provided by an intelligent, networked object which is capable of monitoring its own condition and its surrounding, collecting and processing data, communicating and feedback [4, 20]. Individual customers, customer groups or socio-technical systems like a whole company are targeted as service consumer. Related to the feedback of the networked objects, service providers are able to improve their service offerings by rigorous customization [21].

The latter term is understood as special type of service system "in which smart products are boundary-objects that integrate resources and activities of the involved actors for mutual benefit" [5]. The concept is depicted in Fig. 1. On the left side stands the service consumer as user who operates or integrates the smart product and the underlying capabilities of the smart service system. The consumer may participate in monitoring the smart product and benefits from a certain autonomy of the system since it may adapt itself to the current use case. If the smart product has enough intelligence as well as sensors and if the smart product has a well-defined digital degree-of-freedom, this loop may be closed so that no further interaction between consumer and service provider is necessary.

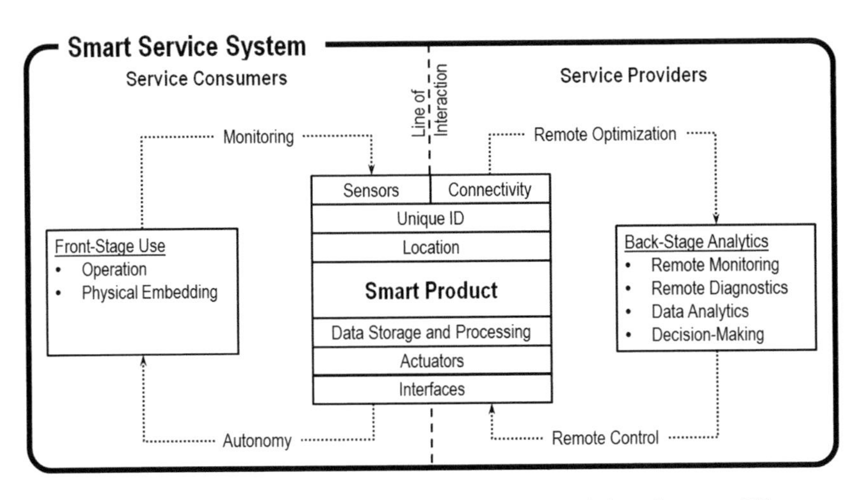

Fig. 1. Conceptualization of smart service systems (adapted acc. to [5])

If customer benefit calls for opening the line of interaction, co-operative and co-creative business models are available where the service provider is allowed to log on the smart product and to remote control or remote optimize the system components [2]. This interaction can be done in two different ways: First, the intervention of the service provider is actively triggered by the consumer. Second, the intervention is part of the committed service and remains hidden in the back-stage [22].

Exemplified on maintenance operations, the first case (without interaction) corresponds e.g. to application of predictive maintenance where the smart product adapts the meantime-between-maintenance to the use behavior of the customer autonomously. The second one (triggered interaction) comes into play when the customer calls for remote support at customer self-service. The third case (back-stage interaction) may be data analytics and the development of the next generation of product components with better performance, tailored for the individual use case of the service consumer.

2.2 Solution Space Development

The definition of all options and degrees-of-freedom in a mass customization offering belongs to the activity of solution space development [8, 9, 23]. As discussed in [24], the solution space itself may be characterized from four different views:

1. **External variety**: The solution space is the set of all existing product variants that either fulfill specific functions or defined requirements.
2. **Internal variety**: Solution spaces are composed by the set of all existing product components and the laws-of-creation, how these are connected to the final product.
3. **Exploration**: The solution space is the search space for an artifact to be developed, which is converged as the development process progresses.
4. **Degrees-of-Freedom**: A Solution space is set up by variation of a known design.

Usually, a solution space is related to a requirement space as a set of all development goals and required product characteristics [25]. In the product development process, the required product properties are compared with the properties of the designed artefact and approximated by synthesis analysis loops. But due to restrictions, not all areas of the solution space are accessible during the development process. Such restrictions originate from various circumstances like manufacturability or logistics, and ensure that a product variant, which follows all of them, is valid [26].

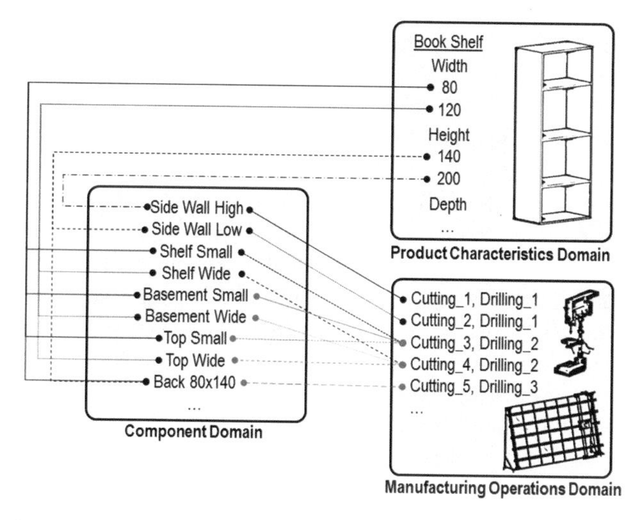

Fig. 2. Combined product and process configuration (adapted acc. to [27])

Another important issue in solution space development is the consideration of the later manufacturing network and its portfolio of capabilities. Here, Aldanondo and Vareilles [27] proposed a domain concept that relates selectable product characteristics, product components as well as their corresponding manufacturing processes (Fig. 2). In turn, a process chain can be assigned resources such as production equipment and processing time. Properties, components and process chains are formulated as a constraint network.

2.3 Product Configurators

In order to make it easier for customers to adapt a product to their individual needs as part of mass customization, many authors see product configuration systems as a necessary building block for the economic success of a corresponding business model (refer e.g. to [8–10, 13]). Generally, two types of configurators have to be distinguished: (1) Sales configurators allow the guided capturing of requirements, their translation into a technical specification and checking for consistency and integrity [9]. (2) Design configurators, also named as CAD configurators or 3D product configurators, use this specification as the basis for the determination of the product variant [17]. Depending on the degrees-of-freedom the customer can choose from, CAD configurators either alter the assembly by replacing components, like e.g. in car configurators or they change the dimensions of a parametric design, e.g. in furniture configuration [9].

Configurators are not only a model of a solution space. They implement techniques from knowledge-based systems and artificial intelligence in order to efficiently explore the solution space and find the best variant for a given set of requirements. As such, different reasoning techniques are available, like rule-based, constrained-based or case-based reasoning [17, 28–30].

3 Smart Service System and Smart Product Solution Space

In the above mentioned concept, smart service systems were introduced as interaction between a service consumer and a service provider where the smart product is the operating resource. This explanatory model is effective for clarifying the concept of smart service systems in general. But in order to explain the control flows and (self-) adaptation of the smart product and to make its portfolio of capabilities visible, the resolution of the model is not sufficient. In order to overcome this, the author proposes to join the concepts of smart service systems and solution space development. First, the smart product is decomposed into a network of components that have multiple, perhaps varying interactions. These components are not static but represent a solution space in which all theoretically valid configurations (regarding hardware and software) are consolidated. Second, a distinction is introduced between active and available components. The extended concept is shown in Fig. 3.

In this model, the smart product contains mandatory elements like its unique ID, sensors, data storage and processing, as well as interfaces to the surrounding world. Since the connectivity of the smart product also belongs to the interfaces and realizes the data reception and transmission, the communication hub is introduced as central

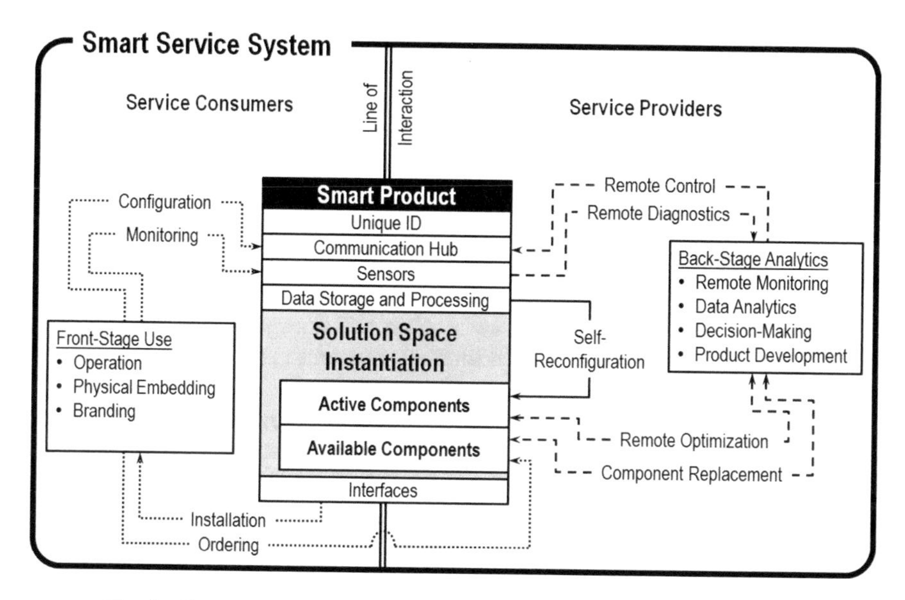

Fig. 3. Smart service system and smart product solution space interaction

interaction platform. This hub manages all access rights, user authentication and contains the (graphical) user interface. Additionally to these elements, the solution space instantiation encompasses all assembled and thus active components as well as information about possible other available ones which may be added or used as replacement for an active component.

The model also shows three different control flows. The first one (dotted line) is the possible interaction between service consumer and smart product. Via interfaces, the consumer is able to install or embed the smart product into his own environment. The sensors allow him to monitor the condition in use, while the communication interface is used to reconfigure the smart product by altering parameters or update software. Internally, the data storage and processing unit takes these inputs and checks for consistency with the solution space. If the new configuration is valid, these changes are sent to the corresponding components, otherwise the reconfiguration is rejected. If the business model allows branding, the communication hub is also used to apply the corresponding user interfaces. Of course, the consumer is able to extend the smart product proactively by ordering additional components.

The second control flow (dashed line) is the interaction between service provider and smart product, which can be classed to four tasks. First, the provider is able to log on the sensors and use the data (plus data from the storage) for remote diagnostics. Second, the access to the communication hub has in principle the same goal as for the consumer, but may additionally be used for remote control. Since the provider (or sub-providers!) may access not only the smart product entirely but also the single active components, he is third capable of remote optimizing and reconfiguring them. The other way round, the component may report about events, i.e. load cases that are relevant for dimensioning, which can be used for product development and improvement of the succeeding component generation. And finally, if the actual solution space instantiation reports limitations of physical components of the smart product, it can trigger component replacement or the provider does this proactively.

The third control cycle (continuous line) is completely internal of the smart product and uses sensed and processed data to self-reconfigure.

4 Application

The model should be able to represent and explain real examples. In the following, two applications are presented, one from business-to-consumer and one from business-to-business domain.

In smart home applications, multiple components like lighting, heating thermostats, ventilation, etc. are connected to a control unit which adjusts the environmental conditions e.g. according to presence of residents (activate corridor lighting when person enters), weather (close windows according to rain forecast) or even use behavior (residents return from work at six p.m., start heating accordingly). Additionally, sensors and surveillance technology detect e.g. a switched-on stove or allow the communication at the front door via smartphone when nobody is inside the building. Especially the last function is enabled by the communication hub in the model above, as well as entering rules and configuring the system to the own feelings, like temperatures etc. New components may be added via the interfaces, as long as their communication protocols and busses match to the communication hub and to the data processing unit. Ideally, a newly installed heating thermostat logs on the hub automatically with its ID and adds its capabilities to the system. In detail, the configuration interface must be extended according to the new component. Its capabilities must be assigned to a room and the reasoning needs information as input variables like e.g. room size etc. for self-adjusting the heating curve. In this scenario, service provider control flows are not necessarily needed but still available.

In an IoT-enabled job shop for autonomous production, single production machines, transport systems etc. are usually orchestrated by an production planning and control system that uses data about the manufacturing operations, lot size etc. From perspective of the service provider, this example is a bit more complex and has to be differentiated with respect to business model and supply chain. E.g., in performance based contracting, the provider may remote control the maximum machine performance of a delivered production machine according to the order situation of the service consumer. In average times, the machine setup is set to "preserving" so that life times of the machine components are maximized. When heavy duty is ordered, the provider reconfigures for more power accepting more wear and shorter component life-times. Additionally, on component level, the provider may remote optimize the components' maintenance intervals based upon sensed load collectives. If maintenance is necessary then, the provider is able to give precise instructions to the service technician, schedule the assignment and deliver all needed spare parts just-in-time. The replacement of components is fed back to the smart product that carries out its reconfiguration autonomously. If the customer is willing to participate in self-service, a corresponding component like e.g. smart glasses is triggered and installed. In case of maintenance, the service provider uses sensor data and the communication hub to assist the technician from remote.

5 Discussion

The extended concept of smart service system is able to represent real applications. A closer attention has to be directed towards the linkage of communication hub and data storage and processing. This combination has to fulfill the following functions:

- List of active components and their IDs
- Knowledge-base for component interaction possibilities
- Reasoning mechanism for consistency-check in case of reconfiguration
- Reasoning mechanism for event-handling
- User-interface for input variables
- User-interface for entering (decision) rules
- User-interface for output of sensor data, including filtering

Most of these functions are also performed by product configuration systems. As introduced above, such systems provide not only the model of the available solution space but also mechanisms for its exploration. As a consequence, the enhanced smart service system concept strengthens the role of the underlying information system as hub for customer interaction and as enabler for smart services. Additionally, it visualizes the three challenges at designing smart service systems that were condensed above:

1. With respect to the digital degree-of freedom, two implications result from the extended model: (1) The information system must incorporate the solution space of each installed component in order to check for consistency as described above. (2) The information system must contain information about all other options and components that may be activated for use in the smart product. Since a (physical) reconfiguration usually involves a change in the control flows, the varying interaction between the active components must be modelled. As in mass customization, the internal processes are more robust the more the solution space is predefined.
2. Regarding modularity: The solution space instantiation calls for modularity the more plug and playability is needed (e.g. if the control flow on side of the service consumer has to be closed loop).
3. Regarding multi-perspective development: The model strengthens the need to design smart products holistically, not only on side of the different domains like product-, service- and software engineering, but also considering the time axis and the future alterations of the smart product.

6 Conclusion

In the present article, the concepts of smart service systems as well as solution space development and product configuration were joined into a common model which is able to represent three different control flows between service consumer, service provider and a smart product as operating resource.

The enhanced model strengthens the role of the involved information systems, namely the communication hub and data storage and processing. As discussed, the necessary functionalities are similar to those of product configuration systems, like checking for consistency, modelling of options and implementing reasoning.

An interesting future extension is the consideration of multiple service providers. In complex smart products like whole production machines, the machine supplier may be only one provider who adjusts and remote supports the equipment on system level. But single components may be maintained by other actors in the supply chain. In this case, the information systems of the smart product might take the role of a coordinating platform for single services and responsibilities of the different service providers as well.

References

1. Porter, M.E., Heppelmann, J.E.: How smart, connected products are transforming competition. Harvard Bus. Rev. **92**(11), 64–88 (2014)
2. Demirkan, H., Bess, C., Spohrer, J., Rayes, A., Allen, D., Moghaddam, Y.: Innovations with smart service systems: analytics, big data, cognitive assistance, and the internet of everything. Commun. Assoc. Inf. Syst. **37**(1), 733–752 (2015)
3. Barile, S., Polese, F.: Smart service systems and viable service systems: applying systems theory to service science. Serv. Sci. **2**(1–2), 21–40 (2010)
4. Byun, J., Park, S.: Development of a self-adapting intelligent system for building energy saving and context-aware smart services. IEEE Trans. Consum. Electron. **57**(1), 90–98 (2011)
5. Beverungen, D., Müller, O., Matzner, M., Mendling, J., vom Brocke, J.: Conceptualizing smart service systems. Electron. Markets **29**(1), 1–12 (2017)
6. Kammler, F., Gembarski, P.C., Brinker, J., Thomas, O., Lachmayer, R.: Kunden-und kontextabhängige Konfiguration Smarter Produkte: Digitales Potenzial jenseits physischer Grenzen? HMD, pp. 1–12 (2019)
7. Boynton, A., Victor, B., Pine, B.J.: New competitive strategies: challenges to organizations and information technology. IBM Syst. J. **32**(1), 40–64 (1993)
8. Fogliatto, F.S., da Silveira, G.J.C., Borenstein, D.: The mass customization decade: an updated review of the literature. Int. J. Prod. Econ. **138**(1), 14–25 (2012)
9. Forza, C., Salvador, F.: Product Information Management for Mass Customization: Connecting Customer, Front-Office and Back-Office for Fast and Efficient Customization. Palgrave Macmillan, Basingstoke (2007)
10. Gembarski, P.C., Lachmayer, R.: Degrees of customization and sales support systems - enablers to sustainability in mass customization. In: Proceedings of the 20th International Conference on Engineering Design (ICED 15) Vol 1: Design for Life, pp. 193–200. The Design Society, Milan (2015)
11. Häubl, G., Trifts, V.: Consumer decision making in online shopping environments: The effects of interactive decision aids. Market. Sci. **19**(1), 4–21 (2000)
12. Salvador, F., De Holan, P.M., Piller, F.: Cracking the code of mass customization. MIT Sloan Manag. Rev. **50**(3), 71–78 (2009)
13. Gembarski, P.C., Lachmayer, R.: Designing customer co-creation: business models and co-design activities. Int. J. Ind. Eng. Manag. **8**(3), 121–130 (2017)
14. Pine, B.J.: Mass Customization: The New Frontier in Business Competition. Harvard Business Press, Boston (1993)

15. Abdelkafi, N.: Variety Induced Complexity in Mass Customization: Concepts and Management. Erich Schmidt, Berlin (2008)
16. Koren, Y.: The Global Manufacturing Revolution: Product-Process-Business Integration and Reconfigurable Systems. Wiley, Hoboken (2010)
17. Hvam, L., Mortensen, N.H., Riis, J.: Product Customization. Springer, Heidelberg (2008)
18. Felfernig, A., Hotz, L., Bagley, C., Tiihonen, J.: Knowledge-Based Configuration: From Research to Business Cases. Morgan Kaufman, Waltham (2014)
19. Gembarski, P.C., Lachmayer, R.: A business typological framework for the management of product complexity. In: Managing Complexity, pp. 235–247. Springer, Heidelberg (2017)
20. Allmendinger, G., Lombreglia, R.: Four strategies for the age of smart services. Harvard Bus. Rev. **83**(10), 131–145 (2005)
21. Wünderlich, N.V., Heinonen, K., Ostrom, A.L., Patricio, L., Sousa, R., Voss, C., Lemmink, J.G.: "Futurizing" smart service: implications for service researchers and managers. J. Serv. Mark. **29**(6/7), 442–447 (2015)
22. Wünderlich, N.V., Wangenheim, F.V., Bitner, M.J.: High tech and high touch: a framework for understanding user attitudes and behaviors related to smart interactive services. J. Serv. Res. **16**(1), 3–20 (2012)
23. Steiner, F.: Solution Space Development for Mass Customization: Impact of Continuous Product Change on Production Ramp-Up. Dr. Kovac, Hamburg (2014)
24. Gembarski, P.C., Lachmayer, R.: Solution space development: conceptual reflections and development of the parameter space matrix as planning tool for geometry-based solution spaces. Int. J. Ind. Eng. Manag. **9**(4), 177–186 (2018)
25. Ponn, J.: Systematisierung des Lösungsraums. In: Handbuch Produktentwicklung, pp. 715–742. Carl Hanser, Munich (2016)
26. Pahl, G., Beitz, W.: Engineering Design: A Systematic Approach. Springer, Heidelberg (2013)
27. Aldanondo, M., Vareilles, E.: Configuration for mass customization: how to extend product configuration towards requirements and process configuration. J. Intell. Manuf. **19**(5), 521–535 (2008)
28. Sabin, D., Weigel, R.: Product configuration frameworks-a survey. IEEE Intell. Syst. **13**(4), 42–49 (1998)
29. Gembarski, P.C., Lachmayer, R.: KBE-modeling techniques in standard CAD-systems: case study – autodesk inventor professional. In: Managing Complexity, pp. 215–233. Springer, Heidelberg (2017)
30. Hopgood, A.A.: Intelligent Systems for Engineers and Scientists. CRC Press, Boca Raton (2012)

An Architecture for Distributed Explorable HMD-Based Virtual Reality Environments

Jakub Flotyński(✉), Anna Englert, Adrian Nowak, and Krzysztof Walczak

Poznań University of Economics and Business,
Niepodległości 10, 61-875 Poznań, Poland
{flotynski,walczak}@kti.ue.poznan.pl, anna.englertt@gmail.com,
adrian.nowak700@gmail.com
https://www.kti.ue.poznan.pl/

Abstract. Interaction is one of the critical elements of VR environments. The use of VR in different domains by domain experts and average users, who are not IT-specialists, requires new approaches to represent interaction in either a domain-specific or a generic way, which is neither directly related to 3D graphics nor animation. In this paper, we propose an architecture for distributed VR environments with an explorable representation of interaction based on the semantic web. The architecture enables creation of logs of interactions between users and objects. Exploration of such logs can be especially useful in VR environments intended to gain knowledge about users' and 3D objects' behavior, e.g., in training and education. The architecture is suitable for lightweight VR devices with limited computational power, such as headsets and smartphones. We present an implementation of the architecture for an immersive service guide for household appliances, which is based on Unity and Samsung Gear VR.

Keywords: Distributed VR · HMD · Semantic web · Ontology

1 Introduction

Virtual reality (VR) environments enable immersive presentation and interaction with complex 3D objects in various application domains, such as training, education, marketing, merchandising, tourism, design, engineering, and medicine. The use of VR in diverse domains has been enabled by the increasing computational power of hardware equipment and network bandwidth as well as the availability of multiple presentation and interaction systems, such as glasses, headsets, haptic interfaces, and motion capture devices. A crucial element of the vast majority of VR environments is interactive animated 3D content. Users and 3D content objects typically interact, which is followed by content evolution over time, introducing changes to objects' geometry, structure, presentation, and high-level semantics, which may be expressed by general or domain-specific knowledge representation.

L. Borzemski et al. (Eds.): ISAT 2019, AISC 1050, pp. 38–47, 2020.
https://doi.org/10.1007/978-3-030-30440-9_5

Various application domains could benefit from logging and further exploration of interactions between users and 3D objects as well as temporal changes of objects' properties affected by the interactions [5]. This would be especially useful in VR environments intended to gain knowledge about users' behavior as well as 3D objects' behavior. For instance, interaction logs collected from virtual guides could be used to train new users and servicemen by explaining the semantics of what is being presented at a particular time in the environment. Second, collected information about activities of customers interacting with products and salesmen in virtual shops could provide knowledge of their interests and preferences for marketing and merchandising purposes. This could be used to present personalized offers to the customers and to arrange spaces in real shops. Third, information about virtual guided tours could be used to join the most interesting ones and address customers' interests in the program of the tours. Next, in urban design, collected information about states of a design process could enable tracking of the evolution of cities. In engineering, collected information about the states and behavior of running machines could be used to analyze how the machines work and to identify their possible faults. Finally, in medicine, collected information about the steps of virtual treatment could be analyzed in the process of teaching students.

Logging interactions can be especially useful in distributed VR environments, which encompass multiple objects and users, thus providing object-to-object, object-to-user and user-to-user interactions. On the one hand, such environments can be successfully built with the use of mobile VR devices, such as wireless headsets with smartphones (e.g., Samsung Gear VR and Google Cardboard) due to their availability and reasonable prices. On the other hand, such devices typically have limited computing power, which is an important obstacle for creating, storing and managing interaction logs, and processing queries to such logs. Nonetheless, exploration of interactions in VR still gains little attention in the research community.

In this paper, we propose a new client-server architecture for distributed explorable VR environments based on lightweight presentation and interaction devices. The architecture permits collecting semantic interaction logs from different clients by a server. The logs are created in a form suitable for further exploration of temporal 3D objects' and users' properties and relationships. Interaction logs leverage a temporal representation of interaction for 3D content, which is based on the semantic web. The logs include information about interactions as well as modifications of 3D content, which typically follow interactions.

The semantic web gets increasing attention in the VR domain. Applications include content description and retrieval [15], design of industrial spaces [13], archaeology [4], and molecular visualization [16,17]. Due to the use of the semantic web in our approach, the collected logs can be expressed with general or domain knowledge, thus being comprehensible to average users and domain experts who are not IT-specialists. Ontologies, which are shared terminological and assertional assets, foster collaborative creation and the use of VR by multiple users. Moreover, ontology-based logs can be processed with standard reasoning engines to infer implicit (tacit) content properties on the basis of explicit properties.

Finally, the logs can be queried by users and services for temporal (past and current) 3D content properties at different moments and periods of time.

The remainder of this paper is structured as follows. Section 2 provides an overview of the current state of the art in semantic representation and modeling of behavior in VR, with the focus on animation and interaction. Next, we outline our approach in Sect. 3. The proposed architecture of explorable VR environments is presented in Sect. 4. It is followed by the description of an immersive service guide that illustrates our approach in Sect. 5. Finally, Sect. 6 concludes the paper and indicates possible future research.

2 Related Work

3D content behavior typically consists of two main elements: interactions between 3D objects and users, and animations which are triggered by such interactions. Multiple 3D modeling tools and game engines permit implementation of interactions and animations, e.g., Blender, 3ds Max, Motion Builder and Unity. The available environments cover various features of 3D content, such as geometry, structure, and appearance, all of which may be animated. The tools permit scripting in different programming languages (e.g., Python and C#) as well as the use of state diagrams and key frames with interpolation of content properties. The environments also support numerous 3D content representation formats, e.g., FBX, OBJ, VRML and X3D. However, the available formats do not enable 3D content exploration with queries.

Several solutions have been devised for representing metadata and semantics of interactions and animations. In [1], an approach to describing metadata for interaction of 3D content has been presented. It is based on XSD schemes and covers events, conditions, and actions. A number of approaches enable modeling of 3D content behavior with ontologies and the semantic web standards. The approach proposed in [2] uses ontologies to build multi-user virtual environments and avatars. Stress is put on representing the geometry, space, animation and behavior of 3D content. In [12], temporal operators are used to design primitive and complex objects' behavior. In addition, upon this solution, a graphical behavior modeling tool has been implemented, which allows for implementation of complex behavior using diagrams. The implemented behavior can be exported to X3D scenes. In [3], primitive actions (move, turn, rotate, etc.) can be combined to represent complex behavior in a way comprehensible to users who are not IT-specialists.

The ontology proposed in [18] represents animations using key frames. The ontology of virtual humans [10] consists of geometrical descriptors (describing vertices and polygons), structural descriptors (describing levels of articulations), 3D animations of face and body and behavior controllers (animation algorithms).

The simulation environment presented in [11] combines a few tools. The Unreal game engine enables rigid-body physics and content presentation. An inference engine enables reasoning and updating the scene representation when events occur. A behavioral engine enables action recognition and changes of conceptual objects' properties. The use of different engines within one environment permits division of responsibilities between users with different expertise.

An approach to spatio-temporal reasoning on ontology-based representations of evolving human embryo has been proposed in [14]. The ontlologies are encoded in RDF [20] and OWL [21], and they describe stages, periods and processes. Representation of 3D molecular models with ontologies has been proposed in [17]. The approach combines different input (e.g., interaction using various haptic and motion tracking devices) and output (e.g., 2D and 3D presentation) modalities to enable presentation and interaction suitable for different kinds of devices, content and tasks to be performed.

The approach proposed in [8,22] uses ontologies to represent 3D content at different abstraction levels: specific to 3D modeling and specific to an application domain. Furthermore, it leverages reasoning to liberate content authors from determining all content details. Finally, it enables semantic queries to generate particular 3D scenes from generalized 3D templates (meta-scenes).

An extensive review of the state of the art in ontology-based modeling and representation of 3D content has been presented in [9]. So far, semantic representation of animations has gained more attention than semantic representation of interactions. However, the available approaches in both areas are still preliminary and lack solutions of multiple important problems, in particular—content exploration.

3 Overview of the Approach

The overall concept of explorable VR environments is depicted in Fig. 1. In such environments, *explorable 3D content* is being generated and used. Interactions between users and 3D content objects are logged (registered) to enable further exploration and visualization of what happened during a particular session of interaction with the environment. To achieve this goal, two documents are being generated, while interactive 3D content is being used:

1. *Interaction log*, which is an ontology that includes statements on interactions that occurred during the session and the results of the interactions. Interaction logs enable semantic query-based exploration of what happened during the session. Interaction logs are encoded in the semantic web standards (RDF [20], RDFS [19] and OWL [21]).
2. *Animation descriptor*, which encompasses values of content properties influenced by interactions, including animated properties. Animation descriptors enable visualization of what happened during the session. The described animations are consistent with the appropriate interaction logs in terms of the moments and the periods of time, when particular events happen. Animation descriptors may be encoded in OWL or other formats, also specific to 3D graphics.

Since the animation descriptor covers gradual changes of the 3D content during a session, and the interaction log covers the semantics of the changes, the session can be restored at both the visual and the semantic levels. The animation descriptor is processed by the explorable VR environment to present

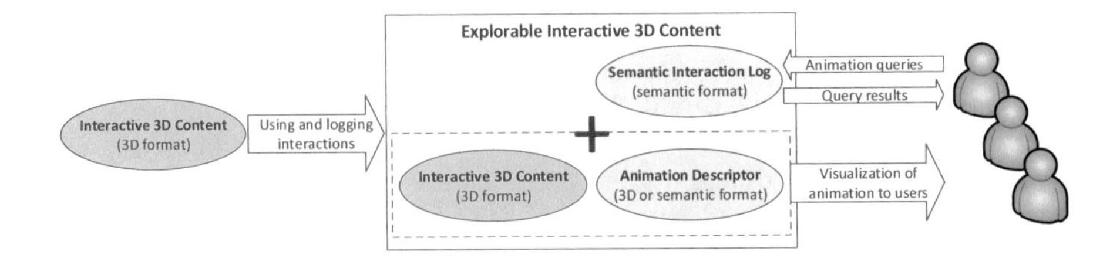

Fig. 1. The concept of semantic 3D content in explorable VR environments.

the 3D content as it was recorded during the session, e.g., servicing household appliances step-by-step. The interaction log is used to query about changes of the 3D content during the session using domain-specific terms. For example, the query: what follows attachment of an indicator to the induction hub at a specific moment, generates the response: getting voltage 200 V at the indicator. The combination of the interactive 3D content used during a session with the generated interaction log and the animation descriptor is referred to as *explorable interactive 3D content.*

4 Architecture of Explorable VR Environments

The main contribution of this paper is an architecture for *distributed explorable virtual reality environments* based on head mounted displays (Fig. 2). The architecture is an extension of the approach presented in [5] with multiple lightweight devices. It uses the interaction representation and the log generation algorithm proposed in [6,7]. A distributed explorable VR environment consists of two parts: HMD-based Clients and the Semantic Server.

1. *HMD-based Clients* enable presentation of 3D content to users and interaction with the content. Every client consists of three elements.
   (a) A head mounted display (HMD), which visualizes 3D content to a user.
   (b) A controller, which enables interaction between the user and 3D content. Controllers may be either separate devices or built-in in the HMD.
   (c) A smartphone or a computer, which renders 3D content and communicates with the *Semantic Server* to store and retrieve interaction logs. Smartphones or computers are selected depending on the HMD used. For instance, Google Cardboard and Samsung Gear VR work with smartphones, whereas HTC Vive and Oculus Rift work with computers.
2. *Semantic Server* creates, stores, manages and provides interaction logs to clients. The Semantic Server consists of several components.
   (a) The *Log Service*, which is a RESTful web service used by clients to create, extend and read interaction logs. Every time a user starts using the system, the Log Service is requested to create a new interaction log for the new customized 3D animation being created. Every time, an interaction between the user and 3D content or between 3D content objects occurs, the service is requested to extend the log with new statements about the

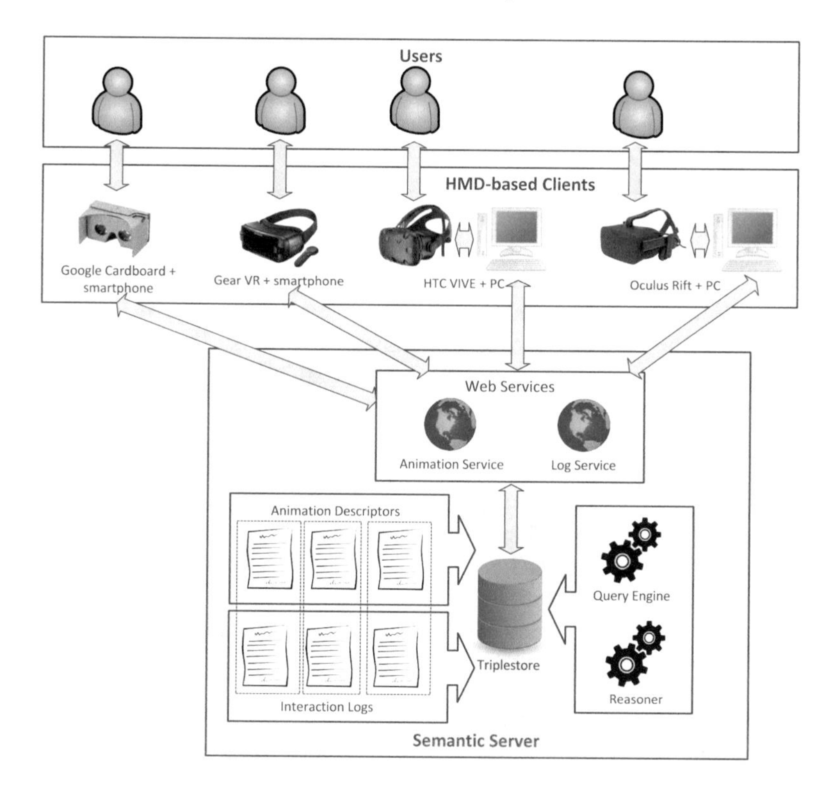

Fig. 2. Architecture for distributed explorable HMD-based VR environments.

user or the content. Only selected interactions are reported in interaction logs, as explained in [6]. Every time a client requests information about the semantics of interactions between users and 3D objects that occurred in a recorded customized 3D animation, the service translates the HTTP request to a SPARQL query, sends the query to the *Triplestore*, and delivers the obtained response to the client.

(b) The *Animation Service*, which is a RESTful web service used by clients to create, extend and retrieve animation descriptors. Every time a user starts using the system, the Animation Service is requested to create a new animation descriptor for the new customized 3D animation being created. Every time, an interaction changes properties of the content, the service is requested to extend the descriptor with the new content properties. Every time, a customized 3D animation is required, the service retrieves the appropriate descriptor from the *Triplestore* and delivers it to the client.

(c) The *Triplestore*, which stores interaction logs and animation descriptors, processes SPARQL queries to logs, which are sent by the Log Service, and performs reasoning. The Triplestore includes an OWL reasoner, which performs reasoning on queries and target logs. The results of reasoning are sent back to the Log Service, which further conveys them to the clients.

5 Distributed Immersive Service Guide

An immersive HMD-based service guide for an induction hob has been developed. The guide presents several scenarios for the repair of the most common defects that can occur while using the appliance. In contrast to a typical paper guide, HMDs enable intuitive user-friendly interactive presentation of the hob at a high level of detail that is appropriate for step-wise training how defects should be repaired. The immersive guide permits to look at different parts of the hob, zoom them in and out, and watch animations that present with high accuracy the activities to be performed. The system can be used to train technicians and home users. Mistakes done when servicing a virtual hob are not followed by costs unlike when using real devices for training.

The main HMD used is Samsung Gear VR with the Galaxy S6 smartphone, which is now a low-cost platform enabling presentation and interaction with reasonable quality. The use of smartphones is enabled by delegating time-consuming computation to the server, which processes queries and performs semantic reasoning. The use of inexpensive hardware makes the platform attainable to small distributors of household equipment as well as average users. The guide is a Unity3D based application that implements the following scenarios of repairing an induction hob. For every scenario, the guide presents interactive animated 3D models as well as movies.

1. Checking the power connection (Fig. 3). This scenario represents one of the most common failures. It includes an animation of measuring power connection, by identifying appropriate terminals and demonstrating how they should be connected.
2. Checking coils and transistors. In this scenario, a user verifies the state of coils and transistors (Fig. 4), which are crucial for the correct work of the hob.
3. Checking the overall work of the hob. This scenario allows a user to verify the entire process of switching on the hob and cooking with it. The user is notified about possible problems that may occur and proper solutions to the problems.

Examples of queries to interaction logs created during sessions of using the guide are:

1. At what time is connecting of terminals presented in the demonstration?
2. What tools are used to check coils and transistors, and at what time in the demonstration is it done?
3. What activity directly precedes cooking with the hob?

Fig. 3. Checking the power connection of an induction hob.

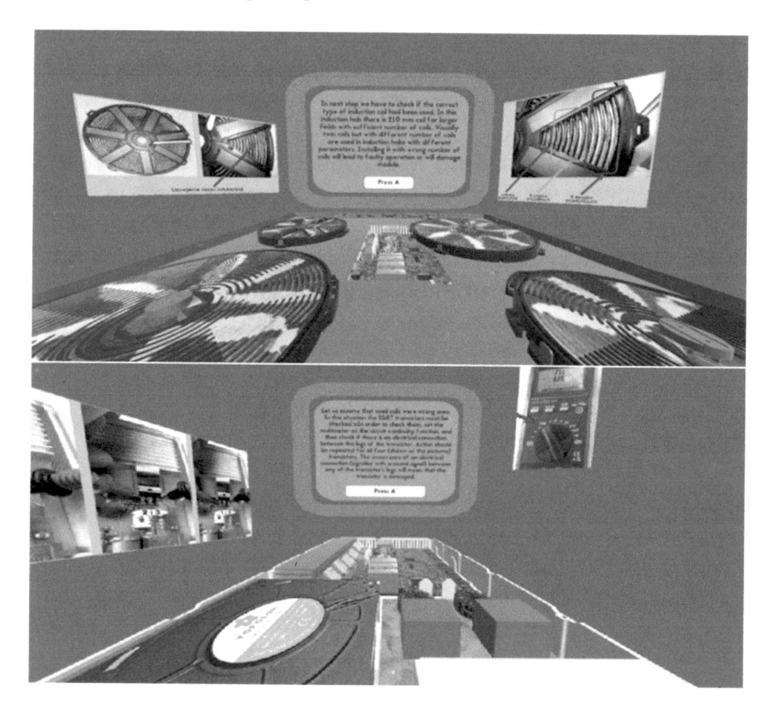

Fig. 4. Checking coils and transistors of an induction hob.

6 Conclusions and Future Work

The use of semantics for building VR/AR applications gains increasing attention in the research community. However, the available methods and tools do not enable creation of temporal representations of interactive animated 3D content that could be explored on-demand with queries constructed by content users. The use of VR and AR in different domains requires new methods and tools of 3D content creation and analysis that are intelligible to average users and domain experts who are not IT-specialists. The development of the current web towards the semantic web offers new technologies that can be used to solve this problem.

In this paper, we have proposed an architecture for distributed explorable VR environments based on lightweight interaction and presentation devices. The primary type of devices used are HMDs equipped with smartphones, which constraints possible solutions to those oriented on centralized computing. The approach is based on the semantic web standards and offers several advantages over the previous solutions. First, it enables on-demand query-based exploration of semantically described animated interactive 3D content, including its past and current states. Second, 3D content can be represented using concepts related to an arbitrary domain, thus being intelligible to users without technical skills. Third, the approach permits inference of tacit knowledge through reasoning. Finally, due to the use of ontologies, which provide common terminology, the approach is suitable for collaborative modeling and exploration of 3D content. This opens new opportunities to develop network- and web-based tools for content creation.

The possible directions of future research encompass several elements. First, the architecture can be evaluated in terms of the performance of logging interactions in different configurations of the distributed VR environment. Another possible evaluation covers the performance of reasoning and query execution as well as the size and complexity of the generated interaction logs and the queries. Also, the environment can be tested in terms of network latency when multiple collaborating client devices are used.

Acknowledgments. We would like to thank Amica S.A. for supporting the project and providing 3D content models.

References

1. Chmielewski, J.: Describing interactivity of 3D content. In: Cellary, W., Walczak, K. (eds.) Interactive 3D Multimedia Content, pp. 195–221. Springer, London (2012)
2. Chu, Y., Li, T.: Realizing semantic virtual environments with ontology and pluggable procedures. In: Applications of Virtual Reality (2012)
3. De Troyer, O., Kleinermann, F., Pellens, B., Bille, W.: Conceptual modeling for virtual reality. In: Grundy, J., Hartmann, S., Laender, A.H.F., Maciaszek, L., Roddick, J.F. (eds.) Tutorials, Posters, Panels and Industrial Contributions at the 26th International Conference on Conceptual Modeling - ER 2007, CRPIT, Auckland, New Zealand, vol. 83, pp. 3–18 (2007)
4. Drap, P., Papini, O., Sourisseau, J.C., Gambin, T.: Ontology-based photogrammetric survey in underwater archaeology. In: European Semantic Web Conference, pp. 3–6. Springer (2017)
5. Flotyński, J., Krzyszkowski, M., Walczak, K.: Semantic composition of 3D content behavior for explorable virtual reality applications. In: Proceedings of EuroVR 2017, Lecture Notes in Computer Science, pp. 3–23. Springer (2017)
6. Flotyński, J., Nowak, A., Walczak, K.: Explorable representation of interaction in VR/AR environments. In: Proceedings of AVR 2018, Lecture Notes in Computer Science, pp. 589–609. Springer (2018)

7. Flotyński, J., Sobociński, P.: Logging Interactions in explorable immersive VR/AR applications. In: 2018 International Conference on 3D Immersion (IC3D), Brussels, 5–6 December 2018, pp. 1–8. IEEE (2018)
8. Flotyński, J., Walczak, K.: Multi-platform semantic representation of interactive 3D content. In: Proceedings of the 5th Doctoral Conference on Computing, Electrical and Industrial Systems, 7–9 April, Lisbon, Portugal, 7–9 April 2014
9. Flotyński, J., Walczak, K.: Ontology-based representation and modelling of synthetic 3D content: a state-of-the-art review. Comput. Graph. Forum, 1–25 (2017)
10. Gutiérrez, M., García-Rojas, A., Thalmann, D., Vexo, F., Moccozet, L., Magnenat-Thalmann, N., Mortara, M., Spagnuolo, M.: An ontology of virtual humans: incorporating semantics into human shapes. Vis. Comput. **23**(3), 207–218 (2007)
11. Lugrin, J.L.: Alternative reality and causality in virtual environments. Ph.D. thesis, University of Teesside, Middlesbrough, United Kingdom (2009)
12. Pellens, B., Kleinermann, F., De Troyer, O.: A development environment using behavior patterns to facilitate building 3D/VR applications. In: Proceedings of the 6th Australasian Conference on International Entertainment, IE 2009, pp. 8:1–8:8, ACM (2009)
13. Perez-Gallardo, Y., Cuadrado, J.L.L., Crespo, Á.G., de Jesús, C.G.: GEODIM: a semantic model-based system for 3D recognition of industrial scenes. In: Current Trends on Knowledge-Based Systems, pp. 137–159. Springer (2017)
14. Rabattu, P.Y., Massé, B., Ulliana, F., Rousset, M.C., Rohmer, D., Léon, J.C., Palombi, O.: My corporis fabrica embryo: an ontology-based 3D spatio-temporal modeling of human embryo development. J. Biomed. Semant. **6**(1), 36 (2015). https://doi.org/10.1186/s13326-015-0034-0
15. Sikos, L.F.: Description Logics in Multimedia Reasoning. Springer, Cham (2017)
16. Trellet, M., Férey, N., Flotyński, J., Baaden, M., Bourdot, P.: Semantics for an integrative and immersive pipeline combining visualization and analysis of molecular data. J. Integr. Bioinf. **15**(2), 1–19 (2018)
17. Trellet, M., Ferey, N., Baaden, M., Bourdot, P.: Interactive visual analytics of molecular data in immersive environments via a semantic definition of the content and the context. In: Immersive Analytics, 2016 Workshop, pp. 48–53. IEEE (2016)
18. Vasilakis, G., García-Rojas, A., Papaleo, L., Catalano, C.E., Robbiano, F., Spagnuolo, M., Vavalis, M., Pitikakis, M.: Knowledge-based representation of 3D Media. Int. J. Softw. Eng. Knowl. Eng. **20**(5), 739–760 (2010)
19. W3C: RDFS (2000). http://www.w3.org/TR/2000/CR-rdf-schema-20000327/
20. W3C: RDF (2004). http://www.w3.org/TR/2004/REC-rdf-concepts-20040210/
21. W3C: OWL (2012). http://www.w3.org/2001/sw/wiki/OWL
22. Walczak, K., Flotyński, J.: On-demand generation of 3D content based on semantic meta-scenes. In: Lecture Notes in Computer Science; Augmented and Virtual Reality; First International Conference, AVR 2014, Lecce, Italy, 17–20 September 2014, pp. 313–332. Springer (2014)

Evolutionary Approach Based on the Ising Model to Analyze Changes in the Structure of the IT Networks

Andrzej Paszkiewicz[1(✉)] and Kamil Iwaniec[2]

[1] The Faculty of Electrical and Computer Engineering, Department of Complex Systems, Rzeszow University of Technology, Rzeszow, Poland
andrzejp@prz.edu.pl
[2] Ropczyce, Poland

Abstract. The paper presents a solution based on the Ising model considering the long-term nature of changes taking place in the network infrastructure. It takes into account the individual nature of separate nodes expressed through the parameter of their susceptibility to possible changes. This parameter directly affects the conviction of a given node about the need to change the state in which it is currently to another one, dominating in its environment. Thanks to the introduced modification of the algorithm, modeled processes occurring in IT networks are more peaceful and changes have an evolutionary character.

Keywords: Ising model · Complex networks · Social networks · Computer networks · IoT · IoE · Phase transitions

1 Introduction

IT networks are the basis for the exchange of information between people, systems and devices. Their proper design and modeling allow to increase the efficiency of their operation and can contribute to the understanding of processes occurring in them. Currently, networks of this type include in their functioning both small environments (e.g. production, IoT, infrastructure of Industry 4.0), as well as global communication (e.g. Internet, social networks, IoE etc.). Therefore, the phenomena occurring in them have the features characteristic of complex systems [1] such as, for example, self-similarity, feedback, power law, additivity, self-adaptation, phase transitions, etc. [2–8]. Of course, at present, it is not possible to take into account all these features. However, one should try to include at least some of these features in created models and algorithms.

Currently, there are many methods and means to support the design and modeling of network structures [9–11]. However, few of them take into account the dynamics of processes occurring in them, which is characteristic of complex systems. Usually, these are methods based on creating a static network structure based on basic, initial information. This is how the physical topology or logical structure is created. However, this approach does not reflect the actual relationships that occur, for example, in social networks, infrastructure of Industry 4.0 and increasingly also in the so-called SDN (Software Defined Networking) networks [12]. The model that is suitable for analyzing

L. Borzemski et al. (Eds.): ISAT 2019, AISC 1050, pp. 48–57, 2020.
https://doi.org/10.1007/978-3-030-30440-9_6

the behavior of IT networks, and in particular social networks, is the Ising model [13]. It was originally used to model ferromagnetic structures. However, over time, it has found applications in other areas of science and engineering [14]. The foundation of its operation is based on the principle that individual nodes take on the state that dominates in their environment. Such an approach enable modeling phenomena related to the creation of opinions in social networks, the spread of viruses or the creation of network congestion. Thus, the classical algorithm based on this model takes into account only the current states of individual nodes, i.e. it is adequate for simple systems. On the other hand, in the case of complex systems, the process of evolution itself should be taken into account, as well as reaching the final decision on changing the state of a given node. Therefore, this paper proposes a modified Ising model, which takes into account the phenomenon of feedback and individual properties of individual nodes when changing their state/opinion.

2 Ising Model

The Ising model allows analyzing the dynamics of changes in the network structure. This model was originally used to analyze behavior in ferromagnetic systems. Currently, it is also used in the area of IT networks. Its basic assumptions determine the principle of changing the state of a given node under the influence of the states of nodes located in its vicinity. Typically, the speed of these changes depends on the degree of domination of a given state in the environment of the considered node. The standard approach assumes that even a small advantage of one of the acceptable states can cause a change in the state of a given node. The introduced threshold model assumes that a possible change of a state may take place only after exceeding the assumed threshold value [15]. This limitation slows down the process of changing individual states and secures a given network structure (system) against possible long-lasting unstable work.

The solution proposed in this paper is aimed at taking into account the factor which determines the individual character of each of the network nodes. Such a factor, presented in the form of a conviction parameter, decides about the readiness of individual nodes to accept the change of their state (or e.g. opinion) under the influence of the state prevailing in its environment. Such an assumption corresponds to real circumstances, in which e.g. different people in a social network need different time to convince themselves to change their opinion to one that dominates in the surrounding environment. Similarly, in an IT network infrastructure, different systems have different levels of security that must be exceeded before a given system is affected by e.g. malware. The concept of this approach is shown in Fig. 1.

In order for the algorithm based on the Ising model to take into account the individual parameter of the changeability of the state, its value should be defined. Thus, the *node's susceptibility* is $P(i) = \{p_j \in \langle 1, \ldots, M \rangle; M \in N\}$. This parameter will directly affect the conviction parameter $C(i)$ node i to change its state. After introducing both parameters, the algorithm can be modified to:

1. Defining the considered population and relations between individual nodes (individuals).

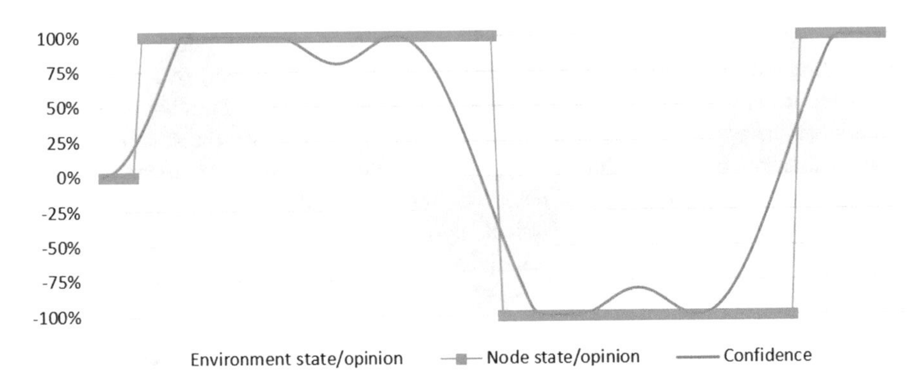

Fig. 1. The idea of changing the node states (opinions) depending on the distribution of states (opinions) in the environment for evaluation Ising model

2. Random generation of status/opinion among the studied population.
3. Counting reviews for each node.
4. Calculation of the current conviction based on its prevailing value and current values for particular considered states/opinions in its vicinity.
5. Changing the state of a given node when the conviction reaches a certain level.
6. Repeating steps 3–5 until the network structure stabilizes.

According to the above algorithm, the considered network will stabilize when no more changes take place in it. As a result, the process of changing states (opinions) will be more gentle. This is shown in Fig. 2, where the example change characteristics for the standard algorithm and for individual conviction are compared. As can be seen in the presented schemes, the system stabilizes much faster in the case of the algorithm based on the standard Ising model (Fig. 2a). This is slower in the case of the evolutionary model (Fig. 2b), in which the nodes need much more time to gain an inner conviction about the necessity of changing their state to the dominant state in their environment. In addition, it can be seen that the classical algorithm has greed characteristics, i.e. some states begin to dominate in a given network infrastructure, causing even the elimination of some initial states. The process itself in the case of the evolutionary algorithm is much more gentle and contributes to a more even distribution of individual states in the network. The threshold approach has been presented in detail in the publication [15] stating that the number of changes is much smaller than in the case of the classical algorithm. Therefore, it is not explained in more detail in this paper.

According to the algorithm presented above, the decision to convert the state of a given node takes place based on the change in conviction of a given node on the grounds of the following expression:

$$C(i) = C'(i) - \frac{\left(\sum_{j \neq i} S_j^k(i) - \sum_{j \neq i} S_j^i(i)\right)}{\sum_j n_j(i)} \cdot P(i), \tag{1}$$

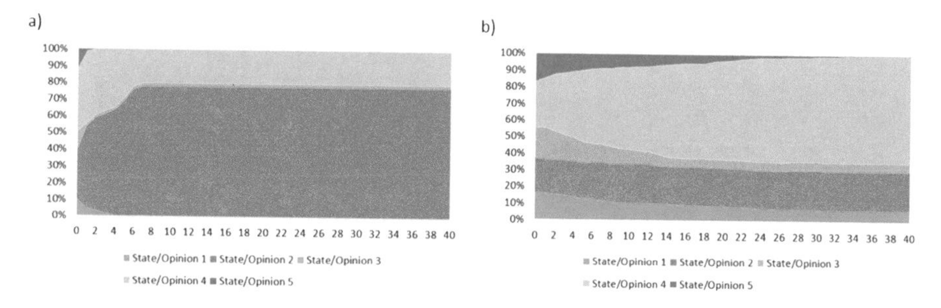

Fig. 2. The course of changes in the distribution of states in subsequent simulation steps: (a) standard Ising algorithm (for 5 initial states); (b) evolutional Ising algorithm (for 5 initial states).

where:

$C'(i)$ – the previous conviction value of the i node,
$P(i)$ – susceptibility to convince the i node,
$\sum_{j \neq i} S_j^k(i)$ – the number of all nodes that are in the node's i environment and are in k state, and this state is different from the current state of i node,
$\sum_{i \neq j} S_j^i(i)$ – the number of all nodes that are in the node's i environment and those that are in the same state that the i node,
$n_j(i)$ – the neighbor of the i node.

The appropriate action expressed in the formula (1) means that the conviction to stay in the current state is gradually reduced in case the given diverse state (opinion) prevails in the environment of the given node. Conversely, the conviction in staying in the current state (opinion) increases when the given diverse state (opinion) is in the minority around the given node in relation to the current state of the node.

Assuming that the susceptibility to changes of a given node can change within the range from 0 to 100%, Fig. 3 presents an example of change in the conviction of a given node for different values of the susceptibility parameter: 100%, 50%, 30%, 10%. Of course, the same distribution of 5 different states (opinions) was assumed in the analyzed network structure in the vicinity of the examined node. At the same time, the node at the time presented in Fig. 3 of the analysis was in the state 1. In addition, the distribution of states in the vicinity of the tested node was as follows: for Fig. 3a: state 1–10 nodes, state 2–25 nodes, state 3–8 nodes, state 4–35 nodes, state 5–22 nodes; for Fig. 3b: state 1–25 nodes, state 2–10 nodes, state 3–15 nodes, state 4–18 nodes, state 5–32 nodes.

In the first case, there is a situation in which the number of nodes with different states from the current state for the tested node prevails in the network infrastructure (Fig. 3a). Then, the value of the conviction parameter of the given node to stay in the current state decreases. Only in step 3 state no. 3 covers less nodes in the vicinity of the analyzed node. Thus, the value of conviction temporarily increases slightly. Another example is demonstrated in Fig. 3b. There, the number of nodes with different states

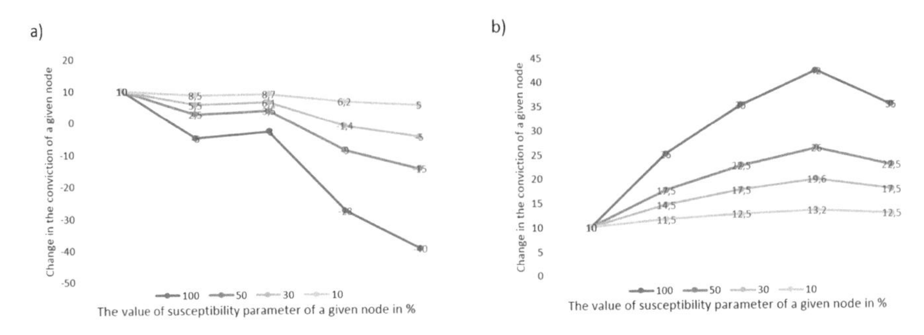

Fig. 3. (a) The first example; (b) The second example

from the current state for the tested node is smaller. This increases the conviction of a given node to stay in its current state. Only in step 5 considered state No. 5 causes a slight decrease in the conviction value of a given node to stay in the current state. Of course, the rate of decrease or increase of this value depends on the susceptibility of a given node to changes and the number of nodes with a different state than the current state of a given node. Analyzing the above results, one can notice how the conviction of the analyzed node evolves under the influence of the surrounding environment.

Taking into account the above relationships, the point at which the final decision about the change of state by a given node should be made:

$$C(i) < 0. \tag{2}$$

According to the expression (2), the given node and in the network changes state from $S^k(i)$ for $S^l(i)$ in the moment, when the susceptibility parameter drops below the value of 0. The initial value of this parameter can be shaped in any way depending on the characteristics of the modeled environment, and in particular the nodes forming them.

Figure 2 presents the course of state changes for 5 different initial states occurring in the network structure. In order to better illustrate the difference between the operation of an algorithm based on a standard model and the algorithm working on the basis of the evolutionary model in Fig. 4, its operation is shown for exemplary distributions when the number of states is equal to 8 and 10. For more states, the charts would be unreadable. Of course, each course has an individual character and is dependent not only on the number of initial states, but also the number of neighbors of each node and the characteristics of the initial distribution of individual states around each node. Nevertheless, Figs. 4a and c reveal the aforementioned feature of the classical approach, which causes the appropriation of structure by some of the states, proceeds in a rapid manner and can be described as violent. This type of changes, however, has few processes occurring in the IT network structures. Most of them, such as changing opinions in social networks, spreading in an evolutionary manner depend on the individual character and vulnerability of individual nodes. This process better illustrates the course of the evolutionary algorithm (Figs. 4b and d).

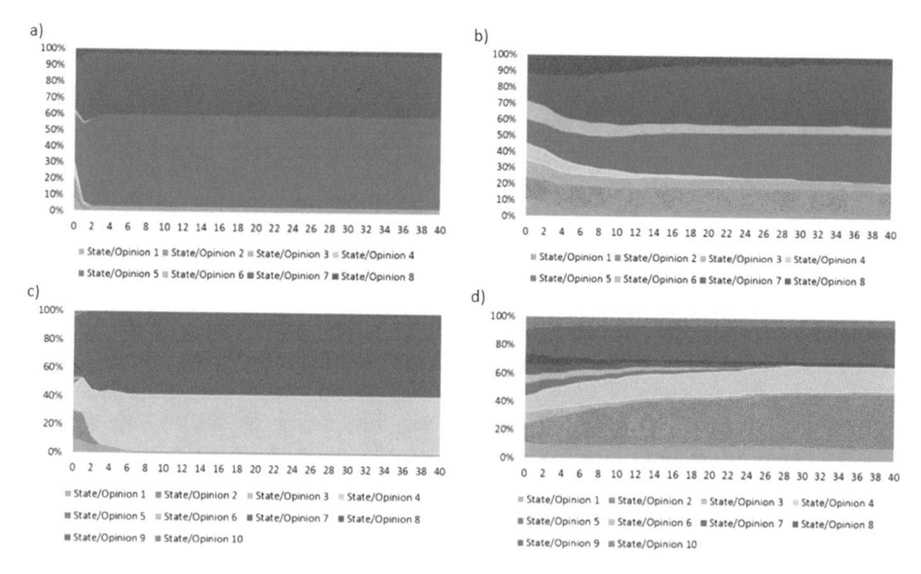

Fig. 4. The course of changes in the distribution of states in subsequent simulation steps: (a) standard Ising algorithm (for 8 initial states); (b) evolutional Ising algorithm (for 8 initial states); (c) standard Ising algorithm (for 10 initial states); (d) evolutional Ising algorithm (for 10 initial states).

3 Analysis of the Dynamics of Changes in the Network Structure

Network structures, especially IT networks (including social networks, computer networks, IoT etc.), are characterized by a large dynamics of changes taking place in them. This applies to both network traffic, relationships in social networks, or the structure of connections in wireless networks, e.g. Ad-Hoc. The study of this type of phenomena can be implemented in various ways. For example, the analysis of network traffic can be carried out on the basis of statistical parameters or the analysis of the Hurst coefficient [16]. In the case of the dynamics of changes in the connection structure, changes in node degrees, network coherence, or the betweenness parameter can be studied. However, for the purposes of this paper the order measure was used [17]. This parameter has one of the special advantages, namely the ability to determine also the phase in which the system is located, in this case the IT network. The paper [15] presents its application in the case of the classic Ising algorithm. Thanks to this, it was possible to analyze the phase change process taking place in such a network infrastructure. Now, according to the evolutionary approach to the Ising algorithm, it will be possible to model long-term processes characteristic of complex systems. The process of evolutionary change itself can contribute to the network structure remaining in a long-term unstable state. Therefore, the period between the phase can be much longer than in the classical approach.

The ordering parameter has been described using the theorem:

$$M = \sum_{S=1}^{q} \left| N_S - \frac{N}{q} \right|, \tag{3}$$

where: N is the number of all nodes in the network structure, N_S is the number of nodes being in the S state, and q is the number of all states in the network. Of course, for a special case where $q = 2$. Eq. (3) can be reduced to a form $|N_1 - N_2|$.

This parameter can be considered in two cases. In the first, when during the whole simulation, we take into account all the initial states present in a given network structure and in the second case, when we take into account only the currently available states at individual stages of the simulation. As a result of the tests and simulations carried out, the indicated difference has a significant impact on the results obtained, and thus the analysis of processes taking place in the network structure. For a better understanding of its meaning, Fig. 5 presents the results of simulations for the classic Ising algorithm.

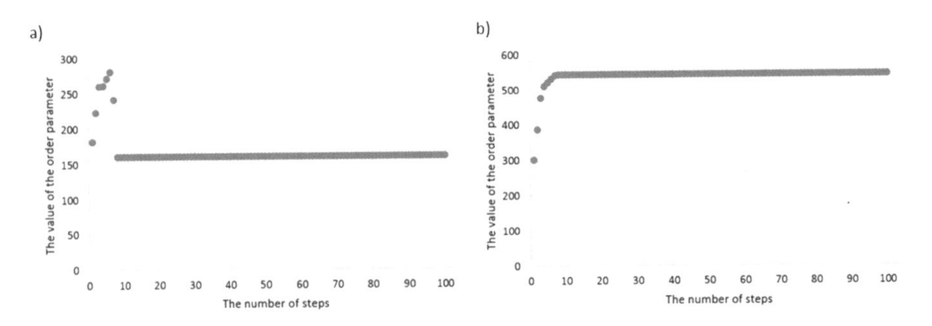

Fig. 5. The values of the order parameter for the classic Ising algorithm: (a) taking into account only existing states in the network; (b) taking into account all initial network states throughout the whole simulation period.

Figure 5a represents a variant in which only currently existing states are taken into account in the next simulation steps. In contrast, Fig. 5b presents an approach in which all initial states are always included in the calculation of the order parameter value. The difference is visible, i.e. in the first case, the value of the order parameter increases, and then after the transitional period, when the network goes into stable operation, this value decreases. In the second case, the parameter value increases to the maximum value, after which the network also goes into an ordered state. However, from the perspective of the phase transitions analysis, both results are similar, i.e. the phase change process is visible, i.e. the system transition from the unsteady state (phase) to the stable state (phase).

Figure 6 refers to the analysis of the order parameter value for the operation of the proposed Ising evolutionary algorithm. At the same time, Fig. 6a, c and e present a situation in which only existing states in a given network infrastructure are taken into

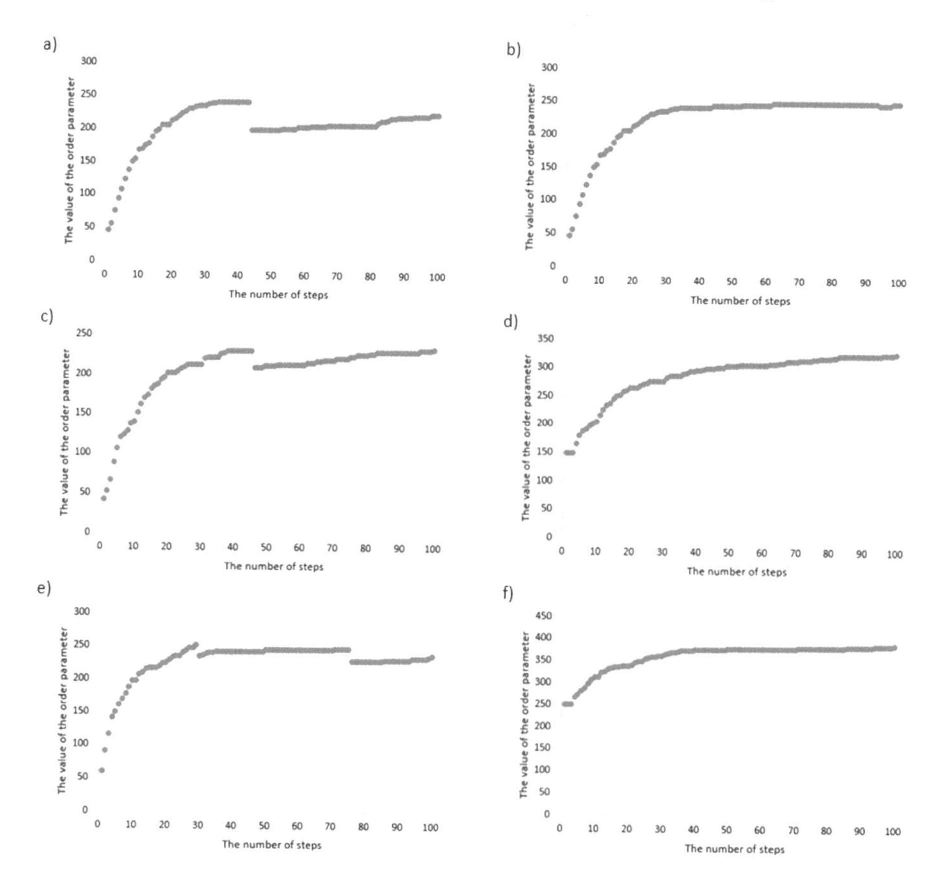

Fig. 6. The values of the ordering parameter for the evolutionary Ising algorithm - including only existing states: (a) 5 initial states, (c) 8 initial states, (e) 10 initial states; taking into account all initial states of the network throughout the simulation period: (b) 5 initial states, (d) 8 initial states, (f) 10 initial states

account, and in Fig. 6b, d and f present the variant when all initial states are included in each simulation step. Considering the operation characteristics of this algorithm, the ordering state can be achieved much later than in the case of a classic or threshold algorithm. Thus, the network may be permanently in an unstable state. Of course, the phase change process is strictly dependent on the number of initial states, the structure of connections and accepted values of conviction parameters and susceptibility to conviction. Thus, it can happen that the process will once go faster or slower. However, the situation presented in Fig. 6a, c and e is much more interesting, in which the reduction of the current number of available states in the simulated environment is clearly visible by a sudden decrease in the value of the order parameter, different from the current trend. Thus, in Fig. 6a and c, this situation occurs once, and in Fig. 6e - twice. In the case of the classical algorithm (see Fig. 5) this situation was not visible, because the possible elimination of individual states from the network occurred at the very beginning of the simulation, when the system was in an unstable phase and the distribution of individual states in the whole network changed rapidly. Taking this into account, the analysis of the change in the value of the order parameter in the case of the evolutionary algorithm allows to observe significant changes in the structural system, and at the same time having a macroscopic character.

4 Conclusions

Network information structures are a kind of complex systems. These systems are characterized by high dynamics of processes occurring in them. These processes may be short-term or long-term, and may have only local as well as global impacts on the entire system. Therefore, modeling such systems is a big challenge. In particular, it is an important aspect of the so-called phase transitions processes when the network moves from one state to another. One of the solutions available for modeling this type of phenomena is the Ising model. In the case of the classic Ising model, network processes that are rapid and short-lived can be perfectly modeled. However, there are also many long-term processes - in that case a better solution is to use the evolutionary approach presented in this paper. This is particularly important in the case of processes characteristic of complex systems. In addition, in their case local changes may have a macroscopic impact and affect changes in the entire network structure. These assumptions are fulfilled by the Ising model together with the modifications introduced to it. In addition, in the case of an evolutionary approach, the situation of changing the number of current available states in the network infrastructure was perfectly visible thanks to the analysis of the ordering parameter. On its basis, the process of phase transitions, i.e. moving from a disordered state into an orderly state was also visible. The solution presented in the paper can be used to model processes occurring in social networks and IoE, the impact of congestion on the functioning of network infrastructure (including IoT), the way the so-called "Fake news" impact a network of views and conviction, gradual and evolutionary spread of malware in distributed information systems.

Acknowledgments. This project is financed by the Minister of Science and Higher Education of the Republic of Poland within the "Regional Initiative of Excellence" program for years 2019–2022. Project number 027/RID/2018/19, amount granted 11 999 900 PLN.

References

1. Auyang, S.Y.: Foundation of Complex-System Theories. Cambridge University Press, Cambridge (1999)
2. Song, Ch., Havlin, S., Makse, H.A.: Origins of fractality in the growth of complex networks. Nat. Phys. **2**(4), 275–281 (2006)
3. Stanley, H.E.: Introduction to Phase Transitions and Critical Phenomena. Oxford University Press (1989). ISBN 9780195053166
4. Fronczak, P., Fronczak, A., Hołyst, J.A.: Phase transitions in social networks. Eur. Phys. J. B **59**(1), 133–139 (2007)

5. Strzałka, D.: Selected remarks about computer processing in terms of flow control and statistical mechanics. Entropy **18**, 93–111 (2016)
6. Paszkiewicz, A.: The use of the power law for designing wireless networks. In: Computing in Science and Technology (CST 2018), ITM Web Conference, vol. 21 (2018). https://doi.org/10.1051/itmconf/20182100001
7. Xiao, W., Peng, L., Parhami, B.: On general laws of complex networks. In: Social Informatics and Telecommunications Engineering, Lecture Notes of the Institute for Computer Sciences, vol. 4, pp. 118–124. Springer, Heidelberg (2009)
8. Grabowski, F., Strzałka, D.: Processes in systems with limited resources in the context of non-extensive thermodynamics. Fundamenta Informaticae **85**, 455–464 (2008)
9. Chang, S., Gavish, B.: Telecommunications network topological design and capacity expansion: formulations and algorithms. In: Telecommunications Systems, vol. 1, no. 1, pp. 99–131. Springer, Amsterdam (2005)
10. Liu, J., Li, J., Chen, Y., Chen, X., Zhou, Z., Yang, Z., Zhang, Ch-J: Modeling complex networks with accelerating growth and aging effect. Phys. Lett. A **383**(13), 1396–1400 (2019). https://doi.org/10.1016/j.physleta.2019.02.004
11. Twaróg, B., Gomółka, Z., Żesławska, E.: Time analysis of data exchange in distributed control systems based on wireless network model. In: Analysis and Simulation of Electrical and Computer Systems, vol. 452, pp. 333–342. Springer (2018)
12. Mazur, D., Paszkiewicz, A., Bolanowski, M., Budzik, G., Oleksy, M.: Analysis of possible SDN use in the rapid prototyping process as part of the Industry 4.0. Bull. Polish Acad. Sci. Tech. Sci. **67**(1), 21–30 (2019). https://doi.org/10.24425/bpas.2019.127334
13. Ishii, M.: Analysis of the growth of social networking services based on the Ising Type agent model. In: Pacific Asia Conference on Information Systems, Proceedings, p. 331 (2016)
14. Selinger, J.V.: Ising model for ferromagnetism. In: Introduction to the Theory of Soft Matter, Soft and Biological Matter, pp. 7–24. Springer (2015). https://doi.org/10.1007/978-3-319-21054-4_2
15. Paszkiewicz, A., Iwaniec, K.: Use of Ising model for analysis of changes in the structure of the IT Network. In: Borzemski, L., et al. (eds.) Advances in Intelligent Systems and Computing, Information Systems Architecture and Technology: Proceedings of 39th International Conference on Information Systems Architecture and Technology – ISAT 2018 PART I, vol. 852, pp. 65–77. Springer (2019). https://doi.org/10.1007/978-3-319-99981-4_7. ISBN: 978-3-319-99980-7
16. Bolanowski, M., Cisło, P.: The possibility of using LACP protocol in anomaly detection systems. In: Computing in Science and Technology (CST 2018), ITM Web Conference, vol. 21 (2018). https://doi.org/10.1051/itmconf/20182100014
17. Toruniewska, J., Suchecki, K., Hołyst, J.A.: Unstable network fragmentation in co-evolution of Potts spins and system topology. Physica A **460**, 1–15 (2016)

Models and Method for Estimate the Information-Time Characteristics of Real-Time Control System

Anatolii Kosolapov(✉)

Dnipro National Rail University, Dnipro 49010, Ukraine
kosolapof@i.ua

Abstract. In the work proposes models for describing and calculating the characteristics of automated real-time control systems at the conceptual design stage. The models are based on the proposed notion of a φ-transaction.

Keywords: φ-transaction · Conceptual design · Real-time control systems

1 Introduction

Modern control systems are geographically and functionally distributed systems of large sizes, combining thousands of computers and working in real time. When creating and improving such systems, the role of conceptual or system design increases [1, 2, 4, 6]. Under these conditions, the process of selecting design decisions becomes more complicated due to the lack of basic data and models for evaluating the information and time characteristics of data processing and transmission processes. Special attention should be paid to real-time systems, in which the violation of time constraints can lead to emergency situations [1, 4, 5, 11]. The paper proposes simple graph-analytical models for engineering evaluation of the following characteristics of computer systems: the processors performance, the amount of information flow, the processing time of applications. The basis of the models is the concept of φ-transaction. The developed models and methods are part of Framework Conceptual Design of Complex Real-Time Management System (CoDeCS) [1].

2 Related Work

The massive development of real-time systems began in the 1970s. This is due to the emergence of reliable computers of the third generation. At this time, there is an introduction of the process control system in industry and transport [7]. One of the main characteristics of control computers is their speed, since it is necessary to ensure the execution of functions for a limited time [8]. At this time, a large number of analytical and simulation methods for estimating the speed of control computers and various performance metrics appear: linear regression models, benchmark programs, or mixes, queuing system models, Petri net models and others, MIPS, FLOPS metrics [8, 10–13]. The first systems were centralized systems. But with the advent of industrial

L. Borzemski et al. (Eds.): ISAT 2019, AISC 1050, pp. 58–67, 2020.
https://doi.org/10.1007/978-3-030-30440-9_7

computers and microcontrollers, there has been a transition to hierarchical geographically and functionally distributed systems, where the role of information flows that affect productivity increases [4, 5]. At the same time, the concept of a transaction similar to that used in banking systems was used in some automated process control systems [3, 14]. For the first time, the author systematized the methods for evaluating the performance of processors in their selection under various design conditions (detailed description of the tasks to be solved) [1, 2]. The relevance of this is confirmed by [15, 16]. Another one of the requirements for models was the simplicity of their calculation in the environment of spreadsheets on gadgets. This is important for the conceptual design of computer systems with a lot of variations in the choice of processors.

3 The Concept of the φ-Transaction and Its Properties

For the study and design of real-time information management systems of sorting stations, we introduce the concept of φ-transaction.

φ-transaction is a logically unified sequence of functional-algorithmic and program blocks (FPB) of operations, which is processed completely from the moment an event occurs in the control system that requires processing until the end of its processing with the output of a message and/or control management action.

φ-transaction must own ACID properties [3]. It is an abbreviation of four words, which means atomicity, consistency, isolation and durability.

Atomicity. Atomicity is a unit of work. In relation to the φ-transaction, this means that either the entire unit of work will be successfully completed, or nothing will be changed.

Consistency. The states before the start of the φ-transaction and upon its completion must be correct. During the φ-transaction state may have intermediate values.

Isolation. Isolation means that the φ-transactions that are executed simultaneously are isolated from the state that changes during the φ-transaction A cannot see the intermediate state of the φ-transaction B until it is completed.

Durability Upon completion of the φ-transaction, its result should be recorded on an ongoing basis. This means that if the microcontroller or the control computer fails, the state should be resumed after their restart.

From the known definitions, the concept of φ-transaction differs by the extended description necessary for calculating the characteristics of an ACS at the early stages of its design, which includes for the set of processed by the system of φ-transactions:

- description of the main arrays of the database M ($\forall m_i \in M \ \{O_i\}$, where, $O_i \ i \in \overline{1, |M|}$ is the size of the array or the amount of memory in bytes), that used when performing φ-transactions;
- description of the set of events E that require the response of the control system; this is start-up φ-transactions ($\forall \varphi_j \in \Phi \ \{s_j, \lambda_j\}$, where $s_j \in S_o$ is the set of input signals and messages associated with the corresponding events E, λ_j is the average intensity of the start of the j-th φ-transaction);

- description of a set of conditions and requirements for successful completion of φ-transactions $\forall \varphi_j \in \Phi\ \forall d_k \in D\{op_{kj}, Q_{kj}, K^{\Gamma}_{kj}, t^{\Gamma p}_{kj}\}$, where d_{kj} is the k-th information transceiver from the j-th φ-transaction, $op_{kj} = \{$ a message $msg_{kj} \in Msg^{out}$ or e control action as a signal $s_{kj} \in S^{out} \ldots\}$, Q_{kj} - the amount of information in bytes that is transmitted to the k-th transceiver from the j-th φ-transaction, K^{Γ}_{kj} is the availability factor and the maximum execution time of the φ-transaction j for the kth transceiver t^{nax}_{kj};
- description of each FPB that is part of the system $\forall \phi_l \in \text{FPB}\ \forall m_i \in M\ \forall com_n \in \mathbb{C}\ \{k_{i,l}, op_{i,l}, K_{nl}\}$, where $\text{FPB} = \{\phi_l | l \in \overline{1, |\text{FPB}|}\}$ the set of FPB systems, $M = \{m_i | i \in \overline{1, |M|}\}$ - arrays of the database, $\mathbb{C} = \{com_n | n = \overline{1, |\mathbb{C}|}\}$ - a set types of commands in a mix of commands of the designed system, $0 \leq k_{i,l} \leq 1$ - coefficient array utilization (what part of the array i is used for one implementation of the FPB l); $op_{j,l,i} = \{\text{rd,wt}\}$ - operations read and write with memory i when executing the l-th FPB in the j-th φ-transaction, K_{nl} - number of commands of type n in the l-th FPB.

A general description of the designed system with the help of φ-transactions is shown in Fig. 1. In the figure the following notation is used: buffering disciplines (BD), service disciplines (SD), device input/output signals (SIOD).

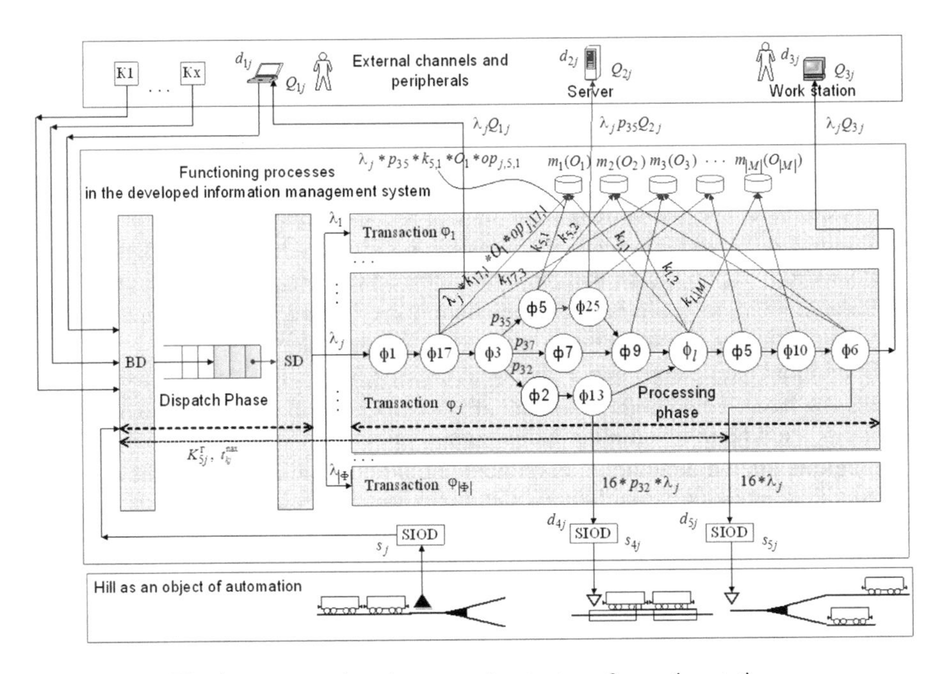

Fig. 1. φ-transactions in automation systems for sorting stations

In the FPB set that is included in the φ-transaction, vertices are singled out in which multiple branching of the data processing process is possible (for example, $\phi 3$ has three options for continuing calculations). In this case, the probability of choosing a direct branching is given, the sum of which is 1 (in the example $p_{35} + p_{37} + p_{32} = 1$).

All described characteristics of φ-transactions are used in the design process of automation systems for sorting stations in accordance with the proposed approach of CoDeCS [1]. In fact, the set of φ-transactions (Φ) is the M-model of the system [6].

4 Methods for Calculating the Characteristics of Real-Time Computer Systems During the Conceptual Design Process

The initial data for calculating the characteristics of the designed control system are:

- description of functional program blocks, which includes a list of FPB, the arrays used by them and the coefficients of their use during each FPB, the number of operations of the selected types in each FPB (Table 1);

Table 1. Description of functional program blocks (FPB).

Options	Functional program blocks					
	ϕ_1	ϕ_2	…	ϕ_l	…	$\phi_{\lvert FPB \rvert}$
Mix command of system	Number of operations in FPB, thousand commands					
com_1	K_{11}	K_{12}		K_{1l}		$K_{1\lvert FPB \rvert}$
com_2	K_{21}	K_{22}		K_{2l}		$K_{2\lvert FPB \rvert}$
…						
com_n	K_{n1}	K_{n2}		K_{nl}		$K_{n\lvert FPB \rvert}$
…						
$com_{\lvert C \rvert}$	$K_{\lvert C \rvert 1}$	$K_{\lvert C \rvert 2}$		$K_{\lvert C \rvert l}$		$K_{\lvert C \rvert\lvert FPB \rvert}$
Arrays and their volumes	Array utilization and type of operation					
$m_1(O_1)$	$k_{1,1}$, $op_{1,1}$	$k_{1,2}$, $op_{1,2}$		$k_{1,l}$, $op_{1,l}$		$k_{1,\lvert FPB \rvert}$, $op_{1,\lvert FPB \rvert}$
$m_2(O_2)$	$k_{2,1}$, $op_{2,1}$	$k_{2,2}$, $op_{2,2}$		$k_{2,l}$, $op_{2,l}$		$k_{2,\lvert FPB \rvert}$, $op_{2,\lvert FPB \rvert}$
…						
$m_i(O_i)$	$k_{i,1}$, $op_{i,1}$	$k_{i,2}$, $op_{i,2}$		$k_{i,l}$, $op_{i,l}$		$k_{i,\lvert FPB \rvert}$, $op_{i,\lvert FPB \rvert}$
…						
$m_{\lvert M \rvert}(O_{\lvert M \rvert})$	$k_{\lvert M \rvert,1}$, $op_{\lvert M \rvert,1}$	$k_{\lvert M \rvert,2}$, $op_{\lvert M \rvert,2}$		$k_{\lvert M \rvert,l}$, $op_{\lvert M \rvert,l}$		$k_{\lvert M \rvert,\lvert FPB \rvert}$, $op_{\lvert M \rvert,\lvert FPB \rvert}$

- description of φ-transactions in the form of oriented vertex and edge-weighted graphs with a description of their parameters (Fig. 1);

- description of the pre-selected type of microprocessor and the execution times of the operations of the types mentioned above (examples of microprocessor descriptions are in Tables 2 and 3 [10]).

Table 2. Arithmetic operations time, ns.

Argument type	int				double		
Operation	^	+	*	/	+	*	/
Pentium 4 2,5 GHz	0,78	0,86	5,6	19	2	2,8	15
CoreDuo 1,8 GHz	1,1	1,1	2,2	4,2	1,7	2,7	18
Pentium M 1,8 GHz	1,1	1,1	2,2	8,7	1,7	2,7	17
Athlon 64X2 2,5 GHz	1,4	1,4	1,6	17	1,8	1,8	8,3

All initial data are presented in tabular form for further processing.

It should be noted that the FPB are either programs (if this block is taken from the developer's program library), or algorithms (for known control algorithms), or functions (for new tasks). This determines the accuracy of the estimate of K_{nl} and $k_{i,l}, op_{i,l}$.

To select a set of commands $\mathbb{C}$ and get K_{nl}, you can use the "POET" technique [7]. For new functions, there are characteristics of typical algorithms that are recommended for the IMS (information management system) design stages (Table 4) [4].

To calculate the speed of computing systems in order to compare and select them, a large number of various models, methods and benchmarks are used (benchmark - Linpack, SPEC, SPECfp95, SPECweb97, TPC, WinMark, Winstone 97, etc.), but most of them focused on already developed systems and they are almost impossible to use in the early stages of design [4, 5, 10].

The developed methodology proposes two methods for evaluating the performance of microprocessors—$\tilde{W}_i$ the average performance of the i-th model for a mix of tasks of the system and $\tilde{W}_i^{\varphi}$ the average performance of the I model for a mix of φ-transactions of the control system that is designed.

For the automated control systems of the hump yard, the author has already identified frequency mix of commands, first described in [7]. To take into account new algorithms and control programs, the necessary adjustment of the frequency characteristics of the mix. It is recommended to use the command sets $\mathbb{C}$ from Tables 2 and 3.

The performance on the frequency mix of commands when designing a system for a pre-selected I of type of microprocessor (MCP) is calculated as follows:

$$\tilde{W}_i = 1 / \sum_{n=1}^{|\mathbb{C}|} \xi_n \tau_{ni}, \tag{1}$$

where ξ_n is the frequency or probability of command occurrence in the process of system operation; τ_{ni} is the average execute time the n command on the microprocessor of the i type; the unit of measurement of performance is the number of operations on the mix of system tasks per second – op/s.

Table 3. The number of cycles required to execute instructions on Intel P4 +.

Instruction	XOR	ADD	MUL	DIV	FADD	FMUL	FDIV (double)
Processing delay	0,5-1	0,5-1	3-18	22-70	5-6	7-8	38-40
Time of processing	0,5	0,5	1-5	23-30	1	2	38-40

Table 4. Characteristics of typical algorithms.

Class of algorithms (tasks)	Input parameters of algorithms	The resulting characteristics		
		Internal connectivity	The number of middle operations	Specific volume
Initial processing of scale measurements	n - number of measurements, m - size of the calibration table	$n + m$	nm	m
Statistical analysis	n - number of measurements, m - number of factors, number of coefficients	nm	nm^2	n
Optimization (linear prog., Etc.)	m - number of variables, n - number of restrictions	$n(m + n)$	$n^2(m+n)$	n
Calculation of technical and economic characteristics	n - number of data, m - the number of constants of the formula	$n + m$	nm	m

The value of ξ_n is calculated as

$$\xi_n = K_n / \sum_{h=1}^{|\mathbb{C}|} K_h. \tag{2}$$

Here K_h is the number of h-type commands that are processed by the system in the process of performing φ-transactions.

K_h is calculated from formula (3) taking into account the data K_{nl} for each FPB from Table 1 and the structure of all φ-transactions processed by the system, taking into account the intensities of their launch, described in Fig. 1.

$$\forall h \in |\mathbb{C}| \quad K_h = \sum_{\forall \varphi_j \in \Phi} \lambda_j * \sum_{\forall \phi_l \in \varphi_j} k_{lj} * K_{hl}. \tag{3}$$

In formula (3), $0 < k_{lj} \leq 1$ is the coefficient that takes into account the decrease in the number of operations in the φ-transaction j in the FPB l due to the previous ramifications and cycles (in the weighted graph). This coefficient is calculated in the process of transforming the structure and calculating the characteristics of the φ-transactions.

The obtained average performance of the i-th model on a mix of tasks of the future system $\tilde{W}_i$ is used in the proposed method of CoDeCS [1] to compare different types of MCPs in their choice, as well as for resource-saving methods for searching rational structures of decentralized control systems for sorting hill [6, 7].

For service-oriented real-time systems that have a limit on the reaction time or the maintenance time of each φ-transaction, it is proposed to calculate the average performance of the i-th model on a mix of φ-transactions, or $\tilde{W}_i^{\varphi}$, which is measured in the number of weighted φ-transactions per second ("φ – tps" - φ-transaction per second) or per minute – "φ – tpm". To do this, we use the expression (4).

$$\tilde{W}_i^{\varphi} = \sum_{j=1}^{|\Phi|} \lambda_j / \sum_{j=1}^{|\Phi|} \lambda_j T_{ji}. \tag{4}$$

In this formula, since the condition for the selected i-th type of the MCP (or clock frequency) must be met

$$\forall j \in |\Phi| \;\; T_{ji} \leq t_j^{\max}, \tag{5}$$

then according to (4) it is possible to make a choice for a project of this type of MCP minimum performance (and cost) at which (5) is performed. The T_{ji} is the time of the implementation of this j-th φ-transaction on i-th MCP.

The developer's knowledge of the values of $\forall j \in |\Phi| \;\; T_{ji}$ in the process of designing a real-time control system allows for the selected i-th type of microprocessor to estimate what part of the processor load is provided by providing the j-th service (j-th φ-transaction), formula (6).

$$\rho_{ji} = \lambda_j T_{ji} \;\; \text{and for} \;\; \forall j \in |\Phi| \quad \rho_{ji} \leq 0,6. \tag{6}$$

Carrying out the distribution of processed φ-transactions on the subsystems obtained as a result of structural optimization [1]. It is necessary to ensure that the total service load of each subsystem in the project $f_p \in \mathbb{F}$ does not exceed the permissible limit (for processors it is 0.6), i.e.

$$\forall f_{pi} \in \mathbb{F}_i \;\; R_{pi} \leq 0,6. \tag{7}$$

Sometimes the lower limit of loading (8) is also established in order to efficiently use computational resources and reduce the total cost of IMS.

$$\forall f_{pi} \in \mathbb{F}_i \;\; 0,1 \leq R_{pi} \leq 0,6. \tag{8}$$

The construction of φ-transactions allows you to evaluate some of other characteristics that are very important for the design of the ASC, in particular, those related to the information flows in the future system. In the ASC we will distinguish three types of information flows: message flows (msg), control action (signals, s_k) flows, and reading (rd) and writing (wt) flows from/to the arrays of the future database.

The information flow of messages $\pi_{jd_k}^{msg}$, which is formed by each φ-transaction j, and transmitted to the peripheral device d_k:

$$\pi_{jd_k}^{msg} = \sum_{\forall \phi_l \in \varphi_j \wedge \exists d_k} \lambda_j \widehat{p}_{jl} Q_{ld_k j}. \quad (9)$$

The total flow of messages to the transceiver d_k during operation of the system will be:

$$\Pi_{d_k}^{msg} = \sum_{j=1}^{|\Phi|} \pi_{jd_k}^{msg}. \quad (10)$$

Flows of control actions, or s_k signals are output through the SIOD on the devices of the control object. For one j-transaction and when encoding one action with 2 bytes (16 bits) they will be:

$$\pi_{s_{kj}} = \sum_{\forall \phi_l \in \varphi_j \wedge \exists s_k} 16 \lambda_j \widehat{p}_{jl}. \quad (11)$$

In the system as a whole, the signal flow s_k will be:

$$\Pi_{s_k} = \sum_{j=1}^{|\Phi|} \pi_{s_{kj}}. \quad (12)$$

Information flows associated with reading/writing data from/to array i, when performing the j-th transaction, can be calculated as follows:

$$\pi_{ji}^{\mathrm{rd/wt}} = \sum_{\forall \phi_l \in \varphi_j} \lambda_j \widehat{p}_{jl} k_{j,l,i} O_i op_{j,l,i}. \quad (13)$$

The total information flow to the array i in the process of the system can be calculated by the formula:

$$\Pi_i^{\mathrm{rd/wt}} = \sum_{j=1}^{|\Phi|} \pi_{ji}^{\mathrm{rd/wt}}. \quad (14)$$

The total information flow to all arrays or the flow to the central database of the ASC will be:

$$\Pi_M^{\mathrm{rd/wt}} = \sum_{i=1}^{|M|} \Pi_i^{\mathrm{rd/wt}}. \quad (15)$$

In addition, φ-transitions allow determining the priorities for processing applications, the average waiting time for applications in the queue, its length and other parameters of the organization of the system's functioning.

5 Conclusions

The model of the φ-transactions proposed in the article and the methods of approximate calculation of the temporal and informational characteristics of distributed real-time computer systems at the stages of conceptual design differ from the known approaches. They are simple and can be implemented on smartphones and tablets in a spreadsheet environment. The proposed models were obtained on the basis of exponential flows of applications and the disciplines of their maintenance, which makes it possible to obtain overestimated values of system characteristics, which is especially important for real-time systems. The considered models are part of the CoDeCS framework, covering the whole complex of tasks of conceptual design of distributed automated control systems [1]. Almost all tasks are solved in a spreadsheet environment or using mobile applications. The developed framework CoDeCS is constantly being improved with the development of information technologies and systems. It has found application in the creation of a number of automation systems for control and management for the railways of Russia and Ukraine.

References

1. Kosolapov, A., Loboda, D.: Framework conceptual design of complex real-time management system (CoDeCS). Eur. J. Adv. Eng. Technol. **5**(8), 559–566 (2018)
2. Kosolapov, A., Loboda, D.: Semiotic-agent-ontological approach to design intellectual transport systems. Am. J. Eng. Res. (AJER) **7**(7), 205–209 (2018)
3. Nagel, C., Evjen, B., Glynn, J., Skinner, M,. Watson, K.: Professional C# 2005 with .NET 3.0, Wrox, p. 1416 (2005)
4. Valkov, V.: Microelectronic Control Computing Systems. System Design and Engineering, Mechanical Engineering. Leningrad Department, p. 224 (1990). (published in Russia)
5. Kosolapov, A., Kosolapova, M.: To the question of the structural design of automated systems. Math. Modevian Sci. J. **2**(5), 124–128 (2000). Dniprodzerzhinsk, DDTU (published in Russia)
6. Kosolapov, A., Shafit, E., Safris, L.: Some characteristics of microprocessors and microcomputers and technological algorithms in control systems for composition dissolution on a hill, Trudy. Interuniversity Thematic Collection, RIZHT, Rostov-on-Don, no. 168, pp. 20–23 (1982). (published in Russia)
7. Kosolapov, A.: Development and application of mathematical modeling methods in the analysis and design of microprocessor control systems for technological processes of dissolution of compositions on hills. PhD Dissertation, VNIIZhT (Central Research Institute of Ministry of Railways), 24 s. (1984). (published in Russia)
8. Mayorov, S., Novikov, G., Alive, T.: Fundamentals of the Theory of Computing Systems, Proc. Textbook for Universities, High School, p. 408 (1978). (published in Russia)
9. Trifonov, P.: Computer Science. Construction and Analysis of Algorithms, Textbook for Universities, St. Petersburg, Peter, p. 95. (2007). (published in Russia)

10. Tsilker, B., Orlov S.: Computer Organization and Systems, Textbook for High Schools, St. Petersburg, Peter, p. 654 (2006). (published in Russia)
11. Heidelberger, P., Lavenberg, S.S.: Computer performance evaluation methodology. IEEE Trans. Comput. **c-33**(12), 1195–1219 (1984)
12. Lilja, D.J.: Measuring Computer Performance: A Practitioner's Guide. Cambridge University Press, p. 15 (2000)
13. Zelkowitz, M., Basili1, V., et al.: Measuring productivity on high performance computers. Presented at International Symposium on Software Metrics, Como, Italy, September, pp. 113–123 (2005)
14. Vtorin, V.A.: Automated process control systems. Basics of PCS. St. Petersburg, St. Petersburg State Forestry Academy named after S.M. Kirov, p. 182 (2006). (published in Russia)
15. Alford, R.S.: Computer Systems Engineering Management. CRC Press, p. 372 (2017)
16. Blanchard, B.S., Blyler, J.E.: Sestem Engineering Management, 5th edn. Wiley, p. 568 (2016)

A Large-Scale Customer-Facility Network Model for Customer Service Centre Location Applications

David Chalupa[1], Peter Nielsen[1(✉)], Zbigniew Banaszak[2], and Grzegorz Bocewicz[2]

[1] Department of Materials and Production, Aalborg University, Aalborg, Denmark
{dc,peter}@m-tech.aau.dk
[2] Department of Business Informatics, Faculty of Electronics and Computer Science, Koszalin University of Technology, Koszalin, Poland
{zbigniew.banaszak,bocewicz}@tu.koszalin.pl

Abstract. We propose a large-scale sparse customer-facility network model that allows a customer to be assigned only to facilities within the vicinity of a customer. In this model, customer-facility distances are integer values representing zones. Experimental results are presented for large instances with up to 100,000 customers and 100 potential facility sites. A mixed-integer linear programming solver reveals large gaps in suboptimal solutions and lower bounds provided, even with a considerable computational effort. Two simple but scalable local search heuristics are computationally investigated, revealing their potential for solving such large-scale problems in practice.

Keywords: Facility location problem · Customer-facility networks · Sparse problems · Customer service centre location

1 Introduction

Facility location problems are among the fundamental problems in operations research and manufacturing [15], with wide applications such as supply chain [8] or manufacturing network design [17]. Many of these applications require solving of large-scale instances of the problem. This is a computationally hard task, given the NP-hardness of the problem [19]. Likewise the problem has many similarities to a number of well-known assignment problems found in literature such job-shop scheduling [2], project selection problems [20] and container placement problems [9].

For applications such as customer service centre location, the problem can get very large in terms of the numbers of customers to assign. This type of scale is also often found in transportation problems. Facility location problems have been tackled for moderately sized but structurally challenging instances by a number of stochastic algorithms [3]. These include including local search [4],

L. Borzemski et al. (Eds.): ISAT 2019, AISC 1050, pp. 68–77, 2020.
https://doi.org/10.1007/978-3-030-30440-9_8

tabu search [1,16,22], genetic algorithms [12], evolutionary simulated annealing [23] or particle swarm optimisation [10]. Related problems have been tackled for relatively large numbers of customers, a recent study solved the competitive facility location problem efficiently for instances with around 7,000 customers [13].

Yet, to the best of our knowledge, algorithmic requirements for instances with tens or even hundreds of thousands of customers remain largely unexplored. This is interesting as the nature of the problem lends the results to have implication for other related problems. In this paper, we develop a sparse customer-facility network model, in which the distances represent zones and a customer can be assigned to a facility only within a limited zone. This has a simple real-world interpretation: it makes sense to restrict the set of potential facility sites to serve the customer only to the ones within their vicinity. This keeps the problem instance sparse and is likely to improve on scalability of problem solving techniques for the problem in practice. It also opens the way to relating the problem closely to other location problems in complex networked systems, such as the k-reachability problems [6].

Contributions. Our sparse customer-facility network model works with randomly generated facility costs and integer distances that represent zones. A zone threshold is introduced and an edge between a customer and potential a facility site is established only if the zone is within the threshold. This leads to a large but sparse networks, for which specific computational techniques are available [21].

Experimental results are presented for large-scale implementations of the model with 10,000 customers, 100,000 customers and 100 potential facility sites. The facility costs are generated uniformly at random and the customers are uniformly assigned to 10 zones for each facility, limiting the connections only to zones 1 and 2. It is worth noting that we treat zones as symmetric distances, i.e. assigning a customer to a zone for a facility, and assigning a facility to a zone for a customer essentially have the same meaning.

We solve the problem using scalable local search (LS) and randomised local search (RLS) algorithms that we propose for problems of this scale. Performance of these local search algorithms in solving various problems is widely studied in evolutionary computation theory [18]. Note that these algorithms were chosen, since simple algorithms usually have good scaling properties.

Compared to a mixed-integer linear programming solver, LS and RLS provided solutions of very good quality much faster, and were the only techniques producing solutions to the instances with 100,000 customers. The gap between the average and mean performances of LS and RLS seems to narrow down with growing instance size. However, while LS seems to be a technique suitable for finding a good local optimum quickly, RLS exhibits high variance in its results. Using these instances and algorithms allowed us to explore the properties of large-scale but sparse instances of general assignment problems. As the problem studied is chosen as an example of such an assignment problem, these implications reach far beyond the facility location problem.

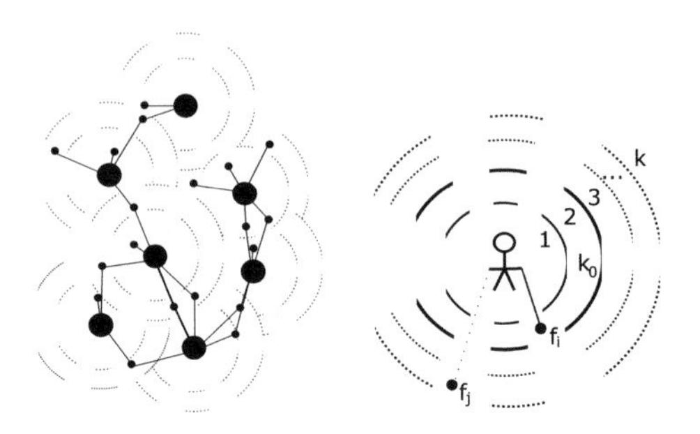

Fig. 1. Illustration of a simple customer-facility network according to our model (on the left-hand side) and distances represented by k zones (on the right-hand side). Facilities are depicted as the larger circles. A customer can be assigned only to a facility within the corresponding zones.

2 Customer-Facility Network Model for Large-Scale Applications

In order to represent the facility location problem as a sparse problem, one needs to slightly adapt the traditional integer linear programming (ILP) formulation of the problem. The following is an adaptation of the most widely used formulation of the problem [1]:

$$\min \sum_{i=1}^{n} f_i y_i + \sum_{\{v_i^f, v_j^c\} \in E} c_{ij} x_{ij}, \tag{1}$$

s.t.

$$\sum_{\{v_i^f, v_j^c\} \in E} x_{ij} = 1, \quad j = 1, ..., m, \tag{2}$$

$$x_{ij} \leq y_i, \quad i = 1, ..., n, \quad \{v_i^f, v_j^c\} \in E, \tag{3}$$

where f_i is the cost of facility i and c_{ij} is the integer zone value for customer j if assigned to facility i. The difference between this and the classical formulation is that one now operates within a network of customers and facilities that can be formalised as a graph $G = [V, E]$. $V = [\mathcal{C}, \mathcal{F}]$ represents the vertices, dividing customers and potential facility sites into separate sets, and E is a set of connections of customers to potential facility sites. Note that $\{v_i^f, v_j^c\} \in E$ if zone c_{ij} is within a bounded interval for customer j and facility site i. This creates a sparse bipartite networked structure, as illustrated in Fig. 1 (on the left-hand side). Constraint (2) traditionally requires that a customer is assigned to a single facility. Here we restrict this so that the customer is assigned locally, i.e. the set of feasible solutions in restricted only to assignments within E.

Figure 1 (on the right-hand side) illustrates the principles of zones within our model. For a customer j and facility i, the facility is situated within zone c_{ij}. There are k zones overall, i.e. $1 \leq c_{ij} \leq k$. Then, we say that $\{v_i^f, v_j^c\} \in E$ if $c_{ij} \leq k_0$, where k_0 is the zone threshold. This threshold limits the possible assignment of customers only to the closest facilities and ensures that the problem instance becomes sparse.

In summary, our model represents large-scale sparse instances of the facility location problem with the following properties:

- n potential facility sites and m customers, with m potentially being a very high number;
- uniform distribution of facility costs $f_i \in \{1, 2, ..., f_{max}\}$ for each facility i;
- uniform distribution of zone assignments $c_{ij} \in \{1, 2, ..., k\}$ for each customer j and facility i;
- zone threshold $1 \leq k_0 \leq k$ to regulate local properties and sparseness of the instance.

This slight modification of the formulation of the facility location problem allowed us to solve it using LS and RLS. These algorithms are expected to be able to deliver reasonable results even for very large-scale problem instances. Importantly, we assume that the attributes exhibited in this problem are somewhat representative of also other types of assignment problems. This highlights the relevance of our study for practice.

In addition, we are interested in how the performance of LS and RLS scales from the smaller instances to the very large instances. We are especially interested in whether the performance scales proportionally. In the following, we discover that this is not true, but rather we see that LS and RLS behave somewhat differently. This uncovers the opportunities not only to exploit these algorithms in facility location problems, but other related problems as well.

3 Scalable Local Search Algorithms

To solve the problem efficiently for the instance sizes discussed, we will use two simple but efficient and scalable local search algorithms [7,18]. Both of the algorithms operate in the search space of binary strings $y \in \{0,1\}^n$, where element y_i represents a decision variable determining whether facility i is open or closed. The values of x_{ij} are computed by assigning each customer to the closest facility. In our design, both algorithms will start with binary string $1, 1, ..., 1$, i.e. all facilities will be open at the start. The reasons for this decision are two-fold. Firstly, both algorithms will always start at the same point in the search space, which eliminates one source of uncertainty in our computational investigation. Secondly, this is a solution that will likely be feasible also in further constrained variants of the problem, since we aim also for an easy generalisation to related problems. The stopping criteria for both algorithms will depend on the number of potential facility sites, following the increasingly popular fixed budget computation paradigm [11].

Local search (LS). Let $y \in \{0,1\}^n$ be the current solution. The neighbourhood of y consists of all binary strings obtained by flipping a single bit, i.e. $\mathcal{N}(y) = \{y' \mid y' = y + c, \ \sum_{i=1}^{n} c_i = 1\}$. Let the objective function in (2) be denoted by $J(y)$. Then, LS picks $y' = \arg\min_{y' \in \mathcal{N}(y)} J(y)$, i.e. it chooses the move that minimises the objective within the neighbourhood. In our further experiments, LS will be stopped after n iterations, where n is the number of facilities.

Randomised Local Search (RLS). RLS works with the same neighbourhood structure $\mathcal{N}(y)$ as LS. However, in each iteration, it always chooses a facility site i and attempts to flip bit y_i, effectively opening or closing the corresponding facility. This leads to a new candidate solution y'. This solution is accepted simply if $J(y') \leq J(y)$, i.e. the new solution is at least as good as the current solution. In our further experiments, RLS will be stopped after n^2 iterations. This procedure effectively evaluates the same number of candidate solutions as LS, since LS scans all n solution in the neighbourhood in each iteration, while RLS attempts to evaluate only a single one.

To implement the checking of the potential moves efficiently, one can use a collection of efficient data structures, similarly to the design of tabu search algorithms [22]. For each facility, we explicitly store the list of customers assigned to it. Conversely, we also store the assignment vector determining the corresponding facility for each customer. This way, the objective value can first be recalculated, before a move is accepted. These structures are then updated after accepting the move.

4 Experimental Results

In order to evaluate our model, we have performed experiments for two sets of large-scale problem instances generated for 10,000 and 100,000 customers. We generated 10 problem instances of size 10000×100 and 100000×100, with a maximum facility cost $f_{max} = 100$ and a maximum of $k = 10$ zones. For a connection between a customer and a potential facility site to be established, each customer has to be within zone $k_0 = 2$ (or lower).

The code of LS and RLS was written in C++ using Qt and compiled using the 32-bit MinGW compiler under Windows 10. We also used the mixed-integer linear programming branch-and-cut solving utility CBC from the COIN-OR package [5,14]. This was to see how difficult it is solve the problem into optimality at this scale and obtain worst-case bounds on the quality of the solutions obtained by LS and RLS. All experiments were performed on a machine with Intel Core i7-6820 CPU @ 2.70 GHz and 32 GB RAM. We used pre-compiled binaries of the COIN-OR package compiled by the 64-bit Intel 11.1 compiler.

In Fig. 2, we present the box-wisker plots obtained for LS and RLS for the corresponding 10000×100 (on the left-hand side) and 100000×100 (on the right-hand side) instances. These are obtained for 10 independently generated instances and 100 runs of both LS and RLS for each of the instances. The results show that LS outperforms RLS in terms of solution quality for the smaller instances. However, for larger instances, the performances of LS and RLS are much more even. Note that the difference between these two instance types is only in the number of customers, all other parameters were the same. For some

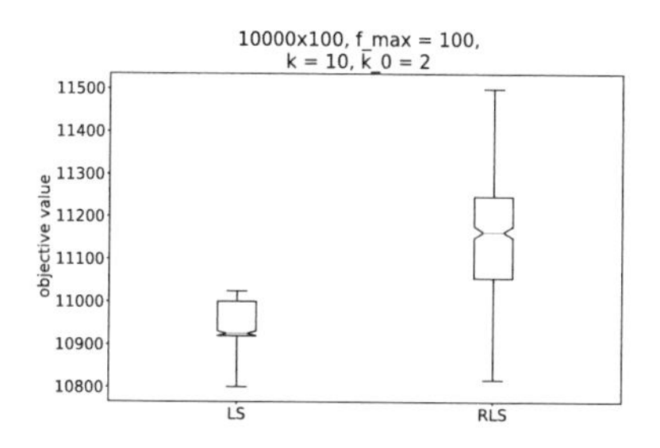

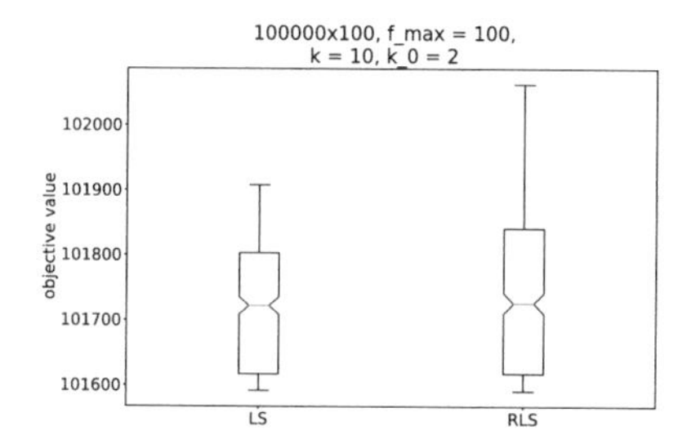

Fig. 2. Box-whisker plots depicting the results obtained by LS and RLS for the instances of size 10000×100 (on the left-hand side) and 100000×100 (on the right-hand side).

of the instances, RLS actually performed better than LS in the best runs, even though it exhibited weaker average and median performances.

Table 1 (upper part) provides detailed results for the 10000×100 instances. These show that both LS and RLS took between 12 and 15 s to solve the problem. One can see that the performance of LS is much more stable, while the variance of results obtained by RLS is much higher. This is likely related to the random sampling of the moves performed by RLS. LS emerged as the better of the two strategies for these instances, even though it is worth mentioning that RLS sampled a better solution for instance 9.

Regarding the performance of CBC in solving these instances exactly, one can see that for all apart of one instance, LS or RLS found a better solution than CBC. All the above solutions found by the CBC are found only by terminating the search after 24 h, still leaving a wide gap between the solution found and the lower bound. This shows that heuristics are indeed very useful especially for such large-scale sparse instances of the problem. It is also worth noting the relatively distant optimal solutions for the LP relaxations and a very limited progress of CBC in improving the lower bound. Given the relatively generous time limit given to CBC, it seems that these instances are not only large but also quite intricate in terms of the search landscape.

This finding complements a large body of work that focused on moderately large but structurally highly intricate instances [1,3,10,12,16,22,23].

In Table 1 (lower part), we also provide the detailed results obtained for the 100000×100 instances. For this instance scale, we already observed very high computational demands of CBC. While the LP relaxation of problem could be solved within hours, already the preparation phase for the branch-and-cut procedure took more than 24 h.

This renders heuristics to be the only practical choice for this instance scale. Both LS and RLS produced solutions in two to four minutes, with LS being slightly quicker, which is likely related to less randomness within the process of LS and less frequent solution updates. One can also see that the performance of LS is quite stable in different runs, often leading to matching average and best upper bounds. However, we also observed that it at times converged to solutions that can be easily identified as suboptimal. This is observed particularly for instances 1, 4 and 10, for which RLS found better solutions than LS in the best runs. However, the inverse was observed for instance 9. This shows that while LS is better both in its average and median performances, the variance in the behaviour of RLS can have the advantage of avoiding local optima or escaping them.

It is also worth noting that the performance gap between the average and median performances of LS and RLS seems to shrink with growing number of customers. While the gap between the performances is quite wide for 10,000 customers, the average and median performances seem much closer for 100,000 customers. We have tested this further for 200,000 customers and the gap seems to close further. However, while LS seems to behave quite predictably, often converging to a limited number of local optima, RLS preserves a much higher variance in its behaviour. This indicates that LS and RLS have complementary qualities for very large-scale instances. While LS is a good strategy to sample promising local optima of these instances quickly, RLS can at times sample solutions of better quality and be used as a multi-start algorithm or as a procedure for escaping local optima.

These results sketch quite a complex picture of the performance of different strategies in our model. The problem becomes increasingly difficult for the branch-and-cut exact solver not only with the number of facilities but also with the number of customers in the model. For 100,000 customers, the exact solver with a 24 h time limit was only able to compute the solution to the LP relaxation of the problem. This highlights the demand for efficient scalable heuristics for solving the problem at such a scale.

The performance of heuristics themselves is also quite complex. LS outperformed RLS for the model with 10,000 customers but with a growing number of them, the gap between the performances of LS and RLS is gradually closing. While the gap between average objective values obtained by LS and RLS was 228.1 in favour of LS for 10,000 customers, it becomes only 18.3 for 100,000 customers. This can also be seen in the gap between the best objective values that was 40.7 in favour of LS for 10,000 customers. However, this gap became 0.6 in favour of RLS for 100,000 customers. In contrast to this, the gap between the best values obtained by the better of the two heuristics and the lower bound seems to widen with growing number of customers. This gap was 132.1 for 10,000 customers and 411.6 for 100,000 customers. LS is therefore a good choice for finding a good solution rapidly, while RLS preserves variability in its performance in large-scale, possibly leading to improved results. A multi-start or a population-based local search heuristic may therefore be a good direction for future algorithm design for this scale.

Table 1. A detailed breakdown of the results obtained for instances with 10,000 customers (the upper part) or 100,000 customers (the lower part) and 100 potential facility sites. The results obtained by solving the LP relaxation, branch-and-cut (both using CBC), LS and RLS are provided. CPU time measures are provided to quantify the scalability.

Instance	LP relaxation		Branch-and-cut			LS			RLS		
	Lower bound	CPU time	Upper bound	Lower bound	CPU time	Average upper bound	Best upper bound	CPU time	Average upper bound	Best upper bound	CPU time
10,000 customers											
1	10787	196 s	10928	10788	24 h	10920	**10920**	13 s	11156	10955	13 s
2	10811	203 s	10971	10812	24 h	10926	**10926**	12 s	11154	10977	12 s
3	10674	168 s	**10796**	10675	24 h	10801	10801	12 s	10968	10818	13 s
4	10852	175 s	11007	10855	24 h	11001	**11001**	12 s	11272	11066	12 s
5	10848	159 s	11069	10849	24 h	11012	**11007**	12 s	11225	11012	12 s
6	10845	167 s	11072	10846	24 h	11020	**11020**	12 s	11318	11114	12 s
7	10682	173 s	10805	10683	24 h	10797	**10797**	12 s	11038	10875	13 s
8	10820	119 s	**10956**	10822	24 h	10958	**10956**	13 s	11199	11008	13 s
9	10801	109 s	10961	10802	24 h	10944	10944	13 s	1153	**10925**	14 s
10	10713	109 s	**10782**	10719	24 h	10782	**10782**	13 s	10965	10811	15 s
100,000 customers											
1	101293	22381 s	Timeout			101592	101592	140 s	101598	**101591**	174 s
2	101379	23973 s				101803	**101801**	150 s	101823	**101801**	197 s
3	101345	23421 s				101722	**101722**	161 s	101727	**101722**	214 s
4	101420	24373 s				101908	101908	163 s	101949	**101905**	194 s
5	101269	24052 s				101617	**101617**	183 s	101623	**101617**	227 s
6	101180	18749 s				101439	**101439**	159 s	101451	**101439**	190 s
7	101287	26976 s				101632	**101632**	155 s	101649	**101632**	192 s
8	101357	27173 s				101837	**101837**	175 s	101848	**101837**	219 s
9	101524	27467 s				102173	**102167**	154 s	102187	102168	193 s
10	101316	26418 s				101777	101777	162 s	101828	**101774**	194 s

5 Conclusions

In this paper, we proposed a new large-scale sparse customer-facility network model for the facility location problem. This model is especially relevant for applications with high numbers of customers, such as customer service centre location.

The model works with facility costs that are distributed uniformly between 1 and a value f_{max}, and the distances were represented by zones, assigning each pair customer-facility with a zone number between 1 and k. We then introduce a distance threshold k_0 and allow a customer to be assigned only to a facility within zone at most k_0. This leads to large but sparse problem instances, allowing to model scenarios with potentially very high numbers of customers, with potential assignments limited only to the close vicinity of a customer or a facility.

In our experiments, we explored instances with 10,000 customers, 100,000 customers, 100 potential facility sites, and 10 zones, with connections between customers and facility possible only within zones 1 and 2. To the best of our knowledge, the problem has not yet been widely tackled in such a large scale.

Our experimental results were presented for a branch-and-cut mixed-integer linear programming solver, as well as two scalable local search heuristics: local search (LS) and randomised local search (RLS). These have uncovered a complex pattern of efficiency of different techniques to solve the problem in such a scale. The branch-and-cut solver found a relatively wide gap between the solution found and a lower bound for the model with 10,000 customers. For 100,000 customers, it was only able to produce a lower bound within 24 h.

Regarding the heuristics, while LS provided a solution of good quality rapidly, RLS exhibits higher variance in its performance. However, RLS is at times able to sample a better solution than LS.

In addition, the gap between the performances of LS and RLS seems to become more narrow with growing instance size, while the variance in performance of RLS is preserved.

We believe that our findings will lead to future investigations of problem structure, scale and their impact on algorithm design. It seems that there are intricate connections between these properties of a problem instance and the right choice of an algorithm to solve such an instance. The real-world relevance of the scenarios investigated in this paper could also lead to further interesting large-scale practical applications.

References

1. Al-Sultan, K.S., Al-Fawzan, M.A.: A tabu search approach to the uncapacitated facility location problem. Ann. Oper. Res. **86**, 91–103 (1999)
2. Applegate, D., Cook, W.: A computational study of the job-shop scheduling problem. ORSA J. Comput. **3**(2), 149–156 (1991)
3. Arostegui, M.A., Kadipasaoglu, S.N., Khumawala, B.M.: An empirical comparison of tabu search, simulated annealing, and genetic algorithms for facilities location problems. Int. J. Prod. Econ. **103**(2), 742–754 (2006)
4. Arya, V., Garg, N., Khandekar, R., Meyerson, A., Munagala, K., Pandit, V.: Local search heuristics for k-median and facility location problems. SIAM J. Comput. **33**(3), 544–562 (2004)
5. Bonami, P., Biegler, L.T., Conn, A.R., Cornuéjols, G., Grossmann, I.E., Laird, C.D., Lee, J., Lodi, A., Margot, F., Sawaya, N., et al.: An algorithmic framework for convex mixed integer nonlinear programs. Discrete Optim. **5**(2), 186–204 (2008)
6. Chalupa, D., Blum, C.: Mining k-reachable sets in real-world networks using domination in shortcut graphs. J. Comput. Sci. **22**, 1–14 (2017)
7. Chalupa, D., Hawick, K.A., Walker, J.A.: Hybrid bridge-based memetic algorithms for finding bottlenecks in complex networks. Big Data Res. (2018, to appear)
8. Daskin, M.S., Snyder, L.V., Berger, R.T.: Facility location in supply chain design. In: Logistics Systems: Design and Optimization, pp. 39–65 (2005)
9. Do, N.A.D., Nielsen, I.E., Chen, G., Nielsen, P.: A simulation-based genetic algorithm approach for reducing emissions from import container pick-up operation at container terminal. Ann. Oper. Res. **242**(2), 285–301 (2016)
10. Guner, A.R., Sevkli, M.: A discrete particle swarm optimization algorithm foruncapacitated facility location problem. J. Artif. Evol. Appl. **545**(C), 39–58 (2008)
11. Jansen, T., Zarges, C.: Performance analysis of randomised search heuristics operating with a fixed budget. Theor. Comput. Sci. **545**, 39–58 (2014)

12. Jaramillo, J.H., Bhadury, J., Batta, R.: On the use of genetic algorithms to solve location problems. Comput. Oper. Res. **29**(6), 761–779 (2002)
13. Lančinskas, A., Fernández, P., Pelegín, B., Žilinskas, J.: Improving solution of discrete competitive facility location problems. Optim. Lett. **11**(2), 259–270 (2017)
14. Linderoth, J.T., Lodi, A.: MILP software. In: Wiley Encyclopedia of Operations Research and Management Science (2011)
15. Melo, M.T., Nickel, S., Saldanha-Da-Gama, F.: Facility location and supply chain management-a review. Eur. J. Oper. Res. **196**(2), 401–412 (2009)
16. Michel, L., Van Hentenryck, P.: A simple tabu search for warehouse location. Eur. J. Oper. Res. **157**(3), 576–591 (2004)
17. Mourtzis, D., Doukas, M., Vandera, C.: Mobile apps for product customisation and design of manufacturing networks. Manuf. Lett. **2**(2), 30–34 (2014)
18. Neumann, F., Witt, C.: Bioinspired Computation in Combinatorial Optimization: Algorithms and Their Computational Complexity. Springer, Heidelberg (2010)
19. Nimrod, M., Arie, T.: On the complexity of locating linear facilities in the plane. Oper. Res. Lett. **1**(5), 194–197 (1982)
20. Relich, M., Pawlewski, P.: A multi-agent system for selecting portfolio of new product development projects. Commun. Comput. Inform. Sci. **524**, 102–114 (2015)
21. Sun, J., Xie, Y., Zhang, H., Faloutsos, C.: Less is more: sparse graph mining with compact matrix decomposition. Stat. Anal. Data Min.: ASA Data Sci. J. **1**(1), 6–22 (2008)
22. Sun, M.: Solving the uncapacitated facility location problem using tabu search. Comput. Oper. Res. **33**(9), 2563–2589 (2006)
23. Yigit, V., Aydin, M.E., Turkbey, O.: Solving large-scale uncapacitated facility location problems with evolutionary simulated annealing. Int. J. Prod. Res. **44**(22), 4773–4791 (2006)

Development of Parameter Measurement Method of Information Systems

Olesya Afanasyeva(✉)

Technics Institute, Pedagogical University in Krakow,
Podchorunzhyh Street 2, 30-084 Krakow, Poland
olesia.afanasyeva@gmail.com

Abstract. Measurement and analysis of information systems parameters is a necessary element of their research. Unlike technical systems, in information systems parameters need to be selected basing on the analysis of their functional characteristics. In this work are reviewed peculiarities of information systems parameters selection. For such systems it is important is not only to measure some values, but also setting boundaries, scale etc. In this work is researched the concealment parameter. This parameter describes the level of ability to detect the hidden message in digital environment, providing that according to the concealment parameter it is invisible. To solve the task are used statistical methods of digital environment analysis. In the work, the scale of the concealment parameter is determined and method of measurement of its value is reviewed. The research provides tools to use this parameter and illustrate an example of conduction of measurements in information systems.

Keywords: Information system IS · Steganosystem SS · Semantics · Secret message · Invisibility · Steganokey · Digital environment DE

1 Introduction

Work with information systems (IS) requires the identification, evaluation of certain characteristics, features and changes occurring in them, therefore, there is a need for the implementation of appropriate parameters that characterize these phenomena. Definition of parameters and description of measurement systems in the IS, in contrast to the classical measurement systems for physical and other natural objects [1], have a number of features, which is related to the fundamental difference between these systems. But all approaches to the definition of parameters in the IS should occur in accordance with the generally accepted stages: theoretical and practical justification of measurements; their implementation and interpretation of the results. Particularly important is the measurement of parameter values and their compliance with the expected value. For IS, in most cases, there are no universal and generally accepted measurement system of parameters, due to the fact that such systems are heterogeneous in their description, and each of the IC has its own peculiarities of operation. An example illustrating this circumstance can be the steganographic system (SS), since the terminology used in it allows for many ambiguities. In connection with this, for the SS there is a situation in that the parameters that characterize it, in different cases, have

L. Borzemski et al. (Eds.): ISAT 2019, AISC 1050, pp. 78–90, 2020.
https://doi.org/10.1007/978-3-030-30440-9_9

different interpretations, do not have unambiguously defined measurement standards of the corresponding parameters, characteristics of the individual identified subprocesses. In the case of SS, the problems of measurement and analysis of parameters characterizing the process of its functioning are of fundamental importance. Examples of such parameters can be the invisibility parameter introduced in the digital environment (DE) message η, a parameter that determines the extent of the message concealment μ in DE and a number of other parameters [2]. The use of these or other parameters necessitates the measurement of their values. Unlike the measurement of technical systems (*TS*), where physical factors are specified whose values of values are to be measured, in IS such factors are the characteristics and parameters of processes that are activated within the IS. In this case, as in technical measurements, the necessary means for measuring parameters, the scale of values of parameters and the unit of measurement of their values, it is necessary to set the values of various selected values, for example, which can serve as the maximum and minimum value.

Measurement tools may be part of the corresponding IS, or may be a separate component of the IS if the latter needs additional resources. An important difference in the measurements in the *TS* from the measurements in the IS is that in most cases, measurements in the IS are implemented within research or the latter test, while in the *TS*, a factor characterized by a physical value can be measured by the appropriate means of measuring it in the process of functioning of the corresponding *TS*.

Means of measuring the parameters characterizing the IS, are oriented on the measurement of each parameter that is defined for this system. Such means of measurement should provide the following possibilities:

- simulation of factors that determine the possibility of occurrence of the parameter to be measured;
- registration of measurable parameters;
- calculation of the value of the parameter values and analysis of the value of the measured parameters.

In most cases, the means of measuring the parameters of IS are implemented in the form of software or software-hardware. In IS, factors that determine certain values of the parameters are information that is passed from the IS to the appropriate means, where the latter is analyzed by the algorithms of the appropriate means.

To simulate the factors that determine the possibility of occurrence of this or that parameter, the IS itself is used predominantly. This is due to the fact that the functional processes that generate and activate the relevant factors are quite complex and interdependent, which leads to the need to separate the processes of the functioning of the IS and the processes of measuring the corresponding parameters. Therefore, the values of the measured parameters, their verification and the results of the analysis may differ in some ways from the parameters of the parameters they accept in the IS during its operation.

In most cases, an IS is a typical situation where the parameter does not manifest itself explicitly in the process of operation. In addition, there are IS as a result of which a certain informational product is formed, the parameters of which are determined by parameters of IS and intended for direct use. In this case, the user is the important parameters of the corresponding product that are formed in the IS. Thus, the IS parameters can be implicitly manifested through the parameters of the corresponding product.

Due to the fact that the IS parameters can be manifested unintentionally, due to product parameters, the task of measuring the values of the IS parameters becomes one of the key tasks, in contrast to the measurements of parameters in the TS, where a separate measuring device is connected to the points of the system, in which the measurable parameters are brought, that Provides a direct measurement of the value of such a parameter with a given accuracy.

The given features of measuring parameters of IS cause a rather large complexity of the solution of the problem of measuring the value of parameters. These features are conditioned by the following factors:

- the establishment of the functional correspondence of the parameters of the IS itself by the selected parameters that appear in the products of the functioning of the IS, which we will call an intermediate;
- establishing the relationship between the selected parameters characterizing the IS and possible factors with the IS, leading to a change in the value of the values of the corresponding parameters;
- determining the accuracy of measuring the values of parameters directly from the IS.

2 Task Statement

Determining the functional correspondence of the IS parameters and the mediated parameters of the product IS is based on the analysis of processes occurring in the IS and lead to the creation of an information product. Such IS will be called closed information systems (CIS). In general, IS can collaborate with individual technical components that result from the operation of the IS process in the manufacture of a particular physical product. Such a case, within the framework of this work, will not be considered.

If the CIS functionality produces some information product (IP), then such product may have some or other data (ID) or isolated information products from the CIS, which, in turn, can implement the information processes of the product (IPP). Then it can be argued that the IP product represents some IPP, or a set of them that can be linked to each other. We restrict one IPP, then you can write: SIS $\rightarrow$ IP $\rightarrow$ IPP. We denote some parameter with the symbol π, we can prove that there is the following relation for the CIS:

$$\left[CIS\,\&\left(\pi_i^S \in PIS\right)\right] \rightarrow \left\{IPP \rightarrow \left[\pi_i^P\,\&\left(\pi_i^P \Leftrightarrow \pi_i^S\right)\right]\right\} \tag{1}$$

where π_i^S is a parameter to be measured, *PIS* is a subsystem of *IS*, where $PIS_i \in CIS$, π_i^P is a parameter of *IPP*, if the following is true: $CIS \rightarrow IPP$ which, in this case, is corresponding to π_i^S and characterizes CIS itself.

The proof will not be considered, since it is closely related to the individual classes CIS, and the corresponding *IPP* products.

Within the CIS system, there must be a relationship between the parameter π_i^S and the variables x_i that affect the value of the value of π_i^S, which is established on the basis of the CIS data as a whole. Such dependencies are formed when designing each of the CIS, which can be written in the form of a relationship:

$$\pi_i^S = f_i(x_{i1}, \ldots, x_{in}) \tag{2}$$

where x_{ij} is a variable that influences the value of the parameter π_i^S, f_i is a function that describes the dependency between variables $\{x_{i1}, \ldots, x_{in}\}$ and the parameter π_i^S.

Since CIS are implemented on the basis of the developed algorithms Al_i, then the function $f_i(x_{i1}, \ldots, x_{in})$ in CIS exists and represents the transformations described by these algorithms Al_i. The accuracy of determining the values of π_i^P in the IPP is determined by the possibilities of the finite measuring means and the accuracy of the transitions $\pi_i^S \rightarrow \pi_i^P$. Since it is about measuring the parameters of the IS, the accuracy of the measurements is determined for the parameter π_i^S. Therefore, the error associated with the transition $\pi_i^S(CIS) \rightarrow \pi_i^P(IPP)$ is taken into account when determining the accuracy of the parameters π_i^S. The need for taking into account the transition $\pi_i^S \rightarrow \pi_i^P$ is conditioned by the fact that the parameters of the system determine the required quality of the product functioning of the IS and directly depend on the methods of implementing the corresponding IS.

In addition to *SS* possible examples of IS, which generate some informational product as a result of their functioning, are information encryption systems, the product of which is the cryptogram formed in IS; systems of protection of IS; etc. In the case of a security system (SecS), the latter is part of the IS, and the product of the operation of SecS is the protection of the IS, which can be activated by SecS algorithms. Such remedies can be expanded. Security systems are some functionally oriented PIS_i and provide management of the security of the main IS. In this case, one of the main parameters of the IS is the security level parameter.

The peculiarity of measuring the parameters in the IS is that the ranges of the values of the main parameters can be changed by changing the algorithms of IS functioning, which are directly related to the selected parameters. The required range of values of the parameters is determined by the consumer of the corresponding IS. The concept of the consumer should be understood in a bit abstract, since this may be a certain task that uses the corresponding IS, and the purpose of solving the problem is determined by the domain of the subject, which is intended to use the results of the solution of the problem. The value of the main characteristics of the parameters depends on the implementation of the necessary algorithms, which represent the information tools. The possible values of the parameters from the standpoint of the IS are not limited to physical or technical factors, which may occur for measurements in technical systems. Such restrictions in the IP are realized only on the basis of interpretation of the corresponding parameters and their values in the subject area, which use the corresponding tasks. Formation of algorithms that analyze the corresponding parameters in the IS are realized on the basis of data on their interpretation in the subject area. A measure that takes into account such an interpretation may be different, which may lead to the need to reconcile the data of such an interpretation with other IS algorithms. In many cases, the interpretation of individual parameters in the process of building an IS and in the process of its use can be expanded or modified. This leads to the need to change the characteristics of the parameters, such as changing the limit values of parameters, changes in values that are defined as key or critical. Therefore, at the general level, the choice of key values of the parameters that determine their critical values are realized on the basis of an analysis of the process of functioning of the IS.

3 Research of a Method of Measuring Values of IS_i Parameters by the Example of a Steganographic System

We will consider the method for measuring the values of parameters by using the example of the μ concealment parameter used in the SS queue. This parameter characterizes the ability to detect the presence of a hidden HM message in the digital environment of DE. The degree of visibility of HM in DE is determined by the possibilities of human vision, which manifest itself in the process of user review of the corresponding image. When the degree of visibility of HM in a graphic image lies beyond the limits of the capabilities of the human vision system, provided that the human's vision meets medical standards, then there is the problem of detecting abnormalities in DE, which would indicate that DE is covered by a message. The concealment parameter determines the possibility of setting when anomalies are detected in DE, or such anomalies are due to the built-in HM message. In addition, this parameter can be used in steganoanalysis systems.

Determining the presence of HM allows, in the event that HM cannot be decrypted or restored, to block the ability to transfer HM to a user, for example by scaling the steganogram SG, either by partially eliminating HM or HM filtering, if it is possible to determine the location of the HM in DE.

Detection of HM in some DE is rather complicated, so the process of detecting HM in DE should be divided into stages: the stage of detecting the fact of the presence of HM in DE; the stage of HM localization in DE; playback stage of the message.

The digital environment, in terms of interpreting the sequence of individual information pixel codes of the information image (II) formed as a result of scanning DE in accordance with the selected trajectory, can be interpreted as a random function of some signal (WFS), which is characteristic of a separate II and can have its own distribution law in DE. This interpretation is permissible because it is rather difficult to associate the semantics of a graphic image with the change in the clarity of individual pixels for a case where the trajectory of the choice of successive pixels is not related to the trajectory of the plot, or to other semantic characteristics of the information image located in DE. The introduction of HM into DE will result in changes in the corresponding pattern of WFS distribution that allow them to be interpreted as occurring anomalies in the corresponding function. If we take into account that the trajectory for introducing HM bits in DE does not correspond to the DE trajectory path leading to the occurrence of WFS with a certain distribution, then the probabilistic nature of the corresponding anomalies that are created when implementing HM can be accepted as changes to the WFS. In accordance with [2], the occurrence of anomalies in WFS compared to the WFS of the entire DE, can be measured using coherence. As shown in [3], the formula for determining the parameter μ can be written as:

$$\mu = A_1\left\{\delta_x\left[\gamma^2_{x,y}(f_N)\right] + \ln k\right\} \tag{3}$$

where A_1 is a possible formula expansion, $\gamma^2_{x,y}$ is a coherence function, δ_x means differentiation of $\gamma^2_{x,y}$ by x, f_N is the frequency of the noise component that interprets the anomaly, k is a coefficient that determines the relation between the DE size and the steganogram (SG) size.

3.1 Forming of Measurement Tools

Proceeding from the relation (2), the value of the parameter μ is measured in units that are determined on the basis of the corresponding interpretation, which is the "measure of concealment". The minimum value of this measure, or $\min \mu$, should correspond to a situation where the visibility parameter $\eta = 0$ and HM in DE cannot be visually detected. Therefore, we assume that $\mu = \min \sigma(DE) \rightarrow \mu = 0$, where σ is the value of the semantic difference between DE i (DE & HM), which determines the degree of visibility of HM in the DE system of human vision. Let's assume the following:

$$[min\Delta\sigma(DE) - \Delta\sigma = 0] \rightarrow [(\min \eta - \Delta\eta) = 0] \tag{4}$$

where $\Delta\sigma$ is a discrete minimal value of the unit of semantic change of an image II, which is represented as $\Delta\sigma(DE)$, $\Delta\eta$ is a value of the unit of change of anomaly visibility in DE. It follows from Eq. (4) that the minimum value of the anomaly visibility $\eta = \Delta\sigma(DE)$ is $\Delta\sigma$ greater than the minimal concealment value, or $\min \mu = \min \eta - \Delta\sigma(DE)$.

In the customary interpretation of the trajectory function as some random function, it can be assumed that the function of the HM embedding trajectory in DE also introduces an incident in the scanning trajectory DE(II), which can be considered as an additional component that results in noise in WFS. The parameter μ cannot have an absolute maximal value, since this would mean that there are no anomalies in the WFS and, consequently, there is no HM.

The peculiarity of measurements in IS is that in many cases the interpretation of a parameter in terms of its use in the IS and the interpretation of the units of measurement of the corresponding parameter may be different. For example, the parameter μ characterizes the measure of the concealment of the HM built-in DE message, which can be measured relative to, for example, the percentages of the range $[\min \mu, \max \mu]$, where $\max\mu = M - \varepsilon$, where ε is an infinite small value, M is a maximum possible large quantity, which is chosen on the basis of the analysis of the domain of interpretation II. From a practical point of view, the measurement of μ in the framework of the IS of the SS type, the value of the parameter μ can be measured as the noise abnormality describing the environment DE. The value of the noise anomaly can be measured by the deviation of the selected noise parameter $P_i[Sh(\text{DE})]$, where Sh means noise, $P_i(Sh)$ determines the value of the parameter characterizing the noise $Sh(\text{DE})$. An example of such a parameter may be the amplitude of the curve that surrounds the noise anomalies, or such anomaly can be measured by the frequency value of the frequency component, which causes the corresponding anomaly.

As means of measuring measurements of the μ parameter, it is necessary to use the generators of HM texts, which must have one or another interpretation. It is also necessary to have DE libraries that can serve as DE for HM.

Since DE, when scanning the clarity of pixels or their groups, is a certain function of WFS, it is possible to use the means of frequency analysis of anomalies in DE, which is conditioned by HM, which is embedded in DE.

3.2 Measuring the Concealment Parameter for *SS*

Measurements can be carried out under the following conditions, or as a type of a measurement method:

- measuring parameters of *IS* with the purpose of determining the capabilities characterizing the process of functioning of the *IS*,
- measuring parameters of *IS* in the process of functioning of the system with the purpose of management of the corresponding process,
- measure the parameters of the product functioning of the *IS* to determine the capabilities of *IS* in terms of system requirements.

The first condition for conducting an experiment from the third is that in the first case parameters are measured to determine the capabilities of the *IS* as a whole, not depending on the characteristics of the product, and in the third condition, the parameters of the product of the operation process of the IS are measured, with the purpose of determining whether the IS can produce products with given characteristics.

Regardless of the type of measurements of the μ parameter in the SS, it is necessary to fulfill the following requirements in order to conduct them:

- for measurements, it is necessary to have information about the DE in which the HM is implemented or the WFS (II) function,
- it is necessary to choose the criteria for determining the fluid value of μ,
- it is necessary to have data on all the critical values of the parameter μ, for example, max μ, min μ, the optimal value of μ in relation to other parameters characterizing IS and others.

In this case, we choose as the measurement criterion the Pearson criterion, or $\varkappa^2$ criterion, which is described by a rather simple and constructive formula [4]:

$$\aleph^2 = n \sum_{i=1}^{k} \frac{(P_i^* - P_i)^2}{P_i} \tag{5}$$

where n is the total number of measurement intervals, P_i^* is the probability of amplitude value of the signal from WFS in case when an HM is embedded in DE, P_i is similar to P_i^*, but for case when there is no *HM* embedded in DE.

The parameter $\mu = \frac{\alpha}{\aleph^2}$, where α is the coefficient of harmonization of the measurement scales $\aleph^2$ and μ [5]. In work, for the measurement, DE is chosen in the form of II. DE has a container for placing $HM_1, HM_2, \ldots HM_{20}$, where each HM_i differs from another by message size.

The digital environment of the DE Steganographic System SS, with which the research was conducted, used the .png format images. To enclosure in a DE environment, information was introduced using a spatial domain method [6], substituting the least significant bits in pixels of color (Quantization Index Modulation) for the RGB color model [7]. For each color channel, 16 bits were used (1 pixel - 48 bits). For the study, an algorithm that implements the controlled requantization method is developed. This allowed to implement the embedding of the selected message in accordance with the specified functions that determine the method of embedding. As a result of the experiment, a weak deviation of the values of the pixels *DE* clarity from the corresponding values of the container is shown, as shown in Fig. 1. When embedding information, the index w of the function δ, which was used for embedding δ, is determined on the basis of the following relation:

$$C^w(n_1, n_2) = \delta(C(n_1, n_2), w(n_1, n_2)) = Q_{w(n_1,n_2)}(C(n_1, n_2)) \tag{6}$$

where Q_w is a quantization function, w is an index of this function that characterizes an amount of step of signal quantization, $C(n_1, n_2)$ is a container, $C^w(n_1, n_2)$ is a data carrier.

The chosen method provides the required stability of the steganograph (steganosystem) to Gaussian noise and therefore it can be recommended for use in *SS* design.

The research used the following types of images: images with a small, medium and large level of detail. In the image of different types, embedded messages of different sizes: from the smallest to the maximum. Limit values of parameters (or image sizes) were determined experimentally.

Fig. 1. An example of insignificant deviations of clarity values of a data carrier from the corresponding values of a container

To increase the level of stability of the steganosystem in relation to unauthorized detection of concealed messages, the algorithm for the implementation of the data that is intended to be concealed, was expanded with the possibilities of noise masking the container, shown in Fig. 2.

Fig. 2. Noise masking the container with the message

The experiments used the following types of data to be implemented: text and images. Due to the fact that for implementation of different data types one implementation algorithm was used, the probability of occurrence of different clarity values in the samples used in the realization of experiments was within the range of small, practically permissible deviations for text data and graphic types of images shown on Figs. 3 and 4.

In experimental studies, it has been shown that the functions of the distribution of the probability of changing the clarity values are close when implementing data in different color channels.

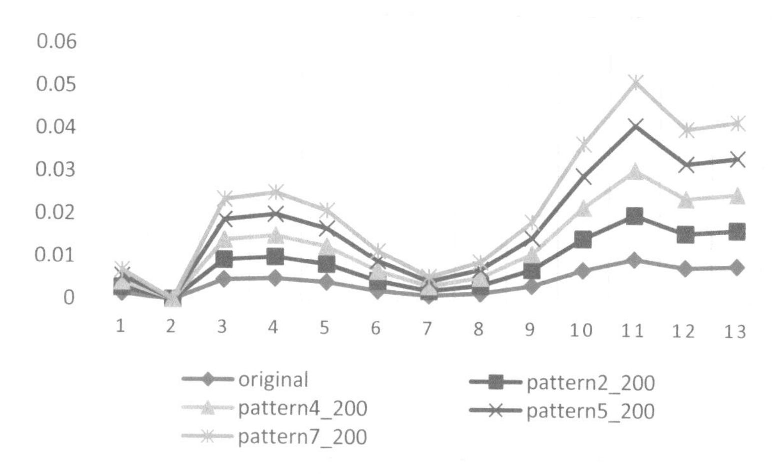

Fig. 3. Probabilistic value of clarity in the samples for textual data with the quantization index 200 in the blue channel

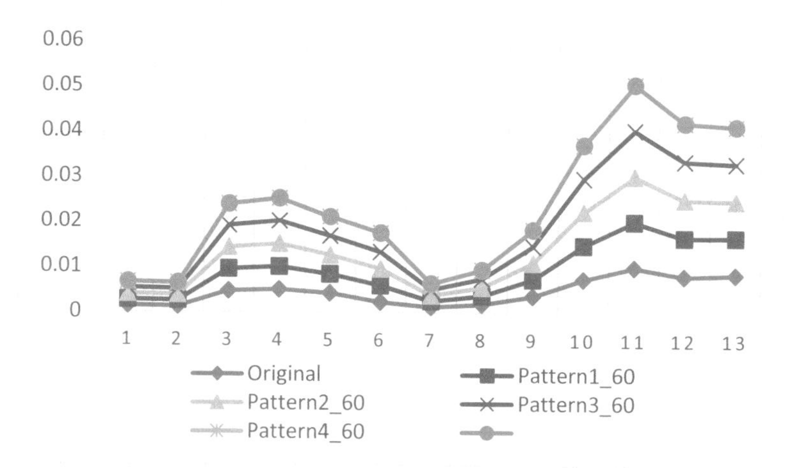

Fig. 4. Probabilistic value of clarity in samples for graphic data with the index of quantization function 60.000 in red, blue and green channels

The difference in relative values of clarity was observed only for different types of images. From experiments conducted for images with an average level of detail, using the different channels of clarity, we get the dependencies shown in Fig. 5.

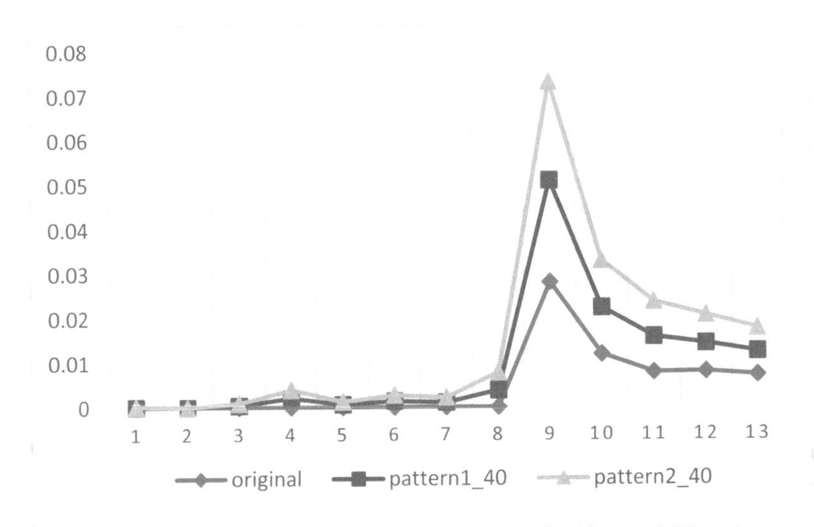

Fig. 5. Probability of clarity value in samples for graphic data with an average level of detail and an index of quantization function of 40.000 in red, blue and green channels

In the case of images with a high level of detail, the nature of changes in the probability values of clarity was observed, which is shown in Fig. 6. To determine the confinement μ, it is necessary to determine the $\varkappa^2$ criterion and analyze its characteristics for all possible cases - different types of images and messages, different quantization indices, and when filling all and different individual color channels. The conducted studies have shown that under different conditions of conducting experiments one can observe the growth of the value of $\varkappa^2$ of the criterion with an increase in the volume of the incoming messages Fig. 7a and b.

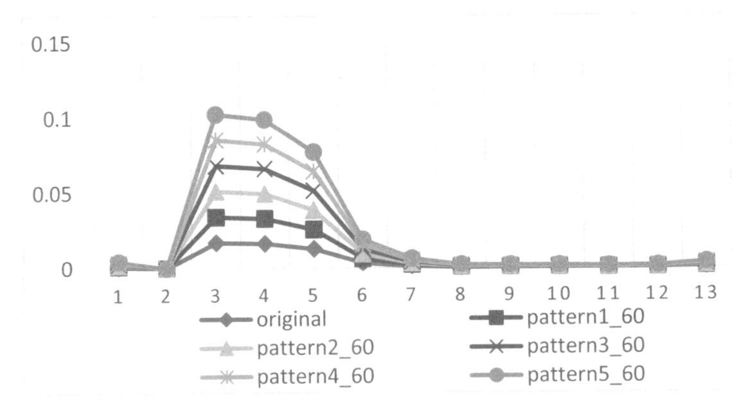

Fig. 6. Probability of clarity value in samples for graphic data with an average level of detail and an index of quantization function of 60.000 in red, blue and green channels

From the research carried out, it appears that the criterion, and hence its related storage parameter, varies depending on the increase in the size of the embedded data, which confirms the possibility of its use in steganography SS systems, when determining the stability of the steganosystem to unauthorized detection of hidden information in a digital environment.

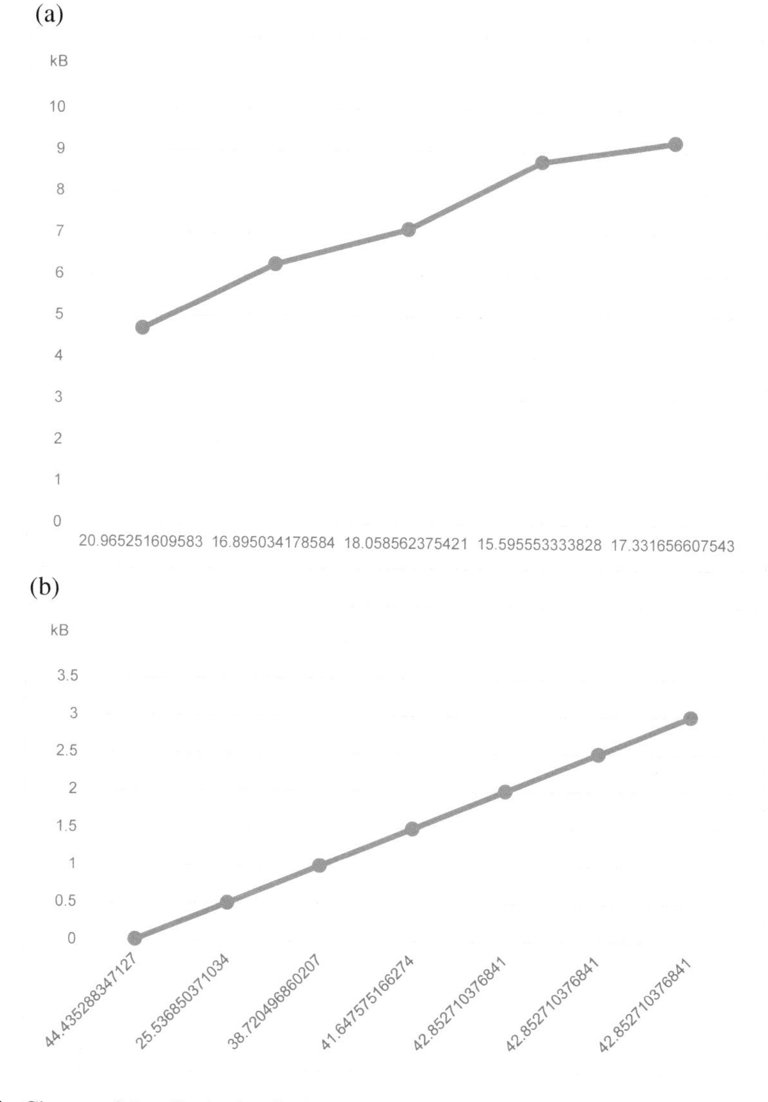

Fig. 7. Change of the $\varkappa^2$ criterion for images of low detail type, various values of the indices of quantization functions 60.000 (a) and 10.000 (b), different types of embedded messages: images and text, from message sizes HM_i

4 Summary

The paper analyzes and researches the main methods for measuring the parameters in the *IS*. The main problems that arise in the implementation of methods for measuring parameters in the *SS* are determined and the ways of their solution are considered. Different types of measurements are determined, which are determined by their purpose. Based on the number of parameters that *SS* characterizes, it is possible to measure the dependence of the selected *SS* parameter on a number of several *SS* parameters, for example, the dependence of μ on the container size, on the parameters of the chosen embedding method HM_i, and others. For example, if the embedding is based on the use of the *DE* cosine transformations, then as parameters you can select the size of the fragment of the image, in which the cosine of the transformation is realized, it is possible to select as the parameter the value of the discrete value of the signal whose magnitude changes the value of the coefficients in the cosine of the decay leading to embedding one message bit.

In the paper, it is shown that for *IS* the method for conducting measurements depends on the selected parameter, which is supposed to measure that, for example, there is a relation with -parameters μ and η for *IS* of type *SS*, since the measurement of the parameter η needs to be used in the process of measuring the data obtained on the basis of using the model of human vision.

References

1. Guide to the Expression of Uncertainty in Measurement. ISO, Switzerland (1995)
2. Fridrich, J., Du, R.: Secure steganographic methods for palette images. In: Proceedings of the Third International Workshop, IH 1999. Lecture Notes in Computer Science, vol. 1768, pp. 47–60. Springer, Heidelberg (2000)
3. Afanasieva, O.: Analysis of parameters of steganography system orientated on graphical digital environments. Sci. J. Marit. Univ. Szczec. **48**(120), 161–170 (2016)
4. Jaynes, E.: Probability Theory The Logic of Science. Cambridge University Press, Cambridge (2003)
5. Karapetoff, V.: Engineering Applications of Higher Mathematics. WENTWORTH Press, Sydney (2016)
6. Cox, I.J.: Digital Watermarking and Steganography, 2nd edn. Morgan Kaufmann Publishers, Burlington (2008)
7. Chen, B., Wornell, G.W.: Quantization index modulation: a class of provably good methods for digital watermarking and information embedding. IEEE Trans. Inf. Theory **47**(4), 1423–1443 (2001)

Social Media Acceptance and Use Among University Students for Learning Purpose Using UTAUT Model

Joseph Bamidele Awotunde[1], Roseline Oluwaseun Ogundokun[2](✉), Femi E. Ayo[3], Gbemisola J. Ajamu[2], Emmanuel Abidemi Adeniyi[2], and Eyitayo Opeyemi Ogundokun[4]

[1] University of Ilorin, Ilorin, Kwara State, Nigeria
Awotunde.jb@unilorin.edu.ng

[2] Landmark University, Omu Aran, Kwara State, Nigeria
{ogundokun.roseline, ajamu.gbemisola, adeniyi.emmanuel}@lmu.edu.ng

[3] McPherson University, Seriki-Sotayo, Ijebo, Ogun State, Nigeria
emmini8168@gmail.com

[4] Agricultural and Rural Management Training Institute (ARMTI), Ilorin, Kwara State, Nigeria
teeposh4u2c@gmail.com

Abstract. Use of Technology and Unified Theory of Acceptance was being utilized in this research study to authenticate the application of a new environment, which is not work associated. The model is used to examine acceptance and willingness of University students' use of social media for learning purposes. Social and peer influence are relevant factors that influence the acceptance of social media for learning purposes. Access to social media might be denied to students by a parent or institutional management if it was used for wrong reasons rather than learning purpose. The results show that all five independents variables have a direct effect on usage for learning purpose. The results also show that social media conditions have a direct hypothesized effect on usage for learning purpose. This may be attributed to the facts that not all social media can be easily used for learning purposes. Another important factor is peer influence that is found to be an important factor. The perception of students would also be affected if social media are perceived only for social activities such as watching films, playing the game and relaxing with family and friends.

Keywords: Acceptance model · Regression analysis · Social media · UTAUT model

1 Introduction

In Nigeria, like many developing countries, Social media have become an important part of a national effort to improve public education. On one hand, Nigeria educationalists hope that e-learning will provide a pathway to education for students who are unable to access higher education; on the other, it is a necessary enhancement for the country to become more competitive among other nations. Social media is built on the

L. Borzemski et al. (Eds.): ISAT 2019, AISC 1050, pp. 91–102, 2020.
https://doi.org/10.1007/978-3-030-30440-9_10

idea of how people interact with each other. There are still various challenges facing the promotion of social media for learning purpose in developing countries. A survey among staff and students from different Nigerian universities shown that unreliable platform and Internet services, low computer literacy level of both staff and students, low acceptance of e-learning was due to the low awareness level, and the high cost of implementation contributed to the low use of social media [2]. There is a need for quantitative analyses of the students' acceptance of social media for the learning purpose, which can offer precise and efficient identification of student willingness for using social media for learning purpose.

A trusted platform and source of information is social media, which is for academician where interaction can take place between them, it is noted to have gained reliability over the years. Educational bodies had tailored this improvement into the system and they depend on collective resources as well as mechanisms to help in enlightening student's life. Furthermore, worthwhile information is gathered by means of the usage of the social network in imparting and acquiring knowledge and at the same time contributes to connecting with studying clusters and other informative systems that will make teaching and learning process convenient. Students and institutions are being offered enough social media tools in order to generate numerous opportunities to advance learning systems.

Technology-based research has become an extensive and emerging interest in academic discipline. The influence and power of the Internet called the World Wide Web (W3) started with computers desktop to laptop computers, moving on to e-learning applications, and now to the dynamic role of social media. These have moved students learning and research into a new technological paradigm, which has helped in the improvement of their academic performances (GPA) [7, 8]. Social media has created a new wave in W3 technologies; this technology has created an intense evolution of the Web, which is daily increased in growth, excitement, and investment [1, 2]. The technology has been used to shares direct kinds of information. It shares a high degree of community formation, likes sharing of academic material, user-level content creation, and a variety of other characteristics within themselves.

There is no doubt Social Media has changed the communication habits of people, provides publicity for various for individuals and organizations and creates opportunities for business. The sharing of information encouraged by social media and also affects life in many other ways. Social media are assumed to encourage and improve university students teaching and learning; therefore, they are the peak latest add-on to the lengthy list of technological novelties.

[3], said the creation and exchange of user-generated content and group of Internet-based applications build on the ideology and technology foundation of Web 2.0 technology is called social media. Therefore, the study reported the acceptance of social media by University students for learning purpose.

This paper proceeds as follows. In Sect. 2, we discussed the literature review of social media and UTAUT model and, Sect. 3 describes the research model and methodology. In Sect. 4, we explain the results and discussion, and finally, in Sect. 5, we conclude the paper.

2 Literature Review

2.1 Social Media to Support Learning Activities

Social media has been identified as a tool in education by many authors for creating personal learning environment [4], open distance education [5], it improves teaching and learning experience [6, 9–15, 32], and many others benefits. It, therefore, served as a tool for improving teaching and a learning experience and an undeniable influence on education.

The significant impact of social media can never be an overemphasis on people's lives and their decisions. The continuous increase of the Internet's penetration rate helps social media to enable communication among people across the world [16]. These broad penetrations allow identifying several communication patterns generated by their use, especially in an academic environment [17]. The used of blogs by schools for advertising for prospective students [18]; and enable live communication on academic events, alerts; positive influence on students' academic; increases interaction between both students and teachers within the academic community, these and others are an instance of the impact and use of social media in an academic environment. From research conducted by [19] on the use of the usage of social media and mobile devices in higher education, their results showed that they contribute and increases the interaction and communication among students and allow them to also engage in online content creation. Social media plugins that will facilitate distribution and communication can be integrated across these networks. There are online tutorials and resources shared through social media network and LMS's, which students can benefit from. There is some appreciated information, which could be derived by means of the usage of social media network such as analytics and understandings on diverse matters or subjects for the drive of learning. In social media, students can ascertain networks that are advantageous for professions. It is fundamental to be effective in several social network platforms as much as possible as an informative institution and this will help to create better student training strategies and also help to shape student culture.

2.2 Unified Theory of Acceptance and Use of Technology (UTAUT)

Research has concluded that there are many user acceptance models in technology research studies. [27], recent improvement and reformation of technology acceptance model studied eight foremost user acceptance hypotheses and these had been a key phase in the study of technology user acceptance and this was named Unified Theory of Acceptance and Use of Technology (UTAUT). Researchers have used UTAUT with other theoretical models to study technology acceptance and use and related issues. [33] studied the impacts of extrinsic motivation and intrinsic motivation on employees' intention to use e-learning in the workplace. They conceptualized performance expectancy, social influences, and facilitating conditions as the components of extrinsic motivation, and effort expectancy as a component of intrinsic motivation. [34, 35] also adopted the same theoretical foundation to examine consumers' purchase intention in the virtual world, but they viewed performance expectancy and effort expectancy as components of extrinsic motivation. Venkatesh et al. (2011) integrated UTAUT beliefs into the two-stage expectation-confirmation model of IS continuance [36] to study citizens continued use of e-government technologies. Other studies have integrated UTAUT with theoretical perspectives such as the equity-implementation model [37], IS success model [38], and task-technology fit [39]. These studies have made some progress. However, here too, there is a lack of integration of the UTAUT moderating variables. Therefore, this paper used UTAUT to study social media acceptance and usage among universities students in Nigeria. UTAUT is said to be a novel model, which inspects this particular model in diverse settings like different technologies as well as user group are necessary. In as much as the primary experiential analysis in UTAUT remained in a recognized setting that is a work context, this study tried to authenticate UTAUT in an extra casual setting such as University, with social media network which is a more informal application as well as with students as subjects nevertheless, the research also attempts to examine the UTAUT background as well. It is additionally deduced that social media was a widespread application among university students [28–31]. This research study, therefore, was divided into two different concepts, which include peer influences, and social influence. Peer-influenced had been considered as an essential element that is regarding the lives of youths and adolescents [20–24]. For illustration, a report discovers that adolescent and youths are 24 times likely to develop to be a smoker if they have 3 or more friends who are smokers [25]. [26], considered user acceptance in the aforementioned studies that user acceptance has also distinguished between the perception of a colleague and public influences. The study utilized UTAUT models for the validation of student acceptance of social media for learning purposes. Yet, as it is exploratory in nature, this study selected measurement items from these theories mainly for the purpose of learning about students' acceptance. The goal of this study does not include the prediction of any future behavior.

Hence, this study furthermore attempted to survey the likelihood that peer influence is likewise an essential element aimed at the usage of social media by youths and adolescents.

3 Research Model and Methodology

3.1 The Proposed Conceptual Framework

Established from the theoretical perspectives of UTUAT model, this study proposes the conceptual framework presented in Fig. 1. The framework depicts the relationships between the independent variables and the dependent variables.

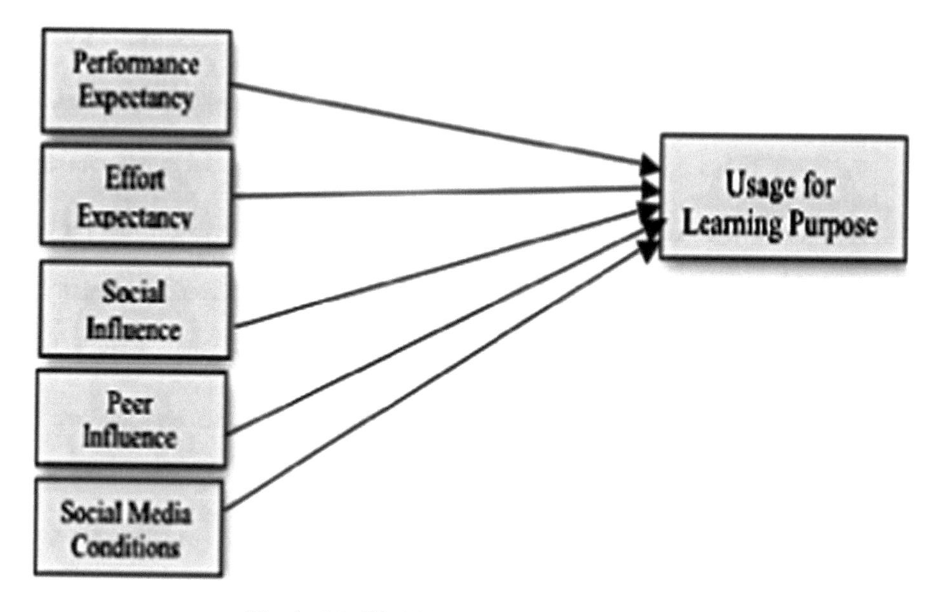

Fig. 1. Modified UTAUT model for the study

4 Research Hypotheses

Since the study group is almost of the same age group, age is taken off from this study. Likewise, since the usage of social media for students under study is highly voluntary, voluntariness' of use is excluded from this study. Dependable with the prior study by [27], the expected relationships among performance expectancy, effort expectancy, social influence, peer influence, and social media condition for learning purposes.

The five (5) option hypotheses for the research questions and the corresponding five null hypotheses are as follows:

H1: There is a significant relationship between performance expectancy and the usage of social media for the learning purpose.
H1a: Performance expectancy will have an effect on social media for the learning purpose.
H1b: Performance expectancy will not have an effect on social media for the learning purpose.
H2: There is a significant relationship between effort expectancy and the usage of social media for the learning purpose.
H2a: Effort expectancy will have an effect on social media for the learning purpose.
H2b: Effort expectancy will not have an effect on social media for the learning purpose.
H3: There is a significant relationship between social influence and the usage of social media for the learning purpose.
H3a: Social influence will have a significant influence on social media for the learning purpose.

H3b: Social Influence will not have a significant influence on social media for the learning purpose.
H4: Peer influence will have a significant influence on social media for the learning purpose.
H4a: Peer Influence will have a significant influence on social media for the learning purpose.
H4b: Peer Influence will not have a significant influence on social media for the learning purpose.
H5: There is a significant relationship between social media conditions and the usage of social media for the learning purpose.
H5a: Social media conditions will have a significant positive influence on social media for the learning purpose.
H5b: Social media conditions will not have a significant influence on social media for the learning purpose.

5 Results and Discussion

5.1 Reliability of the Questionnaire

Cronbach's Alpha is used to test the reliability of the questionnaire.

Table 1. Analysis of Variance (ANOVA)

	Model	Sum of squares	Df	Mean square	F	Sig.
1	Regression	238.022	1	238.022	4958.220	.000[b]
	Residual	57.511	1198	.048		
	Total	295.533	1199			
2	Regression	241.311	2	120.656	2663.622	.000[c]
	Residual	54.221	1197	.045		
	Total	295.533	1199			
3	Regression	243.950	3	81.317	1885.440	.000[d]
	Residual	51.582	1196	.043		
	Total	295.533	1199			
4	Regression	244.661	4	61.165	1436.806	.000[e]
	Residual	50.872	1195	.043		
	Total	295.533	1199			
5	Regression	245.218	5	49.044	1163.845	.000[f]
	Residual	50.314	1194	.042		
	Total	295.533	1199			

a. Dependent Variable: SMU
b. Predictors: (Constant), PI
c. Predictors: (Constant), PI, SMC
d. Predictors: (Constant), PI, SMC, SI
e. Predictors: (Constant), PI, SMC, SI, EE
f. Predictors: (Constant), PI, SMC, SI, EE, PE

Table 2. Stepwise method of regression showing the variables entered/removed

Mode	Variables entered	Variables removed	Method
1	PI	.	Stepwise (Criteria: Probability-of-F-to-enter <= .050, Probability-of-F-to-remove >= .100)
2	SMC	.	Stepwise (Criteria: Probability-of-F-to-enter <= .050, Probability-of-F-to-remove >= .100)
3	SI	.	Stepwise (Criteria: Probability-of-F-to-enter <= .050, Probability-of-F-to-remove >= .100)
4	EE	.	Stepwise (Criteria: Probability-of-F-to-enter <= .050, Probability-of-F-to-remove >= .100)
5	PE	.	Stepwise (Criteria: Probability-of-F-to-enter <= .050, Probability-of-F-to-remove >= .100)

From Table 1, the Analysis of Variance (ANOVA) tests the significance of each of the model to see if the regression predicted by the independent variables explains a significant amount of the variance in the dependent variable. Using the stepwise method, ANOVA for each model. Considering all the five models.

Chi-Square Test of Independence

Pearson Chi-Square test of independence is used here to test if ULP is related to each of PE, EE, SI, PI, and SMC.

Test of independence between ULP and PE
Ho: ULP and PE are independent vs H1: ULP and PE are not independent $x^2 = 972.458$, df = 12, $p < 0.001$.
Ho is rejected. ULP and PE are not independent.
Test of independence between ULP and EE
Ho: ULP and EE are independent vs H1: ULP and EE are not independent $x^2 = 2045.152$, df = 12, $p < 0.001$. Ho is rejected. ULP and EE are not independent.
Test of independence between ULP and SI
Ho: ULP and SI are independent vs H1: ULP and SI are not independent $x^2 = 2135.596$, df = 12, $p < 0.001$. Ho is rejected. ULP and SI are not independent.
Test of independence between ULP and PI
Ho: ULP and PI are independent vs H1: ULP and PI are not independent $x^2 = 2450.189$, df = 12, $p < 0.001$. Ho is rejected. ULP and PI are not independent.
Test of independence between ULP and SMC
Ho: ULP and SMC are independent vs H1: ULP and SMC are not independent $x^2 = 129.093$, df = 12, $p < 0.001$. Ho is rejected. ULP and SMC are not independent.

Table 3. Model summary

Model	R	R Square	Adjusted R Square	Std. error of the estimate
1	.897[a]	.805	.805	.219
2	.904[b]	.817	.816	.213
3	.909[c]	.825	.825	.208
4	.910[d]	.828	.827	.206
5	.911[e]	.830	.829	.205

a. Predictors: (Constant), PI
b. Predictors: (Constant), PI, SMC
c. Predictors: (Constant), PI, SMC. SI
d. Predictors: (Constant), PI, SMC, SI, EE
e. Predictors: (Constant), PI, SMC, SI, EE, PE

From Table 3, the R Square column shows the amount of variance in the dependent variable that can be explained by the independent variable, with the highest value equal to 0.830 as seen in Model 5. This means that PI, SMC, SI, EE, and PE altogether account for 83% of the variance in the ULP.

Table 4. Pearson correlation coefficients for N-1200

Study assumption	The correlation coefficient (r)	The coefficient of determination (r^2)	Verified result on hypothesis
H1: There is a significant relationship between PE and the usage of social media for learning purpose	+0.6413	0.4752	Proved
H2: There is a significant relationship between EE and the usage of social media for learning purpose	+0.6406	0.4475	Proved
H3: There is a significant relationship between SI and the usage of social media for learning purpose	+0.6789	0.5870	Proved
H4: There is a significant relationship between PI and the usage of social media for learning purpose	+0.6721	0.5432	Proved
H5: There is a significant relationship between SMC and the usage of social media for learning purpose	+0. 6721	0.4389	Proved

Hypothesis 1.
Pearson Correlation Coefficients analysis was used to investigate the influence of performance expectancy on social media acceptance in a University. The results showed that performance expectancy positively influenced social media acceptance in a University ($r = +0.6413$, $r^2 = 0.4752$).

Therefore, Null hypothesis 1 was rejected and Hypothesis 1 was supported. This means that when students expect social media to increase their performance, they increase their acceptance to use it.

Hypothesis 2.
Pearson Correlation Coefficients analysis was used to investigate the influence of effort expectancy on social media acceptance in a University. The results showed that effort expectancy positively influenced social media acceptance in a University ($r = +0.6406$, $r^2 = 0.4475$).

Therefore, Null hypothesis 2 was rejected and Hypothesis 2 was supported. This means that when students expect social media to be easy to use, they increase their acceptance to use it.

Hypothesis 3.
Pearson Correlation Coefficients was used to investigate the influence of social influence on social media acceptance in a University. The results showed that social influence positively influenced social media acceptance in a University ($r = +0.6789$, $r^2 = 0.5870$).

Therefore, Null hypothesis 3 was rejected and Hypothesis 3 was supported. This means that when students are suggested to use social media by someone important to them, they increase their acceptance to use it.

Hypothesis 4.
Pearson Correlation Coefficients was used to investigate the influence of peer influence on social media acceptance in a University. The results showed that peer influence positively influenced social media acceptance in a University ($r = +0.6789$, $r^2 = 0.5870$). Therefore, Null hypothesis 4 was rejected and Hypothesis 4 was supported. This means that when students are suggested to use social media by their peers, they increase their acceptance to use it.

Hypothesis 5.
Pearson Correlation Coefficients was used to investigate the influence of social media condition on social media acceptance in a University. The results showed that social media condition positively influenced social media acceptance in a University ($\beta = .619$, $p < .001$).

Therefore, Null hypothesis 5 was rejected and Hypothesis 5 was supported. This means that when students can easily use social media platforms, they increase their acceptance to use it. Finding results of Pearson Correlation Coefficients rejected the five null hypotheses in the research question and support the five alternative hypotheses.

6 Conclusion

Hypothetically, this study provides significant contributions to the body of knowledge to validate the use of the UTAUT model in another environment. The framework work adopted the UTAUT model using five constructs namely: Performance Expectancy, Effort Expectancy, Social Influence, Peer Influence, and Social Media Conditions to measure university students' perceptions towards the use of social media technology for learning purpose.

The findings provided additional evidence to the applicability of the UTAUT model in other related technology adoption. Other trends were also revealed such as the popularity of social media applications. Findings in this study provided insights into social media acceptance by the selected universities. Universities, government, and education practitioners could use the finding results as references in social media application and education innovation. Finally, the UTAUT model is highly recommended for analyzing data on social media acceptance and use among university students for learning purposes in Nigeria.

Acknowledgment. This research is fully sponsored by Landmark University Centre for Research and Development, Landmark University, Omu-Aran, Nigeria.

References

1. Manoj, P., Andrew, B.W.: Social computing: an overview. Commun. Assoc. Inf. Syst. **19**(7), 762–780 (2007)
2. John, G.C., Irene, G.: Students' perceptions and readiness towards mobile learning in colleges of education: a Nigerian perspective. S. Afr. J. Educ. **37**(1), 1–12 (2017)
3. Kaplan, A.M., Haenlein, M.: Users of the world, unite! The challenges and opportunities of Social Media. Bus. Horizons **53**(1), 59–68 (2010)
4. Dabbagh, N., Kitsantas, A.: Personal learning environments, social media, and self-regulated learning: a natural formula for connecting formal and informal learning. Internet High. Educ. **15**, 3–8 (2012)
5. Kulakli, A., Mahony, S.: Knowledge creation and sharing with Web 2.0 tools for teaching and learning roles in so-called University 2.0. In: 10th International Strategic Management Conference, 19–21 June 2014, Elsevier Ltd., Social and Behavioral Sciences, vol. 150, pp. 648–657 (2014)
6. Danciu, E., Grosseck, G.: Social aspects of web 2.0 technologies: teaching or teachers' challenges? In: WCES 2011, 03–07 February 2011, Elsevier Ltd, Social and Behavioral Sciences, vol. 15, pp. 3768–3773 (2011)
7. Ogundokun, R.O., Adebiyi M., Abikoye, O., Oladele, T., Adeniyi, A., Akande, N.: Performance Evaluation: Dataset on the Scholastic Performance of Students in 12 Programmes from a Private University in the South-West Geopolitical Zone in Nigeria. F1000 Research (2018)
8. Lukman, A.F., Adebimpe, O., Onate, C.A., Ogundokun, R.O., Gbadamosi, B., Oluwayemi, M.O.: Data on expenditure, revenue, and economic growth in Nigeria. Data Brief **20**, 1704–1709 (2018)
9. Wagner, R.: Social media tools for teaching and learning. Athletic Training Educ. J. **6**, 51–52 (2011)
10. Oberer, B., Erkollar, A.: Social media integration in higher education. cross-course google plus integration is shown in the example of a master 's degree courses in management. In: CY-ICER 2012, 08–10 February 2012, Elsevier Ltd., Social and Behavioral Sciences, vol. 47, pp. 1888–1893 (2012)
11. Popoiu, M.C., Grosseck, G., Holotescu, C.: What do we know about the use of social media in medical education. In: WCES 2012, 02–05 February 2012, Elsevier Ltd., Social and Behavioral Sciences, vol. 46, pp. 2262–2266 (2012)

12. Thomas, M., Thomas, H.: Using new social media and Web 2.0 technologies in business school teaching and learning. J. Manage. Dev. **3**, 358–367 (2012)
13. Bexheti, L.A., Ismaili, B.E., Cico, B.H.: An analysis of social media usage in teaching and learning: the case of SEEU. In: 2014 International Conference on Circuits, Systems, Signal Processing, Communications and Computers, 15–17 March 2014, Venice, pp. 90–94 (2014)
14. Dear, A.R., Potts, L.: Teaching and learning with social media tools, cultures, and best practices. Programmatic Persp. **6**, 21–40 (2014)
15. Rasiah, R.R.V.: Transformative higher education teaching and learning: using social media in a team-based learning environment. In: TTLC 2013, 23 November 2013, Elsevier Ltd., Social and Behavioral Sciences, vol. 123, pp. 369–379 (2014)
16. Williams, D.J., Gownder, P., Laura, W.: How to Turn Social Media Assets into Social Co-Creation Assets. Forrester Research, Cambridge, MA (2010)
17. Davis, C.H.F., Deil-Amen, R., Rios-Aguilar, C., Gonzalez Canche, M.S.: Social Media in higher education: A literature review and research directions. The Center for the Study of the Higher Education at the University of Arizona and Claremont Graduate University (2012). http://work.bepress.com/hfdavis/2/
18. Violino, B.: The buzz on campus: social networking takes hold. Community Coll. J. **79**(6), 28–30 (2009)
19. Gikas, J., Grant, M.M.: Mobile computing devices in higher education: Student perspectives on learning with cell phones, smartphones & social media. Internet High. Educ. **19**, 18–26 (2013)
20. Brittain, C.V.: Adolescent choices and parent-peer cross-pressures. Am. Sociol. Rev. **28**, 385–391 (1965)
21. Berndt, T.J.: Development Changes in Conformity to Peers and Parents. Dev. Psychol. **15**, 608–616 (1979)
22. Savin-Williams, R.C., Berndt, T.J.: Friendship and peer relations. In: Elliott, S.S.R. (ed.) At the Threshold: The Developing Adolescent, pp. 277–307. Harvard University Press, Cambridge (1990)
23. Sim, T.N., Koh, S.F.: A domain conceptualization of adolescent susceptibility to peer pressure. J. Res. Adolesc. **13**(1), 57–80 (2003)
24. Santor, D.A., Messervey, D., Kusumakar, V.: Measuring peer pressure, popularity, and conformity in adolescent boys and girls: predicting school performance, sexual attitudes, and substance abuse. J. Youth Adolesc. **29**(2), 163–182 (2004)
25. Lloyd-Richardson, E.E., Papandonatos, G., Kazura, A.: Differentiating stages of smoking intensity among adolescents: stage-specific psychological and social influences. J. Consult. Clin. Psychol. **70**(4), 998–1009 (2002)
26. Taylor, S., Todd, P.A.: Understanding information technology usage: a test of competing models. Inf. Syst. Res. **6**(2), 144–176 (1995)
27. Venkatesh, V., Morris, M.G., Davis, G.B., Davis, F.D.: User acceptance of information technology: toward a unified view. MIS Q. **27**(3), 425–478 (2003)
28. Grinter, R.E., Palen, L.: Instant messaging in teen life. In: Proceedings of the 2002 ACM Conference on Computer Supported Cooperative Work, pp. 21–30 (2002)
29. Huang, A.H., Yen, D.C.: The usefulness of instant messaging among young users: social vs work perspective. Hum. Syst. Manage. **22**(2), 63–72 (2003)
30. Nardi, B.A., Whittaker, S., Bradner, E.: Interaction and outeraction: instant messaging in action. In: Proceedings of the 2000 ACM Conference on Computer Supported Cooperative Work, pp. 79–88 (2000)
31. Renneker, J., Godwin, L.: Theorizing the unintended consequences of instant messaging for worker productivity, sprouts: working papers on information environments. Syst. Organ. (3) (2003). http://weatherhead.cwru.edu/sprouts/2003030307.pdf

32. Miloševic, I., Zivkovic, D., Arsic, S., Manasijevic, D.: Facebook as a virtual classroom – social networking in learning and teaching among Serbian students. Telematics Inform. **32**, 576–585 (2015)
33. Yoo, S.J., Han, S.H., Huang, W.H.: The roles of intrinsic motivators and extrinsic motivators in promoting e-Learning in the workplace: a case from South Korea. Comput. Hum. Behav. **28**(3), 942–950 (2012)
34. Guo, Y., Barnes, S.: Purchase behavior in virtual worlds: an empirical investigation in second life. Inf. Manage. **48**(7), 303–312 (2011)
35. Guo, Y., Barnes, S.J.: Explaining purchasing behavior within World of Warcraft. J. Comput. Inf. Syst. **52**(3), 18–30 (2012)
36. Bhattacherjee, A., Premkumar, G.: Understanding changes in belief and attitude toward information technology usage: a theoretical model and longitudinal test. MIS Q. **28**(2), 229–254 (2004)
37. Hess, T.J., Joshi, K., McNab, A.L.: An alternative lens for understanding technology acceptance: an equity comparison perspective. J. Organ. Comput. Electron. Commer. **20**(2), 123–154 (2010)
38. Kim, C., Jahng, J., Lee, J.: An empirical investigation into the utilization-based information technology success model: integrating task-performance and social influence perspective. J. Inf. Technol. **22**(2), 152–160 (2007)
39. Zhou, T., Lu, Y.B., Wang, B.: Integrating TTF and UTAUT to explain mobile banking user adoption. Comput. Hum. Behav. **26**(4), 760–767 (2010)

Data and Process Management

A Generic Approach to Schema Evolution in Live Relational Databases

Anna O'Faoláin de Bhróithe, Fritz Heiden, Alena Schemmert, Dschialin Phan, Lillian Hung, Jörn Freiheit, and Frank Fuchs-Kittowski(✉)

HTW Berlin, Berlin, Germany
{annaofaolain.debhroithe, fritz.heiden, alena.schemmert, dschialin.phan, lillian.hung, joern.freiheit, frank.fuchs-kittowski}@htw-berlin.de

Abstract. Schema evolution is an important theme for many database users across a broad range of fields. This paper introduces a generic data management layer, GeneRelDB, which allows the schema of a relational database to evolve during run time without the need to rewrite database queries in the application code. It is designed to run as an abstraction layer, handling all communication (queries and data exchange) between the user interface and the database back-end. The only restriction to the changes that can be made relate to data type conversion for existing columns in the database. Foreign key constraints are supported and referential integrity is maintained during evolution.

Keywords: Schema evolution · SQL · Relational databases · Database abstraction

1 Introduction

Schema evolution, the ability to alter already-deployed schemata, is an important theme for many database users across a broad range of fields. For example, it is relevant to medical databases, in which the number of possible tests and diagnoses increases every year and the symptoms, tests, and diagnoses relevant to a patient in a single doctor's visit can in no way be predicted or even standardised; project management, in which the size of a project can grow significantly since its start date; scientific collaborations and experiments, in which the volume and type of data that must be stored increases over time and cannot be fully predicted in advance; or even companies who wish to provide a general and reusable database solution to their clients without a large customisation/reimplementation effort for each individual client. It is also a central issue in the modern era of web services and applications, many of which have 24/7 uptime requirements but must still be able to have deficiencies patched and their databases expanded in the background.

Document-oriented databases such as MongoDB [1], Amazon's proprietary DynamoDB [2], or Apache's CouchDB [3] (to list but a few) naturally accommodate flexible data modelling as the schema can vary from document to document. However, relational databases are very well established, remain in broad use, and offer some

L. Borzemski et al. (Eds.): ISAT 2019, AISC 1050, pp. 105–118, 2020.
https://doi.org/10.1007/978-3-030-30440-9_11

advantages over NoSQL databases. For example, database normalisation in relational databases provides a safeguard against data anomalies and inconsistencies that cannot be enforced by a schemaless database, data types ensure that data are well-formed, and ACID requirements have been solved with a good balance between durability and performance. A tool to ease the burden of evolving schema in relational databases could offer similar flexibility to a schemaless database while still retaining the advantages of an RDBMS.

Another issue associated with evolving database schema is handling necessary changes to the application code to deal with the updated data structure [4]. Even the use of schemaless databases does not avoid this problem: as the database schema is free, the act of finding and extracting all relevant attributes and information is shifted to the application code.

The generic data management layer, dubbed GeneRelDB, described in this paper provides a solution to easily managing evolving schema in relational databases. It monitors the database schema by means of two metadata tables and automatically generates SQL queries based on the current database schema, removing the need to rewrite any application code when a schema change occurs. Data records in the database are automatically migrated to satisfy the new schema. GeneRelDB is envisaged for use in diverse management scenarios – from project management to environmental management – in which the user may wish to define Key Performance Indicators (KPIs) as mathematical functions of data contained in the database and monitor their values over time. To easily facilitate this, GeneRelDB manages the creation and calculation of KPIs, thereby enabling the user to quickly and easily define and track KPIs.

The article is structured as follows: Sect. 2 provides an overview of related work on schema evolution in relational databases. Section 3 describes GeneRelDB, the generic data management layer, and a discussion of this concept is provided in Sect. 4. Section 5 is dedicated to an example application of GeneRelDB, and a summary and outlook is given in Sect. 6.

2 Existing Solutions for Relational Databases

Schema evolution has been the topic of research spanning the last few decades as evidenced by the number of entries in the online bibliography of schema evolution [5]. In this section, we present an overview of some of the existing models and tools supporting schema evolution in relational databases.

The Entity-Attribute-Value (EAV) model or vertical database model is one way of achieving flexible schema in a relational database. In this model, a generic attribute table is created with the aim of storing attributes as separate rows, thus easily allowing a varying number of attributes and attribute types for different entities. The table has three columns: entity, usually a foreign key to a parent table of entities; attribute, the name of the attribute; and value, the value of this attribute. By using this model, the number of columns does not have to increase in order to support new attributes – however, this is offset by the disadvantage of not being able to support data integrity checks. For example, there is no guarantee that attributes, which are intended to be the same, have the same name (e.g. start_date vs. start-date vs. startdate vs. date_begin);

referential integrity cannot be enforced; data types cannot be enforced for specific attributes; and common constraints such as UNIQUE or NOT NULL cannot be defined for specific attributes.

When using a more traditional relational model, adding or removing attributes require changes to the database schema. The authors of [6] describe a model-driven approach that provides the adaptability to respond to changes in evolving and expanding clinical and translational data management. This approach requires an ontology of the domain as input – e.g. a medical ontology for the case of clinical data management – from which an entity relationship (ER) model is generated. The ER model is then converted to a relational model, and the relational model is transformed into a MySQL schema script, both conversions performed by means of an XSLT-based code generation technique. Finally, model-based queries are translated into SQL queries based on the information in the ER model. As the model-based query is on a higher level of abstraction than the SQL queries, it can still be used even when the underlying RDB schema evolves; however, it must be translated to SQL anew for every schema evolution.

Another suite of tools, PRISM++ [7], provides support for schema evolution by offering a language for the evolution of schema and integrity constraints, tools that evaluate the impact of the evolution, automatic data migration, and tools for rewriting legacy database queries. The rewriting is achieved by performing a "virtual migration" of the current schema back to that expected by the legacy query. This results in a sequence of modification operators for schema and integrity constraints which can be used to determine the semantics of rewriting the queries.

Some changes set table locks while executing, which can cause interruptions in the current era of web services and applications with around-the-clock uptime requirements. The open-source QuantumDB [8] aims to automate schema evolution while still maintaining zero downtime for continuous deployment environments in the process. It provides a mixed state for every set of schema changes by automatically maintaining a set of synchronised "ghost tables" for those tables affected (either directly or indirectly) by the changes. It thus adopts mixed state by construction, maintains referential integrity, and renders all DDL statements non-blocking. This tool supports database schema evolution at runtime and avoids downtime for client applications.

GeneRelDB aims to solve the problem of schema evolution in a generic way. It monitors the current database schema by means of metadata tables and, based on this information, automatically generates appropriate SQL queries on the fly.

3 Generic Data Management Layer for Relational Databases

The generic data management layer proposed in this paper, GeneRelDB, allows the schema of a relational database to evolve during run time without the need to rewrite database queries in the application code. It is designed to run as a server handling all communication (queries and data exchange) between the user interface and the database backend. The only restriction to the changes that can be made relate to data type conversion for existing columns in the database. For example, it is not possible to convert a date to a boolean and ensure a meaningful output; thus, such a change to the database is forbidden.

In order to more easily present the concept and avoid confusion, let us introduce the following glossary of terms (Table 1):

Table 1. Glossary terms

Generic data management layer	The concrete database schema supplemented by metadata tables and the prepared sql queries utilising this metadata provided by the server
Concrete schema	The schema of the database tables that hold actual data
Concrete table	A table in the database that contains actual data (as opposed to metadata)
Concrete column	A column in a concrete table
Metadata tables	These tables contain information on the concrete schema. The generic data management layer requires two metadata tables: TABLES and COLUMNS
TABLES table	A metadata table containing information on all concrete tables in the database (excluding column information)
COLUMNS table	A metadata table containing information on all concrete columns in the database

3.1 Metadata Tables

GeneRelDB is implemented with the help of two metadata tables: tables and columns. These tables provide information on the concrete schema. Tables holds metadata on all existing tables in the database and columns holds information on all columns in the database.

The table metadata consists of an ID and the table name, i.e., the table name as given by the user. There is exactly one entry in the tables table for each concrete table in the database. The concrete tables are named according to the convention table < table_id> where <table_id> is a unique integer that is incremented with each additional table. This means that if a table is renamed, only the entry in the tables table must be updated and nothing happens with the actual table itself.

The column metadata comprises the column ID, column name (as given by/visible to the user), data type, and the table ID of the table to which it belongs. For each entry in the columns metadata table, there exists a concrete column. Similar to concrete tables, concrete columns are named according to the convention column <column_id>, where <column_id> is a unique integer that is incremented with each newly-added column. Analogous to the procedure for renaming a table, if a column is renamed, only the value of the column name in the columns table must be updated and the actual column itself remains untouched.

The data type of the column is encoded as an integer in the columns table. Negative values represent native data types while positive values represent a reference to another table, i.e., a foreign key. In the case of a foreign key, the table_id of the referenced table acts as the data type. Table 2 shows the chosen datatype encoding for common data types.

Table 2. Integer encoding of common data types

ID	Data type
–1	Integer
–2	Double
–3	String
–4	Boolean
–5	Date
–6	JSON
>0	Reference to another concrete table (foreign key)

3.2 Concrete Tables and Columns

The concrete tables and columns hold the data records. Their structure reflects the information contained in the metadata tables but they themselves provide only a poor overview for users due to the use of the generic naming convention (table <table_id> for tables and column <column_id> for columns). The generic naming convention makes the implementation much easier than working directly with the user-defined table and column names. Database users can still get a full overview of the database schema by querying the metadata tables.

3.3 Changing the Schema

By maintaining an overview of the database schema in the metadata tables and building generic database queries (using the PreparedStatement interface from the java.sql package) that use this metadata information, it is possible to make changes to almost every aspect of the schema during run time without any interruption to database availability and without having to rewrite any SQL requests.

Adding Tables: When a new table is added to the schema, a new entry is created in the TABLES table and a corresponding concrete table is created. The metadata entry must be created first in order to generate the table ID number which is then used as part of the name of the concrete table.

Renaming Tables: The user-given name of a table is pure metadata. The concrete schema, as it exists in the database, is therefore independent of the user-given table names. When a table is renamed, only the value of the "name" column in the TABLES table is changed. The concrete table remains unaffected.

Deleting Tables: When a table is deleted, the concrete table is deleted along with all the data it holds. It is possible that another table can have a foreign key referring to the table to be deleted. There are two ways to resolve this conflict: Firstly, the "DROP CASCADE" SQL command deletes the target table as well as all other objects that depend on it (e.g., views or other tables with a foreign key that references the target table). If the other tables that are now additionally deleted are also referenced by foreign keys in further tables, then the tables containing these foreign keys must also be deleted, and so on. Thus, the DROP command "cascades" through the database based on dependencies. The second option is to change the data type of the columns containing foreign keys to the target table. If the data type is changed to an integer, for example, then the relationship is broken. The column loses its meaning in this case, but additional tables do not have to be deleted.

Once the target table has been deleted, its metadata must also be deleted. This includes all relevant entries in the TABLES and COLUMNS metadata tables. The metadata is only deleted after the target table has been successfully dropped so that in the case that the target table cannot be deleted, the metadata (as well as the table) remain in the database.

Adding Columns: To add a column to an existing table, an entry must first be created in the COLUMNS table. The generated column ID number is then used to name the concrete column. The target table is modified and a column of the requested data type (native data type or foreign key) is added.

Renaming Columns: Renaming a column is an analogous procedure to renaming a table. The user-given column name is pure metadata, meaning that nothing in the concrete schema need change when the column name is altered. Only the value of the "name" column in the COLUMNS table must be changed.

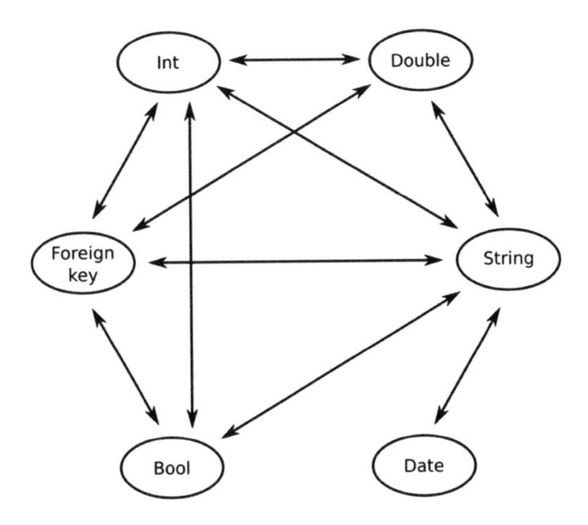

Fig. 1. Data type conversion

Changing the Data Type of Columns: The ability to change the data type of a column depends on a number of factors. If there are no data in the corresponding concrete table, then changing the data type poses no problem. However, even this depends heavily on which database management system is used. If the table already contains data, care must be taken in order to ensure that the existing data can be converted directly to the new data type. Allowed conversions are shown in Fig. 1 Date↔Bool, Date↔Foreign key, Date↔Integer, Date↔Double, and Bool↔Double conversions are not supported.

When a data type conversion is called on a column that already contains data, a check is performed to see if the existing data can really be converted into the target data type. For example, if a column of strings is to be converted to doubles, the server must check if all strings in the column can be properly expressed as doubles and if not, throw and exception and stop the conversion process. If the data type of the concrete column is successfully changed, the corresponding metadata in the COLUMNS table must also be adjusted accordingly.

Deleting Columns: Deleting a column modifies the concrete table to which the column belongs by removing the target column. The corresponding metadata is adjusted by deleting the entry for the target column in the COLUMNS table.

3.4 Key Performance Indicators

The generic data management layer is envisaged not only for flexible data management but also for use in generalised management scenarios – from project management to environmental management – in which the user may wish to define KPIs as mathematical functions of data contained in the database and monitor their values over time. GeneRelDB manages the creation and calculation of KPIs by means of a specialised KPI table, thereby enabling the user to quickly and easily define and track KPIs.

KPIs are characterised by a unique name and a formula saved as text (JSON format). The form of the mathematical formula represents a binary operator tree, corresponding to how the value of the KPI is calculated. The left and right elements on each level of the tree can be of different types: a data set, consisting of data records from the specified table and column; an aggregation of a data set (e.g., sum, mean, min, max, etc.); a data point consisting of a single row from the specified table and column; a constant number; another binary operator expression. The depth of the tree is not limited and can be arbitrarily extended by nesting binary operator expressions. Values of KPIs are not stored in the database, but are calculated on the fly when requested.

3.5 Schema Templates

Database schemata can be imported by means of a schema import function. When preparing the template, tables can be given in any order – the order in which it is necessary to actually create the tables in the database so that foreign keys can be resolved is determined by the import function itself. The only restriction is the recursive foreign keys cannot currently be interpreted by the importer and must be added afterwards.

All changes to this schema must then be performed incrementally – it is not possible to fast-forward the schema evolution by importing another schema on top of the existing one. Schema can be imported at any time, but doing so will delete all data in the database. The current schema (structure only, no data records) of the database can be exported at any time via the schema export function.

3.6 Software Design

GeneRelDB is coded in Java and is therefore fully cross-platform in its implementation. A component diagram of the system is shown in Fig. 2.

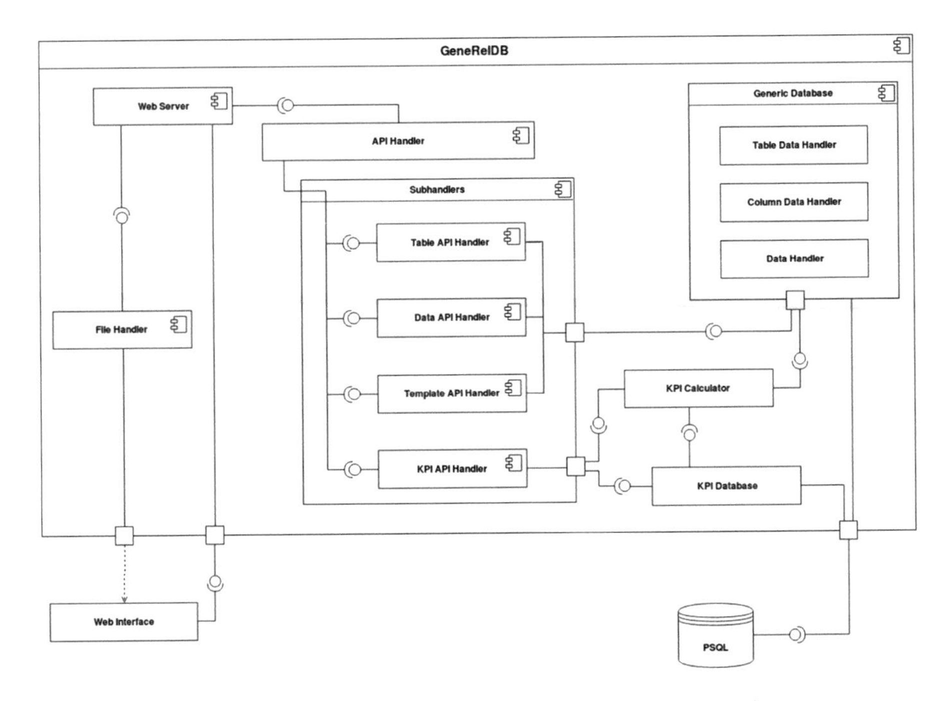

Fig. 2. Component diagram of the GeneRelDB implementation

The File Handler handles all requests for files such as jpg, css, html, javascript and svg.

The Web Server listens on a given port, receives the incoming HTTP requests and makes substitutions in the URIs (e.g., for percent encoding) before forwarding them to the API Handler, receives the responses from the API Handler and forwards them to the requester.

The API Handler forwards the incoming HTTP requests to the appropriate sub-handlers and receives the responses.

- The Table API Handler deals with requests regarding tables and table structure.
- The Data API Handler deals with requests related to data records (including importing from and exporting to csv files).
- The Template API Handler deals with requests for importing and exporting schema templates.
- The KPI API Handler deals with requests regarding the storage and calculation of KPIs.

The Generic Database assembles and executes the generic SQL queries via its three subsidiaries: Table Data Handler and Column Data Handler for DDL and Data Handler for DML.

The KPI Database assembles and executes the SQL DML queries for KPIs.

The KPI Calculator parses the stored declaration of a KPI (described in the Appendix), performs the calculation and returns the result.

GeneRelDB sits as an abstraction layer between the user interface (Web Interface) and the RDBMS (in this case, PostgreSQL). The connection with the database is managed via JDBC, so it is possible to replace the PostgreSQL database with any other JDBC-enabled RDBMS. However, there may be some subtlety in correctly handling exceptions from other RDBMS that requires some extra changes to the code.

4 Discussion

GeneRelDB has been developed with the aim of solving the problem of schema evolution in relational databases in a generic way and so there are two "parts" to the resultant database when using this tool: the generic part, consisting of the metadata tables described in Sect. 3.1 and the realisation of the database schema as concrete tables with generic names assigned by the tool. It is entirely possible to detach the realised schema created within GeneRelDB from the tool in order to use it in a different context. The column data types are used explicitly in the concrete model and the relationships between the tables are also declared as foreign key constraints, so it is only necessary to replace the generic names of the tables and columns with the user-given names stored in the metadata tables. Once this is done, deleting the two metadata tables provides a completely independent database schema.

Compared with the model-driven approach of [6], the database design is left to the user, avoiding multiple steps of code translation. To simplify startup, the initial database schema can be imported into the tool by means of a schema import function. All changes must then be performed incrementally – it is not possible to fast-forward the schema evolution by importing another schema on top of the existing one. Schema can be imported at any time, but doing so will delete all data in the database. The current schema (structure only, no data records) of the database can be exported at any time via the schema export function.

Data can be imported and exported to and from the database in bulk via csv files. For data import, the csv file must correspond to one single table in the database and the names of the columns in the csv file must match the user-given (non-generic) names of the columns in the database table. Otherwise, the tool cannot map the columns of the csv file to their intended counterpart in the database. Before importing the data, GeneRelDB performs a consistency check to make sure that the column names in the csv file exist in the specified database table and that the format of the data matches the data type of that database column. GeneRelDB does not check for duplicates when importing data. Data can also be exported from the database into csv files and, similar to the import, individual tables are written to separate csv files.

In contrast to the query-rewriting approach of PRISM++, GeneRelDB automatically generates the appropriate SQL queries on the fly from the information provided in the HTTP requests. This has the advantage that it does not rely on automatically replicating the exact sequence of modification operators for schema and integrity constraints in order to create the SQL queries. Any well-designed application code can leverage the generic approach of GeneRelDB to ensure that the functions dealing with requesting information from and sending data to the database via the API will never have to be rewritten.

QuantumDB is focussed on the problem of removing downtime due to locking behaviour of schema changes, while GeneRelDB is focussed on removing the need to rewrite SQL queries – and thereby application code – when database schemata evolve. However, due to the exact implementation of GeneRelDB, there is some minor overlap: concrete tables and columns in the database have generic names supplied by the tool – the user-given names are stored only in the metadata tables. This means renaming tables and columns only affects the metadata tables and not, in fact, the concrete tables. In PostgreSQL 9.6, used exclusively in development and testing of GeneRelDB, renaming entities requires an ACCESS EXCLUSIVE lock on the table, thereby blocking all other queries on the table while the rename command is executed. By abstracting the renaming process away from the concrete tables, only a row lock is required on the corresponding entry of the metadata table, and all concrete table remain unaffected.

5 An Example Application of GeneRelDB

Due to its generic approach, GeneRelDB is domain independent and it is possible to deploy the system in many different scenarios. In this section, we will focus on an example database for environmental and energy management in a commercial or industrial setting. This database must be able store measurements from a variety of sensors along with appropriate physical units and a record of the quantity being measured (the "measurand"). It must be possible to associate meters with a physical location or measurement point. It should also be possible to store a readout interval for each meter, as well as if the meter data should arrive automatically or must be entered manually. An example ER model for a relatively simple database capable of the aforementioned points is shown in Fig. 3.

In this model, the physical structure of the enterprise is stored as a hierarchical structure in the table "Organisation Unit". The types of organisation units (e.g., branch, department, building, storage room, etc.) are stored in a separate table "Organisation Unit Type" and referenced via a foreign key. The process for recording meter data (e.g., automatic, manual, etc.) is stored in the table "Measurement Method". All possible measurands are stored in a dedicated "Measurand" table and physical units are stored in the "Unit" table. Different readout intervals (e.g., hourly, daily, weekly, etc.) are stored in the "Measurement Interval" table. Data values are stored in the "Data" table along with a timestamp and a foreign key reference to the meter that recorded them. The meters themselves are stored in the "Meter" table and have foreign keys referencing "Measurement Interval", "Measurement Method", "Measurand", and "Organisation Unit".

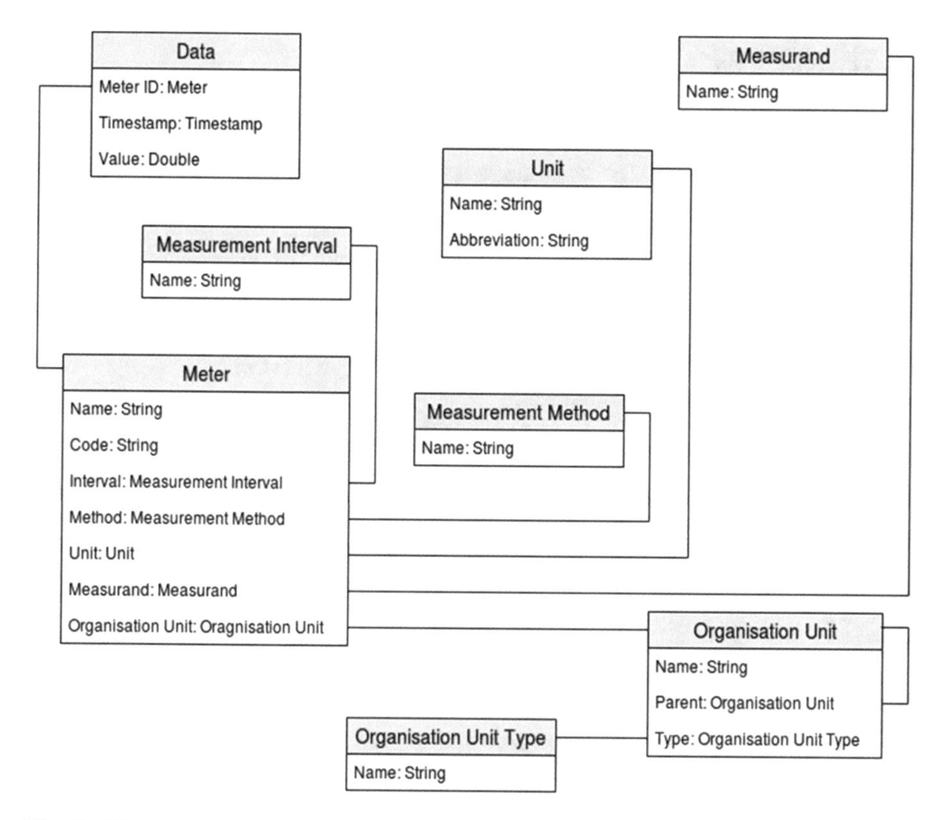

Fig. 3. ER model of a database for environmental and energy management in a commercial/industrial setting

Figure 4 shows both parts of the resulting database (generic metadata tables and realised schema: The generic part is shown to the left of the vertical dashed line and the realised version of the schema - the concrete tables - are shown to the right. The arrows show the corresponding concrete table and column for two example metadata entries) created with GeneRelDB for this scenario. The recursive foreign key in "Organisation Unit" cannot be created at the same time as the table itself: the table must first be created and the the recursive foreign key can be added.

This schema can be easily updated via GeneRelDB as needed. For example, Fig. 5 shows and extension of the ER model from Fig. 3 in which two new columns have been added to the "Unit" table with the intention of facilitating unit conversions between units of the same physical dimension (e.g., volume, mass, energy, etc.). To achieve this, one unit is selected as the "reference" unit for a given dimension (e.g., the cubic meter could be chosen as the reference unit for volume) and other units of that dimension must have an associated conversion formula in terms of the reference unit; this is recorded via a recursive foreign key to the reference unit and a string containing the conversion formula which can be parsed and calculated by the application code.

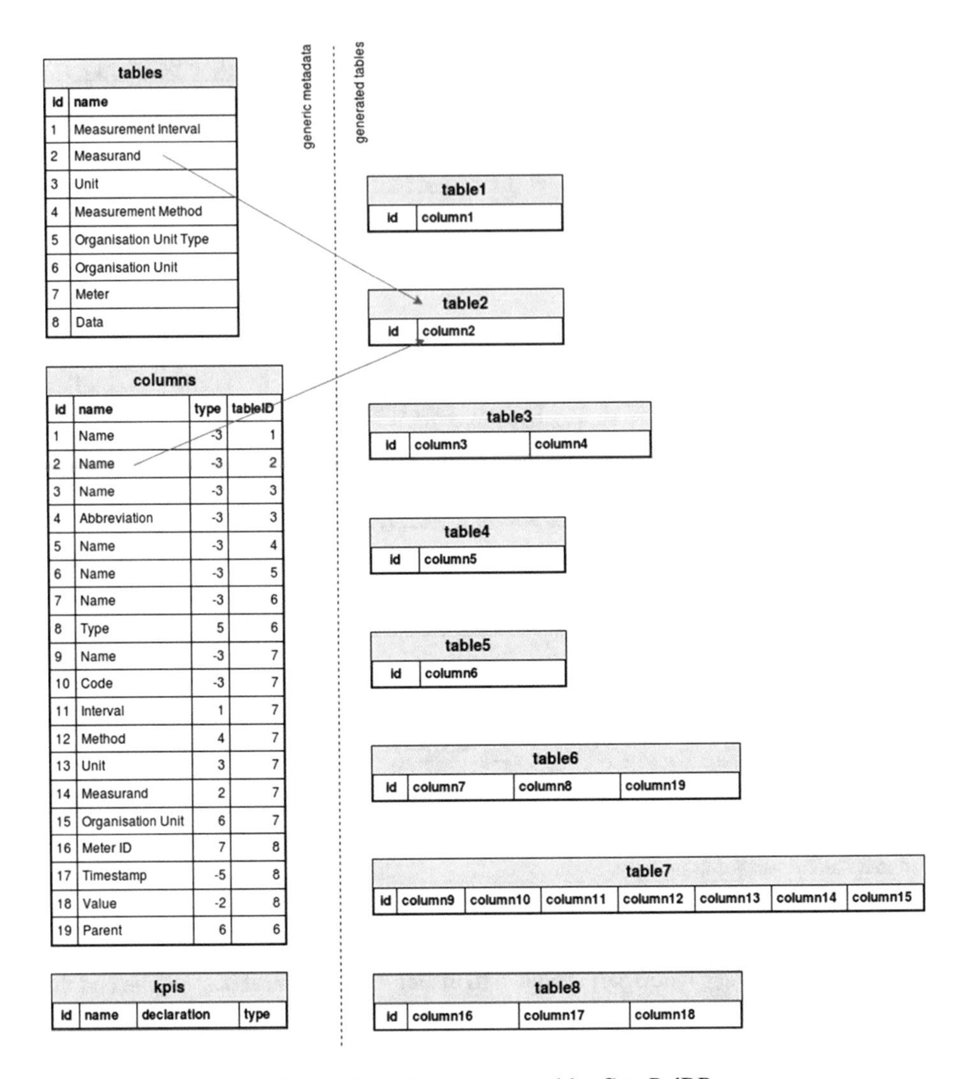

Fig. 4. The database generated by GeneRelDB

In this case, two entries are added to the COLUMNS metadata table (Fig. 6) and the concrete table corresponding to "Unit" is altered (Fig. 7). By leveraging the generic approach of GeneRelDB, it is possible to ensure that application code functions dealing with requesting information from and sending data to the database via the API do not have to be rewritten.

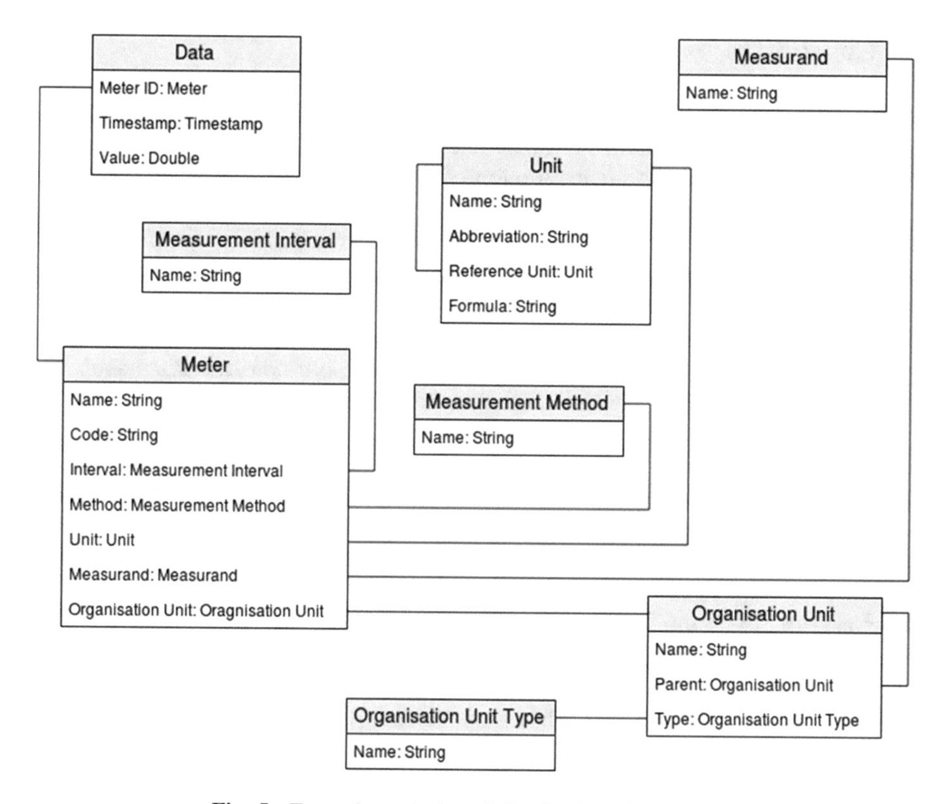

Fig. 5. Example evolution of the database from Fig. 3

columns			
id	**name**	**type**	**tableID**
⋮	⋮	⋮	⋮
20	Formula	-3	3
21	ReferenceUnit	3	3

Fig. 6. Effect of the evolution on the COLUMNS metadata table

table3				
id	**column3**	**column4**	**column20**	**column21**

Fig. 7. Effect of the evolution on the concrete table corresponding to the "Unit" table

6 Conclusion

In this paper we have introduced GeneRelDB, a generic data management layer which allows the schema of a relational database to evolve during run time without the need to rewrite database queries in the application code. It runs as an abstraction layer, handling all communication (queries and data exchange) between the user interface and the database backend. It is designed to be as nonrestrictive as possible, allowing users to build up and evolve arbitrarily complex databases with the tool. The only restriction enforced by GeneRelDB is in relation to data type conversion for existing columns in the database: Date↔Bool, Date↔Foreign key, Date↔Integer, Date↔Double, and Bool↔Double conversions are not supported. Foreign key constraints are supported and referential integrity is maintained during evolution.

The GeneRelDB implementation, while already functional, is not complete – there is room for expansion and improvement. For example, it is currently not possible to add column constraints (such as UNIQUE or NOT NULL) when adding columns via the tool. More complex data retrieval queries requiring SQL JOINS are also not currently supported. Such functionality is not prohibited by GeneRelDB or its design; it is simply a matter of implementing the appropriate methods. Other aspects that could be developed in the future include functionality to support the definition of recursive foreign keys at the time of table creation/schema import and the capability to manage views via the tool. However, the current version of GeneRelDB demonstrates that schema evolution in relational databases can be successfully handled in a generic way by maintaining an overview of the database schema in dedicated metadata tables and building generic database queries that use this metadata information. It is then possible to make changes to almost every aspect of the schema during run time without any interruption to database availability and without having to rewrite any SQL requests.

References

1. MongoDB: The MongoDB 4.0 Manual (2019). https://docs.mongodb.com/manual/
2. Amazon Web Services: Amazon DynamoDB Developer Guide. https://docs.aws.amazon.com/amazondynamodb/latest/developerguide/Introduction.html
3. Apache CouchDB: CouchDB Technical overview (2019). http://docs.couchdb.org/
4. Qiu, D., Li, B., Su, Z.: An empirical analysis of the co-evolution of schema and code in database applications. In: Foundations of Software Engineering, pp. 125–135. ACM (2013)
5. Rahm, E., Bernstein, P.A.: An online bibliography on schema evolution. SIGMOD Rec. **35** (4) (30–31) (2006)
6. Lin, Q., Pu, C., Lee, E.K.: A model-driven approach to manage evolving clinical and translational data in relational databases. In: International Conference on Bioinformatics and Biomedicine. IEEE (2008)
7. Curino, C.A., Moon, H.J., Deutsch, A., Zaniolo, C.: Update rewriting and integrity constraint maintenance in a schema evolution support system: PRISM++. Proc. VLDB Endowment **4**(2), 117–128 (2010). https://doi.org/10.14778/1921071.1921078
8. de Jong, M., van Deursen, A., Cleve, A.: Zero-downtime SQL database schema evolution for continuous deployment. In: 39th International Conference on Software Engineering, pp. 143–152. IEEE (2017). https://doi.org/10.1109/icse-seip.2017.5

The Concept of a Flexible Database - Implementation of Inheritance and Polymorphism

Waldemar Pokuta(✉)

Opole University of Technology, Prószkowska 76 Street, 45-758 Opole, Poland
w.pokuta@po.opole.pl

Abstract. In recent years, a lot of variety of Database Management Systems have been created. Despite of the considerable progress in this area, the most popular are still relational databases (RDB). Designing a relational database requires a lot of experience - not only in computer science. The initial database structure depends on the current customer expectations. These expectations, however, change during operation of the application. Changes in the application usually lead to changes of table structures or field types in the working database. This induce the use of redundant storage space on unneeded data or loss of data archives. The article presents the concept of flexible structures in relational databases. These structures enable the implementation of inheritance, creation of polymorphic tuples within one entity, and the creation of attributes with a variable type of value. Along with the description of structures, a modified version of SQL queries that operate on these structures was proposed.

Keywords: Databases · Flexible databases · Object relational mapping

1 Inheritance as a Method of Making the Database More Flexible

Changes made to the application lead to changes in the definition of data in database and their connections. The computer system evolving, takes on the features, the way of activity and linkage between data, which was not planned at the designing stage. The best solution would be to design the database in such a way that, regardless changing expectations of the client, the database does not require structural changes to existing data. However, the conventional approach to database designing using existing relational mechanisms, where each entity corresponds to a separate table, causes the structure rigidity and changes may turn out problematic. Flexible databases are sometimes defined as databases that do not have a predetermined structure or the structure is easy to change. There should be method to store entities with tuples that differ in the number of attributes or their types are not determined.

One of the ways to make the database more flexible may be to implement the inheritance mechanism.

The implementation of inheritance in relational databases can take place by extending the functionality of the database management system (for example in the

L. Borzemski et al. (Eds.): ISAT 2019, AISC 1050, pp. 119–128, 2020.
https://doi.org/10.1007/978-3-030-30440-9_12

PostgreSQL database) [1] or by object-relational mapping. Generally, there are four approaches to map inheritance in a relational database [2–7]:

- *table-per-hierarchy* – all hierarchy classes are kept in one large table – consecutive records are filled with attribute values of objects of these classes. The fields not used by class are set to NULL;
- *table-per-difference* – each class is mapped to a separate table. For the descendant classes, columns for the fields added during inheritance are only created;
- *table-per-class* – each class is mapped to a separate table. For the descendant classes, tables contain all inherited fields from superclasses up to the root class;
- *generic table structure* – all of the classes are mapped to the same table, but attributes of the classes take up a separate table. A separate table is used to store the object values. The structure may also contain an inheritance table (in the case of multiple inheritance), as well as a table of defined types.

The first and third type of mapping seems the simplest, but also causes unnecessary usage of the tablespace (in the first case by not using space of formed columns in the third case, by the duplication of stored data). The second mapping type seems to be better than the first and the third but is more complex, table corresponding to the subclass should then contain an additional foreign key column to the table corresponding to the superclass. The fourth mapping method is the most complex but offers the highest capabilities, because it does not impose a tabular, rectangular structure for each entity (Each tuple has in its memory only fields that have been assigned to it).

2 The Idea of Flexible Database

2.1 Assumptions of the System

To design a database that can be called flexible (FDB), some assumptions should be made. Database will be flexible when

- the schema allow the implementation of inheritance;
- changes in the schema should not result in excessive use of database storage space or cause loss of archival data (polymorphism of data within collections);
- field value types can be changed without losing archive data.

The proposed flexible database (FDB) schema is presented in Fig. 1.

In the FDB project a relational database was used to its implementation. *"Generic table structure"* was used to implement object-oriented features.

The entire database structure is stored in four tables. The *fdbEntity* and *fdbAttribute* tables describe the database structure. The main table is *fdbEntity*. The *fdbEntity* table contains the names (*fdbName* field) of all entities, inheritance (*fdbSuperID*) and abstractness (*fdbAbstract*) information. The *fdbAttribute* table contains the names (*fdbName* field) of all attributes in the database, foreign key to the collection (*fdbEntityID*) it refers to, information about whether it is a private attribute or not (*fdbPrivate*).

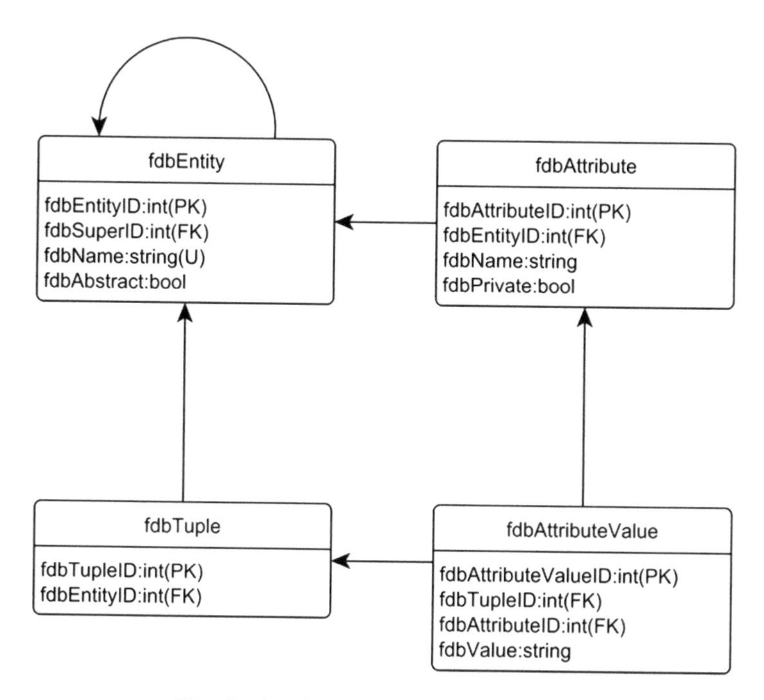

Fig. 1. A schema of the flexible database.

The remaining two tables: *fdbTuple* and *fdbAttributeValue* are used to store values. The *fdbTuple* table is a tuple - contains only foreign key to the Entity it belongs to (*fdbEntityID* field). The *fdbAttributeValue* table contains foreign keys to the attribute definition (*fdbAttributeID*) and to the tuple (*fdbTupleID*) it belongs to. The *fdbValue* field contains all values stored in the database. In order to self-define values in the database, these values are stored in the JSON format. This allows storing data such as logical values, strings, numbers, nulls, arrays, objects or any combination of the above.

3 Using the FDB System

In SQL, creating a new entity requires the *CREATE TABLE* command. The command should contain attribute names and their types. In the FDB system there is no need to enter types, because all values are in JSON format - the type is defined by the value placed in the field. We also do not need to store primary keys. The primary keys are part of the FDB system structure. On the other hand, information about inheritance, abstractness or attribute privacy should be placed.

Due to these differences, the FDB system introduced a modified version [8–10] of the SQL language, namely fdbQL.

3.1 Defining Data Structures

The FDB system has a structure of 4 tables. When we want to design a database, we need to create entities. For example, let's create an entity *Person*. The name of the entity is stored in the table *fdbEntity*, while its attributes in the table *fdbAttributes*. In SQL, creating a new entity requires the *CREATE TABLE* command. In the FDB system this command will be broken down into several instructions. In Table 1 we have a query in the fdbQL, its interpretation using SQL in the FDB system and data stored in tables *fdbEntity* and *fdbAttribute* after executing instructions.

Table 1. Creating entity in FDB system.

fdbQL Command

```
create table Person (fname, lname, pIdentity private)
```

Operations in the FDB system

```
begin transaction
  insert into fdbEntity
    (fdbEntityID,fdbSuperID,fdbName,fdbAbstract)
  values (1,null,'Person',false);
  insert into fdbAttribute
    (fdbAttributeID,fdbEntityID,fdbName,fdbPrivate)
  values
  (1,1,'fname',false),(2,1,'lname',false),
  (3,1,'pIdentity',true);
commit;
```

fdbEntity Table

fdbEntityID	*fdbSuperID*	*fdbNameID*	*fdbAbstract*
1	null	Person	false

fdbAttribute Table

fdbAttributeID	*fdbEntityID*	*fdbName*	*fdbPrivate*
1	1	fname	false
2	1	lname	false
3	1	pIdentity	true

Let's create another entity *Customer* that inherits from the *Person* entity. Table 2 shows the commands needed to create the entity and the contents of the *fdbEntity* and *fdbAttribute* tables.

Table 2. Creating entitity with inheritance in FDB system.

fdbQL Command

```
create table Customer (cIdentity) inherits Person
```

Operations in the FDB system

```
begin transaction
  select fdbEntityID into @fdbE from fdbEntity
  where fdbName='Person';
  insert into
   fdbEntity (fdbEntityID,fdbSuperID,fdbName,fdbAbstract)
   values (2,@fdbE,'Customer',false);
  insert into
   fdbAttribute (fdbAt-
tributeID,fdbEntityID,fdbName,fdbPrivate)
   values (4,2,'cIdentity',false);
commit;
```

fdbEntity Table

fdbEntityID	*fdbSuperID*	*fdbNameID*	*fdbAbstract*
1	null	Person	false
2	1	Customer	false

fdbAttribute Table

fdbAttributeID	*fdbEntityID*	*fdbName*	*fdbPrivate*
1	1	fname	false
2	1	lname	false
3	1	pIdentity	true
4	2	cIdentity	false

3.2 Structure Reading

Having created entities, we can read their structure with the *DESCRIBE* statement (Table 3).

Entities description is simple if they do not inherit from other entities. In case of inheritance, non-private attributes of the parent entity whose names are not hidden in the inherited entity should be added to the entity description. The case of inheritance is shown in Table 4.

Of course, in the general case, where the chain of inheritance is longer, one would have to implement the reading of attributes in the loop.

Table 3. Describing entitity in FDB system.

fdbQL Command

```
describe Person
```

Operations in the FDB system

```
select a.fdbAttributeID, a.fdbName, a.fdbEntityID
from fdbAttribute a, fdbEntity e
where (a.fdbEntityID=e.fdbEntityID) and (e.fdbName='Person')
```

Query result

fdbAttributeID.	*fdbName.*	*fdbEntityID.*
1	fname	1
2	lname	1
3	pIdentity	1

Table 4. Describing entitity with inheritance in FDB system.

fdbQL Command

```
Describe Customer
```

Operations in the FDB system

```
select fdbEntityID,fdbSuperID into @fdbE,@fdbS
from fdbEntity where fdbName='Customer';
create temporary table temptable1
select fdbAttributeID, fdbName, fdbEntityID
from fdbAttribute where fdbEntityID=@fdbE;
create temporary table temptable2
select fdbAttributeID, fdbName, fdbEntityID
from fdbAttribute
where (fdbEntityID=@fdbS) and (fdbPrivate=false) and
  (fdbName not in (select fdbName from temptable1));
select * from temptable1 union select * from temptable2;
```

Query result

fdbAttributeID.	*fdbName.*	*fdbEntityID.*
4	cIdentity	2
1	fname	1
2	lname	1

3.3 Inserting Data

If one want to insert a new tuple into the database, he should first know all the attributes for which he wants to create values. Attributes can be obtained using the describe statement (Table 4). When inserting a larger number of tuples, we can group instructions that operate on the *fdbTuple* and *fdbAttributeValue* tables (Table 5). Values are stored in JSON format, therefore all text data must be enclosed in quotes.

Table 5. Inserting new tuples in the database.

fdbQL Command

```
insert into Person(fname,lname,pIdentity)
values('"Jim"','"Brown"','123456789');
insert into Customer(fname,lname,cIdentity)
values('"John"','"White"','"cus001"');
```

Operations in the FDB system

```
begin transaction
  insert into fdbTuple(fdbTupleID,fdbEntityID)
  values(1,1),(2,2);
  insert into fdbAttributeValue(
    fdbAttributeValueID,
    fdbTupleID,
    fdbAttributeID,
    fdbValue)
  values(1,1,1,'"Jim"'),(2,1,2,'"Brown"'),
    (3,1,3,'123456789'),(4,2,1,'"John"'),
    (5,2,2,'"White"'),(6,2,4,'"cus001"');
commit;
```

Figure 2 shows the scheme of connections after inserting data.

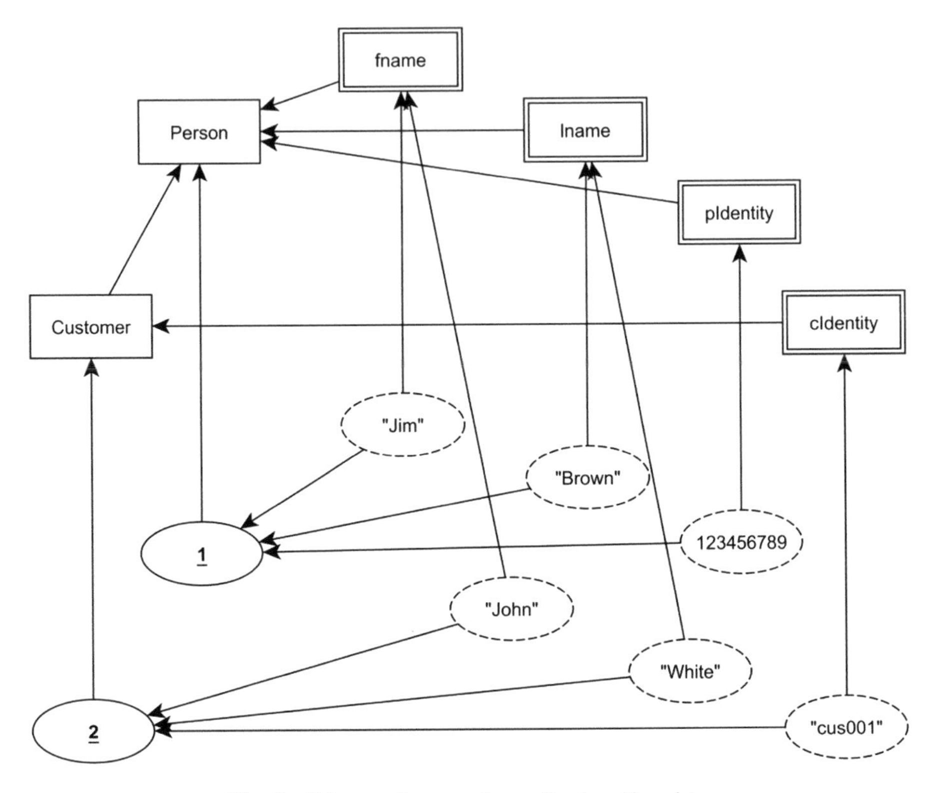

Fig. 2. Scheme of connections after inserting data.

3.4 Receiving Data from the FDB System

The *Person* entity is a generalization of the *Customer* entity. So you can say that the *Customer* entity is included the *Person* entity. By receiving all data from the *Person* entity, we also get all data from child entities (Table 6).

4 Discussion

The presented way of recording the entity may seem difficult to use, due to the lack of an engine enabling efficient navigation on the proposed system. There are, however, some issues where such a structure, due to its versatility, can be very useful. An example of this is the creation of logs. This is especially useful for systems based on databases up to several hundred tables. Saving the dependencies between tables is requested to enter data into two tables (*fdbEntity* and *fdbAttribute*). In other tables (*fdbTuple* and *fdbAttributeValue*) it is possible to save any values that appear when the

Table 6. Receiving polymorphic data from FDB system.

fdbQL Command

```
select * from Person;
```

Operations in the FDB system

```
// get information from 'describe Person'
// (fname,lname,pIdentity)
// get information from child entity 'describe Customer'
// (fname,lname,cIdentity)
// get logical sum of Attribute names
// (fname,lname,pIdentity,cIdentity)
// Construct the sql query for "Person" and "Customer"
select
  (select av.fdbValue from fdbAttributeValue av,fdbAttribute a
   where (av.fdbTupleID=t.fdbTupleID)
   and (av.fdbAttributeID=a.fdbAttributeID)
   and (a.fdbName='fname') limit 1) as 'fname',
  (select av.fdbValue from fdbAttributeValue av,fdbAttribute a
   where (av.fdbTupleID=t.fdbTupleID)
   and (av.fdbAttributeID=a.fdbAttributeID)
   and (a.fdbName='lname') limit 1) as 'lname',
  (select av.fdbValue from fdbAttributeValue av,fdbAttribute a
   where (av.fdbTupleID=t.fdbTupleID)
   and (av.fdbAttributeID=a.fdbAttributeID)
   and (a.fdbName='pIdentity') limit 1) as 'pIdentity',
  (select av.fdbValue from fdbAttributeValue av,fdbAttribute a
   where (av.fdbTupleID=t.fdbTupleID)
   and (av.fdbAttributeID=a.fdbAttributeID)
   and (a.fdbName='cIdentity') limit 1) as 'cIdentity'
from fdbTuple t, fdbEntity e
where (e.fdbEntityID=t.fdbEntityID)
  and ((e.fdbName='Person') or (e.fdbName='Customer'));
```

Polymorphic query result

fname.	*lname.*	*pIdentity.*	*cIdentity.*
"Jim"	"Brown"	123456789	
"John"	"White"		"cus001"

database is filled with data. In order to show the time dimension in changing data, the fdbTuple table can be additionally expanded by two attributes. The first one should indicate the creation date of the entry. The second can describe the type of operation performed (addition, removal, modification).

Another place where one can use such a structure are backup copies of the database. In the presented layout, each database schema can be mapped to 4 tables. If subsequent versions of backups remain on the server, using appropriate procedures, you can compare changes that occurred between the structures of subsequent backups, e.g. new tables, deleted tables, new attributes, deleted attributes. In addition, using SQL, you can also search the contents of data in saved structures. An additional advantage of this solution is the fact that it is a universal mechanism, independent of database providers.

One of the basic features of relational databases is to design integrity constraints and type validation. In the current scheme, the FDB system does not ensure data integrity or type compliance. This will be the subject of further research to develop the FDB system.

References

1. Stonebraker, M., Rowe, L.A.: The POSTGRES data model (PDF). In: Proceedings of the 13th International Conference on Very Large Data Bases, pp. 83–96. Morgan Kaufmann, Brighton (1987). ISBN 0-934613-46-X
2. Ambler, S.W.: Agile Database Techniques: Effective Strategies for the Agile Software Developer, pp. 223–268. Wiley, Indianapolis (2003). ISBN-13: 978-0471202837
3. Dykstra, T., Anderson, R., Schonning, N., Addie, S., Parente, J., Pasic, A., Warren, G., Latham, L.: Implementing Inheritance with the Entity Framework 6 in an ASP.NET MVC 5 Application (11 of 12), (AspNetDocs) (2015)
4. King, G., Bauer, C., Andersen, M.R., Bernard, E., Ebersole, S.: Hibernate ORM documentation (5.0), Chapter 9. Inheritance mapping, Red Hat, Inc (2015)
5. Sun, W., Guo, S., Arafi, F.: Supporting inheritance in relational database systems. In: Proceedings Fourth International Conference on Software Engineering and Knowledge Engineering, Capri, Italy, pp. 511–518 (1992). https://doi.org/10.1109/seke.1992.227910
6. Gottlob, G., Schrefl, M., Stumptner, M.: Selective inheritance of attribute values in relational databases. Discrete Appl. Math. **40**, 187–216 (1992). https://doi.org/10.1016/0166-218X(92)90029-A
7. Awang, M.K.: Transfering object oriented data model to relational data model. Int. J. New Comput. Architectures Appl. (IJNCAA) **3**, 403–410 (2012)
8. Pokrajac, D., Patel, H., Rasamny, M.: Inheritance constraints implementation in PostrgreSQL (2004)
9. Alashqur, A., Su, S.Y.W., Lam, H.: OQL: A query language for manipulating object-oriented databases. In: International Conference on Very Large Data Bases, pp. 433–442 (1989)
10. Assaf, M.A., Badr, Y., Barbar, K., Amghar, Y.: AQL: A Declarative Artifact Query Language, pp. 119–133 (2016). https://doi.org/10.1007/978-3-319-44039-2_9

UAV Detection Employing Sensor Data Fusion and Artificial Intelligence

Cătălin Dumitrescu[1(✉)], Marius Minea[1], and Petrica Ciotirnae[2]

[1] Transports Faculty, TET, University Politehnica of Bucharest, 313 Splaiul Independenței, 060042 Bucharest, Romania
catalin.dumitrescu@upb.ro

[2] Military Technical Academy "Ferdinand I", 39-49 George Cosbuc Avenue, Bucharest, Romania

Abstract. The purpose of this paper is to present a multi-sensorial detection method for discovering and obtaining characteristics of flying Unmanned Aerial Vehicles (UAVs) in restricted areas. Different solutions may be applied for this purpose: radio signals analysis, acoustic patterns analysis, video processing, IR imaging, RADAR, LIDAR etc. The new *Concurrent Neural Networks* (CNN) classification has been introduced as a collection of low-volume neural networks that perform parallel classification. In the present paper the identification and classification of drones is analyzed employing two CNNs, a multilayer perceptron (MLP) for acoustic pattern recognition and a self – organizing map (SOM) to recognize an object from a video stream.

Keywords: UAV detection · Acoustic/video deep learning · Wavelet decomposition

1 Introduction

In the present days, the evolution of Unmanned Aerial Vehicles (UAVs), or Systems (UASs), also known as drones (usually employed as industrial/military equipment, or simple entertainment gadgets) has experienced a rapid growth. The interest of the large public in these remote controlled and partially robotic vehicles is explainable due to their versatility and utility in different actions, such as taking aerial pictures, for postal services deliveries, for inspecting difficult accessible areas, for surveillance in search and rescue actions etc. UAVs may help people working in various places and fields where man operates in agriculture, shipping and delivery, SAR[1] operations, engineering, 3D mapping, providing wireless communications on wide areas, research and nature science, constructions etc. However, this trend also brings some drawbacks, as it became in some cases uncontrolled (despite the measures of declaring to an Aeronautic Authority the flight area or plan, or the interdiction of overflying sensitive areas). Moreover, the impact of not complying with these regulations could prove illegal and

[1] SAR – Search and Rescue.

L. Borzemski et al. (Eds.): ISAT 2019, AISC 1050, pp. 129–139, 2020.
https://doi.org/10.1007/978-3-030-30440-9_13

potentially disastrous in many cases, e.g. flying uncontrolled over a take-off runway in an airport, unauthorized transport of goods or surveillance of prisons or military facilities, drugs and forbidden substances merchandise etc. An uncontrolled flight of a drone, or a deliberately flight over a restricted area, may lead, as exampled before, to potentially dangerous situations, or security violations. This aspect is very important for places where aerial incidents and unauthorized access may produce severe damages, such as in airports, military facilities, prisons, nuclear powerplants, high voltage transport lines etc. In some situations, where the risk of security breaches is high from this point of view, specific detectors have been installed to discover and track unauthorized flight of drones over the controlled area. However, in most of the sensible civilian areas, such as in airports, a large deployment of such protection systems has not been yet implemented.

The purpose of this paper is to present a multi-sensorial detection solution for discovering and collecting information of flying UAVs in restricted areas.

The remaining of the paper is organized as following: the next section makes a brief resume of a literature survey regarding the detection of drones, third section presents the proposed detection method, fourth section some experimental results and the conclusions at the end of the paper.

2 Detection Techniques. Literature Survey

There are a series of scientific papers that focus on applicable technologies in the process of drone detection. Different solutions may be applied: analysis of radio signals, acoustic patterns analysis, video processing, IR imaging, RADAR, LIDAR etc. Each technology usually has its advantages and drawbacks.

In [1], Jeon et al. investigate the use of deep neural network to detect commercial hobby drones in real-life environments by analyzing their sound data. They focus on detecting potentially malicious or terrorist drones. The authors employ Gaussian Mixture Model (GMM), Conventional Neural Network (CNN) and Recurrent Neural Network (RNN) to recognize commercially available drones flying in a typical environment. They obtained an F-score recognition of 0.8009 when employing the RNN methodology, with 240 ms of signal input and short processing time. The authors stated that the most difficult challenge of their work was to train the system, in the presence of environmental noise, in specific restricted areas, where flying of drones was not allowed.

In [2], Vasquez et al. deal with multisensorial 3D tracking for counter small unmanned air vehicles. They show the difficulties arising from the small size and low radar cross section of small drones, when attempting to detect and track these targets with a single sensor such as radar or video cameras. The authors present a multi-sensor architecture that exploits sensor modes including EO/IR[2] cameras, an acoustic array, and future inclusion of a radar. The fundamental approach of their multi-sensor system is to divide the problem into the areas of detect, track, and ID. The sensors were therefore divided into three categories: Wide Field-Of-View (WFOV), Medium

[2] EO/IR – Electro-Optic/Infrared camera sensing.

Field-Of-View (MFOV) and Narrow Field-Of-View (NFOV). They also included an acoustic sensor in their setup and a cooperative subsystem with the radar of the airport. The authors wrote that the preliminary results for EO/IR cameras and acoustic array were very promising.

In [3], Nguyen et al. investigate the cost-effective RF-based detection of drones. These authors propose two technical approaches: in the first approach, they employ active tracking when the system sends a radio signal and then listens for its reflected component. The second approach is passive listening: the system receives, extracts, and then analyzes the acquired wireless signal. They also propose two methods, an active one and a passive one. The active method is performed by observing the reflected wireless signal, while the passive by listening to the communication between the drone and its controller. The solution proposed by the authors rely on three main sources of wireless signals caused by the drone's rotating propellers, the drone's communication and drone's body vibration. The principle used is based on the signature of the signal reflected from the drone's propellers, which is observed on an off-the-shelf wireless receiver (i.e. Wi-Fi receiver or WARP).

In the paper [4], M. Strauss et al. present localization of a sound source employing a microphone array. They do not search for UAV detection, but instead use a drone and the microphone array to locate specific sound sources. Of course, the process being reversible, it is possible to employ the same technology for drone detection.

In [6], Eriksson focuses on how to deal with trespassing and potentially dangerous multi-copters in civil applications in the context of early concept development for a radar company. The author presents in a comparative study all advantages and disadvantages of the most known detection technologies, including simple radar, mono-static pulse radar, FMCW[3] radar, high-frequency FMCW radar, Micro-Doppler UAS[4] radar return, passive radar with commercial signals, WLAN/RF detection, audio sensing, EO[5] sensing etc.

Andrasi et al. [6] have tested the applicability of a low-cost long-wave infrared sensor for detection of various UAVs in flight. For the test they used a longwave infrared sensor produced by FLIR, called Lepton, with spectral range ranging from 8 μm to 14 μm and the spectral response from 9.5 μm to 12.5 μm.

B.H. Kim et al. observe that one common detection method for UAVs or UASs in restricted or sensitive areas is the EO/IR one, but this approach is sometimes strongly affected by adverse meteorological conditions [7]. The authors present a technique for data augmentation with existing LiDAR and LADAR sensors for the development and testing of detection framework. An augmented dataset is generated by fusing the designed target shape and trajectory profiles. The augmented datasets are classified according to different scenarios and a bounding box is added to the absolute target location to calculate the ground truth data. The authors propose a LADAR data augmentation process.

[3] FMCW – Frequency Modulated Continuous Wave.

[4] UAS – Unmanned Aerial Systems.

[5] EO – Electro-Optical.

Also, Guvenc et al. perform in [8] an overall review of techniques that rely on ambient radio frequency signals (emitted from UASs), radars, acoustic sensors, and computer vision techniques for the detection of malicious UASs. They also present some experimental and simulation results on radar-based range estimation of UASs, and receding horizon tracking of UASs.

3 The Proposed Solution

This work performs an investigation on the way the concept of competition in a collection of neural networks can be implemented and, on the other hand, the way the importance of inputs influences the recognition performance of some types of neural networks. The new *Concurrent Neural Networks* (CNN) classification has been introduced as a collection of low-volume neural networks that perform parallel classification. In the present paper the identification and classification of UAVs will be analyzed employing two CNNs, a multilayer perceptron (MLP) for acoustic pattern recognition and a self-organizing map (SOM) – to recognize an object from a video stream. In contrast, SOM-Concurrent consists of modules that are trained by an unsupervised algorithm and the data consist only of positive examples.

It is also described the realization of an "acoustic camera" prototype and the implementation of a *Spiral Beamforming Analysis* designed for the detection of drones along with the realization of the data fusion. The system is composed of a spiral microphone array (log-spiral configuration) using 30 microphones (Bruel & Kjaer - type 4957) and an Axis F1004 video camera installed in the center of the microphone array. The configuration of the acoustic camera system is show in Fig. 1.

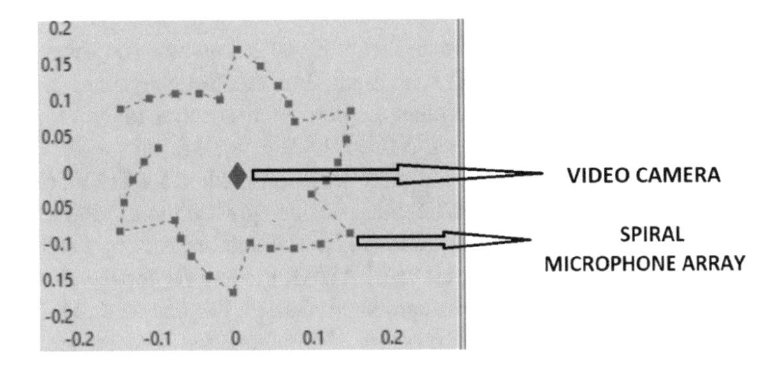

Fig. 1. Array geometry configuration for acoustic camera systems.

The acoustic array is arranged in five spiral arrays, the distance between the microphones being 3.5 cm. Using a beamforming algorithm, the acoustic camera can detect and locate both stationary and moving noise acoustic sources. The resulting high – speed color map-overlay and video are ideal identifying transient sources of acoustic noise.

3.1 Acoustic Signatures Analysis

Extracting the Characteristics of the Acoustic Signal

The process that is common for all types of recognition systems employing acoustic signals is the extraction of characteristic vectors from uniformly distributed segments of time of the sampled sound signal. *Preprocessing the acoustic signal:* before the features are extracted, the acoustic signal is subject of the following processing phases:

(a) *Highlighting*: A high-pass filter is used to emphasize high frequencies and compensate the system's tendency to attenuate these frequencies;
(b) *Segmentation*: the acoustic signal is considered non-stationary on a long-term observation, but quasi-stationary on short term (10–30 ms), therefore the acoustic signal is divided into fixed-length segments, called *frames*. The typical duration of a frame is 20 ms, generating it from 10 to 10 ms so that a 15 ms overlap occurs from one window to the next;
(c) *Attenuation*: Each window is multiplied by a specific function, usually the Hamming window, to mitigate the effect of finishing windows segmentation;
(d) *MFCC*[6] *parameters*: In order to recognize an acoustic pattern, it is important to extract specific features from each frame. The Spectrum-Based MFCC are used and their success is due to the employment of a filter bank using wavelet transforms with a si-milar perceptual scale to the human auditory system, to process the Fourier Transform. Also, these coefficients are robust to noise and flexible due to cepstrum processing;
(e) *Adaptive filtering*: the role of this adaptive filter is to best approximate the value of a signal at a given moment, based on a finite number of previous values. The linear prediction method allows for very good estimates of signal parameters. By minimizing the sum of square differences on a finite interval between real signal samples and samples obtained by linear prediction, a single set of coefficients, called prediction coefficients, can be determined.
(f) *Time-frequency analysis*: the Cohen class method involves the selection of the nea-rest kernel function that corresponds to the fundamental waveform which describes the acoustic signatures specific to drones. Thus, the shape of the nucleus based on the peak values (localization) and the amplitude of a "control" function has to be chosen;
(g) *Performing beamforming algorithms*: DOA (Direction of Arrival) estimation based on MUSIC (Multiple Signal Classification) algorithm;
(h) *Performing classifications*: MLP-Concurrent use supervised trained modules and training instruction sets contain both positive and negative examples from sound signatures of drones.

[6] MFCC – Mel Frequency Cepstral Coefficients - a representation of the short-term power spectrum of a sound, based on a linear cosine transform of a log power spectrum on a nonlinear mel scale of frequency.

3.2 Image Processing

In order to obtain UAV-specific training patterns, inter-band correlation computing is proposed for only two adjacent scales and for each orientation apart (i.e., horizontal, vertical and diagonal). This may be justified by considering that certain contours can be put in evidence only according to specific directions, those being visible only in some images. This fact suggests that different images must be processed separately. Therefore, in this work, the processing procedure is applied separately to sub-images, according to the three orientations. The concept of *hierarchic correlation* is introduced. This concept takes into account both the correlation between the (absolute) values of the neighboring coefficients laying in the same sub-band (but still having the same parent coefficient), and the *inter-band adjacent correlation* along the partial tree, employing or the evaluation a group of five neighboring coefficients (four of which being neighboring and the *parent coefficient* from the adjacent scale) (Fig. 2).

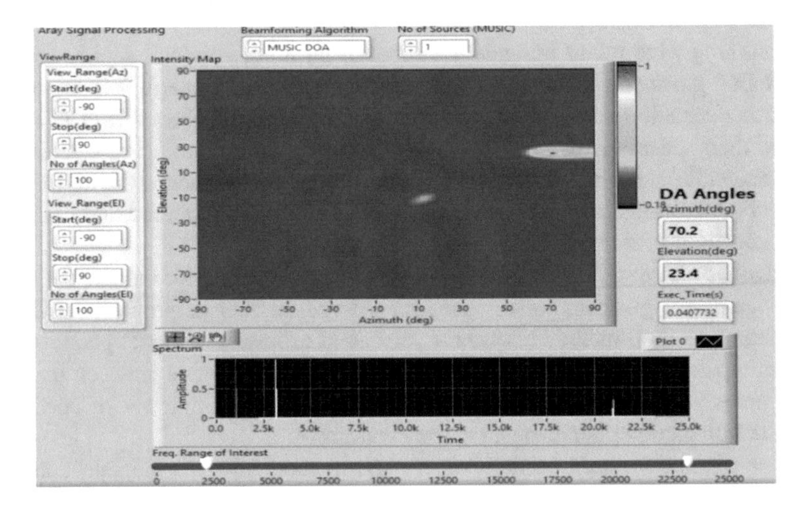

Fig. 2. Acoustic pattern detection specific to drones, fundamental frequency 4.3 kHz and harmonic response of 21.3 kHz (white spectrum), Cohen spectrogram for power engine sound detection and beamforming DOA MUSIC analysis.

In this way, for every pair of sub-bands from the same orientation but belonging to different scales, the hierarchical correlation is computed from:

$$c_{ier}(i,j) = \sqrt{\left|w\left(i_p, j_p\right)\right| \cdot c_l(i,j)}, \tag{1}$$

where i_p and j_p are the coordinates of the parent coefficient, $i_p = \lfloor i/2 \rfloor, j_p = \lfloor j/2 \rfloor, \lfloor \ \rfloor$ being the integer function, and $c_l(i,j)$ representing an intra-band correlation coefficient; this coefficient is differently defined, according to the orientation of the processed sub-band, in the following way:

– For LH-orientation sub-band:

$$c_l(i,j) = max\left\{\sqrt{|w(i,j)\cdot w(i+1,j)|},\sqrt{|w(i,j+1)\cdot w(i+1,j+1)|}\right\} \tag{2}$$

– For HL-orientation sub-band:

$$c_l(i,j) = max\left\{\sqrt{|w(i,j)\cdot w(i,j+1)|},\sqrt{|w(i+1,j)\cdot w(i+1,j+1)|}\right\} \tag{3}$$

– For HH-orientation sub-band:

$$c_l(i,j) = max\left\{\sqrt{|w(i,j)\cdot w(i+1,j+1)|},\sqrt{|w(i+1,j)\cdot w(i,j+1)|}\right\} \tag{4}$$

In such a way, a correlation map is formed having only a quarter of the total pixel count of the original image. The coefficients of this map represent the contours, while the lower values coefficients represent the smooth areas. The specific part of the correlation map that corresponds to the lowest frequency sub-band will not be processed, this sub-band not being taken into consideration. This hierarchical correlation decomposition allows for the prelevation of fine structures from the image, that do not appear as local maximals. In Fig. 3 it can be noticed that on the correlation map, in conditions of a low noise level, there appear points that do not belong to the contours, but may overpass the threshold value which is used for obtaining the contours map from the hierarchical correlation map, processed with a hybrid-meridian filter. Starting from the Concurrent Neural Network model, the Concurrent Self-Organizing Maps (CSOM) has been developed and experimented. That stands out as a new technique with excellent performance.

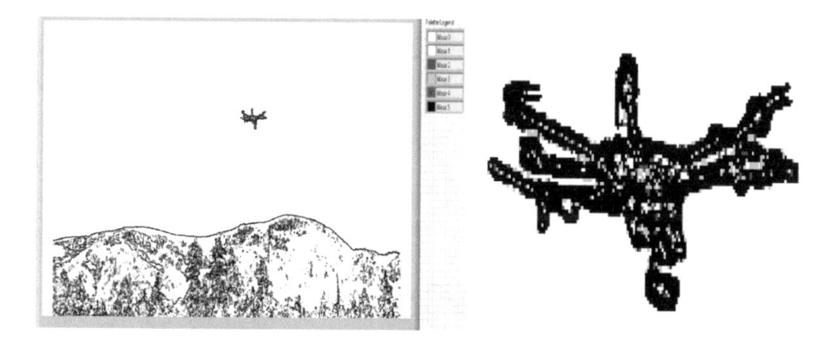

Fig. 3. Spatial analysis in wavelet domain, based on hierarchical correlation map, and the realization pattern template training of the drone.

3.3 Data Fusion

The proposed model, called *Concurrent Neural Networks*, introduces a new neural recognition technique based on the idea of competition between multiple neural networks working in parallel. The number of employed networks equals the number of classes in which the vectors are grouped, and the training is supervised. Each network is designed to correctly recognize vectors in a single class, so the best answers only occur when presented to vectors in the class they were trained with. This model is a frame that provides architecture flexibility because modules can be represented by various types of neural networks. For the experimental tests artificial neural networks have been used together with the Cohen class spectrograms and wavelet decomposition for the system training. Concurrent Self-Organizing Maps (CSOM) are classifiers specially designed to work with images. Their structure facilitates the classification of data with many parameters, as these types of networks allow the efficient processing of such data categories (Fig. 4).

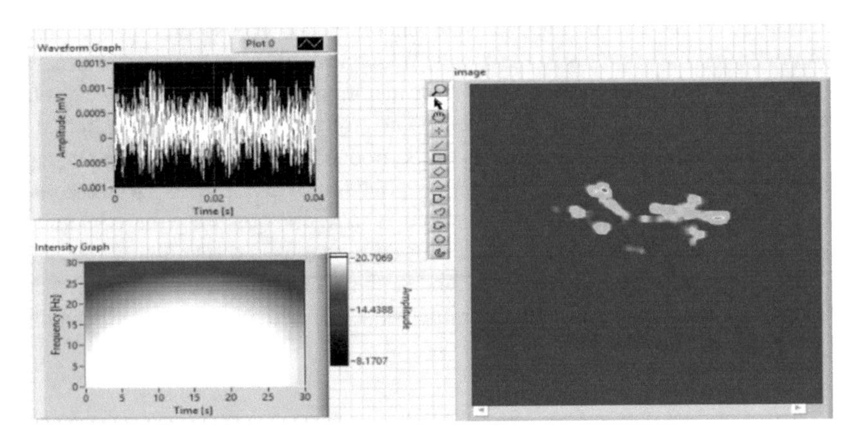

Fig. 4. Data Fusion Interface, acqusition of audio signal from microphone array (top left), detection acustic signature from drone (bottom left) and recognition of the drone in image along with the acoustic emission of the power engine (right).

CSOM do not have a pre-established architecture or structure, it adapts to the type of image being processed or to the nature of the objects to be classified (Fig. 5).

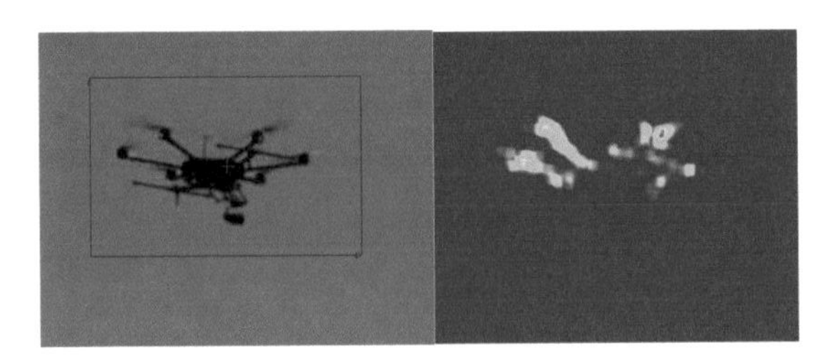

Fig. 5. Drone recognition from video frame (left) and data fusion tracking drone from image data with sound signature of power engine (right).

4 Experimental Results

After observing that the CSOM model, compared to classic models, brings remar-kable improvements to recognition rates of acoustic fingerprints and pattern recognition in image, in this section we focus on recognizing and identifying acoustic signals and shapes contour detection of UAVs. A database has been created by analyzing acoustic signals and shape of contours from a set of 5 drones. Each of the 5 drones was tested 10 times. Every time the recognition system training vectors have been extracted from the first 5 rehearsals, keeping the last 5 rehearsals for the tests. As well as recognizing the drones a set of experiments using a first one single neural network for pattern recognition, and then Concurrent Neural Networks have been tested.

This time only the Kohonen network has been tested, given the results that were achieved in recognitions and comparation of their behavior to that of a Concurrent Neural Network. For the variant that uses a single SOM, the network has been trained with the whole sequence of vectors obtained after preprocessing the selected shapes contour. A Kohonen network with 10×15 nodes through the Self-Organizing Feature Map (SOFM) algorithm has been trained. The first stage, the organization of clusters, took place along 1000 steps and the neighborhood gradually declined to a single neuron. In the second stage, the training was done in 10000 steps and the neighborhood remained fixed to the minimum size. Following training and calibration of the neural network with the training vectors, a set of labeled (annotation) prototypes whose structure is that of Table 1, has been obtained. The applied technique for recognition is described below. The acoustic frequencies identified in the test signal are preprocessed by means of a window from which a vector of the component parts is calculated.

Table 1. Set number of labeled (annotation) prototypes structure has been obtained

UAV types	1	2	3	4	5
Training	472	616	439	625	553
SOM	45	38	14	25	25
CSOM	27	27	27	27	27

The window moves with a step of 50 samples and a collection of vectors whose sequence describes the evolution of the acoustic signal specific to the drones is obtained. Experimentally, a maximum threshold for quantization error, to eliminate elements that are supposed to not belong to any class, was set. Through this process, a sequence of class labels that show how the patterns specific to the drones were recognized by the system has been obtained. In Table 2 below are presented the experimental results of recognition and identification of the drones with SOM and CSOM.

Table 2. Experimental results [%] of recognition and identification of the drones

Drones types	SOM	CSOM
1	73.65%	95,21%
2	60.47%	84.95%
3	50%	67.34%
4	5,89%	50.43%
5	15.03%	82.37%

5 Conclusion

The paper investigates the possibility to implement the concept of competition in a collection of neural networks and the way the differing importance of inputs influences the recognition performance of some types of neural networks. The idea of competition is employed in a collection of neural networks that are independently trained to solve different sub-problems. The authors' contribution consists in: (i) The realization of an "acoustic camera" prototype and the implementation of a *Spiral Beamforming Analysis*; (ii) A spatial analysis in the wavelet domain, based on a hierarchical correlating map for image processing, in order to obtain a training pattern; (iii) A newly developed Concurrent Neural Networks classification as a collection of low-volume, audio and video neural networks that work in parallel and perform classification for data fusion.

Development perspectives: it is foreseen that a radio detection and analysis module will be added to the two modules audio and video in the future. The module will be implemented based on software defined radio, digital modulation recognition and classification, and DOA MUSIC beamforming localization.

Acknowledgement. This work was supported by a grant of the Ministry of Innovation and Research, UEFISCDI, project number 9SOL/12.04.2018 within PNCDI III.

References

1. Jeon, S., Shin, J.-W., Lee, Y.-J., Kim, W.-H., Kwon, Y.H., Yang, H.-Y.: Empirical study of drone sound detection in real-life environment with deep neural networks. Computer Science, Sound (2017)
2. Vasquez, J.R., Tarplee, K.M., Case, E.E., Zelnio, A.M., Rigling, B.D.: Multisensor 3D tracking for counter small unmanned air vehicles (CSUAV). In: Acquisition, Tracking, Pointing, and Laser Systems Technologies XXII Proceedings of Society of Photo Optical Instrumentation Engineers 2008, vol. 6971, pp. 697107 (2008). https://doi.org/10.1117/12.785531
3. Nguyen, P., Ravindranathan, M., Nguyen, A., Han, R., Vu, T.: Investigating Cost-effective RF-based Detection of Drones. DroNet 2016, June 26, Singapore (2016). https://doi.org/10.1145/2935620.2935632
4. Strauss, M., Mordel, P., Miguet, V., Deleforge, A.: DREGON: dataset and methods for UAV-embedded sound source localization. In: IEEE/RSJ International Conference on Intelligent Robots and Systems (IROS 2018), October 2018, Madrid, Spain, pp. 5735–5742. IEEE (2018). https://doi.org/10.1109/iros.2018.8593581hal-01854878

5. Eriksson, N.: Conceptual study of a future drone detection system. Countering a threat posed by a disruptive technology. Department of Industrial and Materials Science, Chalmers University of Technology, Gothenburg, Sweden (2018)
6. Andrasi, P., Radisic, T., Mustra, M., Ivosevic, J.: Night-time detection of UAVs using thermal infrared camera. Transp. Res. Procedia **28**, 183–190 (2017)
7. Kim, B.H., Khan, D., Bohak, W., Choi, C., Lee, H.J., Kim, M.Y.: V-RBNN based small drone detection in augmented datasets for 3D LADAR System. Sensors **18**, 3825 (2018). https://doi.org/10.3390/s18113825
8. Guvenc, I., Ozdemir, O., Yapici, Y., Mehrpouyan, H., Matolak, D.: Detection, Localization, and Tracking of Unauthorized UAS and Jammers. https://ntrs.nasa.gov/search.jsp?R=201700094652019-04-08T07:39:27+00:00Z

Heuristic Algorithm for Recovering a Physical Structure of Spreadsheet Header

Viacheslav Paramonov[1(✉)], Alexey Shigarov[1], Varvara Vetrova[2], and Andrey Mikhailov[1]

[1] Matrosov Institute for System Dynamics and Control Theory of Siberian Branch of Russian Academy of Sciences, Irkutsk, Russia
{slv,shigarov,mikhailov}@icc.ru
[2] School of Mathematics and Statistics, University of Canterbury, Christchurch, New Zealand
varvara.vetrova@canterbury.ac.nz

Abstract. Tables in electronic documents (spreadsheets) contain large volumes of useful information about different domains. Efficient extraction of data from document tables plays a crucial role in its further usage including analysis and integration. The visual or logical structure of table elements might differ from its physical structure. Such differences cause difficulties for automated table processing and understanding. Automated correction from physical form to visual allows to simplify tables processing operations. In this paper, we propose a heuristic approach for transformation of tables' header cells. The main goal of the proposed approach is to provide an algorithm and software tool for recovering a physical structure of a spreadsheet header. The proposed approach is illustrated by application to the Statistical Abstract of the United States (SAUS) dataset.

Keywords: Spreadsheets · Table structure · Cells · Heuristics · Table layers

1 Introduction

Tables are a very common form of data representation. It is possible to include and present different types of information (i.e. numerical, textual, graphical) in a tabular form. Some common examples of data such as calendars, different schedules, statistical reports, experimental results, and grade reports usually presented in the form of tables [5]. If a table is generated in electronic form is commonly referred to a spreadsheet. Extraction, integration, and processing of information presented in spreadsheets allows gathering huge volumes of data

This work was supported by the Russian Science Foundation, grant number 18-71-10001.

L. Borzemski et al. (Eds.): ISAT 2019, AISC 1050, pp. 140–149, 2020.
https://doi.org/10.1007/978-3-030-30440-9_14

from different domains. The collocated data could, in turn, be used for further research. Thus, tabular document is a valuable source of semi–structured information. Presently, a large part of such documents is presented on the Internet as open–access data including the form of Open Data. The number of spreadsheets on the Internet is currently estimated as more than 14 billion [2], about 41% of which are Excel documents [6]. The main reasons for such a high proportion of documents in Excel format is a widespread distribution of office software which supports these documents coupled with fast data processing, and sufficient simplicity of use. These spreadsheets contain useful information such as statistical, analytic reports, business intelligence data, etc. and therefore could be used for further analysis. One of the important tasks for further data reuse is automatic extraction, normalisation and integration. However, different ways of data presentation, schemes, and models (for example, reports prepared by the statistical agencies of different countries) greatly complicate the integration processes. Thus, automated processing and preparation of heterogeneous data for integration present a degree of challenge.

The reason for such complexities lies in a fact that tables are commonly generated in a form that is easily readable by people. First of all, people create spreadsheets for use by other people rather than computers. There are no standards of data presentation schemes. Therefore, each author of a document generates it in a form most suitable for their needs. In this regard, sometimes spreadsheet visual presentation is different from its physical form. By a visual structure of the header, we mean here its representation for human, when visually it is possible to determine borders of the cells. By a logical structure, we mean spreadsheets cells that form the table structure. Thus one visual cell may be presented by a set of several logical cells. In detail, it is presented in Sect. 4. It makes automatically data processing, integration, and sharing more difficult. Often researchers use their own implementation approaches for casting tables to a suitable form for integration. One of such methods is a representation of the table in a canonical [15] view. In order to represent a table in a canonical view, methods of table understanding are required. For example, the correct representation of cells in a table header allows improving the quality of table understanding processes.

This paper presents an approach for automatically correcting physical structure of table header in Excel documents. The approach proposed in this paper is based on heuristics. We evaluate the accuracy of this approach and illustrate it by utilising SAUS dataset[1].

2 Related Work

Spreadsheet and systems of their creation and editing are very common. Huge amount of spreadsheets (more than 55 million in the US only) are created every day by specialists and managers. A spreadsheet is a way of organising data into rows and columns for simplifications of reading and manipulation by

[1] https://catalog.data.gov/dataset/statistical-abstract-of-the-united-states.

people. Methods automated information extraction from semi-structured tabular documents (such as Excel, PDF, HTML) is a very active area of development [4,11,13].

Unfortunately, a lot of tabular documents contain errors. These errors could be in a form of incorrect tabular structure, various kinds of typos, errors in calculations, presence of redundant data, etc. Some studies report that 90% of documents contain errors [1,9]. For the most part, these are errors which are introduced by users.

Human errors, occurred in document preparation (including spelling, composition, design and optical recognition errors) process is a well-known problem. Research on human errors in many fields has shown that the problem lies in rather fundamental limitations in human cognition [12]. In the [9], noted that "we do not think the way we think we think". Human cognition is built on complex mechanisms that inherently sacrifice some accuracy for speed of information processing. Spreadsheets are also not an exception.

Some of the errors are not significant for human interpretation. However, these errors may be a huge problem for automated reading, interpretation, and analysis. These errors lead to one of the challenges of the classification problems when there is a need to identify individual table's section. These sections are building blocks of the tabular document. If it is possible to identify these sections correctly, it gives an opportunity for more effective automated document processing. Here we define the following sections of spreadsheets:

- **Header** – cells that represents the label(s) of a column. Their structure could be flat or hierarchical;
- **Attributes** – these cells might be found in the left-most or right-most columns of a table. Attributes can be seen as instances from the same or different (relational) dimensions placed in one or multiple columns in a way that conveys the existence of a hierarchy.
- **Metadata** – cells that provide additional information regarding the worksheet as a whole or specific sections.
- **Data** – cells that form the actual content of the table.
- **Derived** – cells, that may be presented in the Data block. They contain derived values of other cells. However, derived cells can have a different structure from the core data cells, therefore we need to treat them separately.

An example of the table sections is shown in Fig. 1. Types of section's elements are written vertically on the right–hand side.

A person could visually correctly identify all the section's elements. Normally, it does not matter whether these elements consist of one or more cells and whether the cells are merged or split. People identify tabular structure by analysing the relationship between items. However, elements of the physical structure might be different from the logical one. This discrepancy may have an affect on the results of automated table understanding. In order to get a correct result of table processing, there is a need to analyse elements of sections layout and classify all links between blocks. Suggestions about table elements layout are based on empirical analysis of classification results of neighbouring cells.

Table 4. Components of Population Change

See Notes

Period	Population as of beginning of period (1,000)	Net increase \1 Total (1,000)	Percent \2	Births (1,000)	Deaths (1,000)	Net international migration \3 (1,000)	Population as of end of period (1,000)
April 1, 2000 to July 1, 2000 \4	281,425	747	0.3	989	561	319	282,172
July 1, 2000 to July 1, 2001	282,172	2,868	1.0	4,047	2,419	1,200	285,040
July 1, 2001 to July 1, 2002	285,040	2,687	0.9	4,007	2,430	1,078	287,727
July 1, 2002 to July 1, 2003	287,727	2,484	0.9	4,053	2,423	822	290,211
July 1, 2003 to July 1, 2004	290,211	2,681	0.9	4,113	2,450	986	292,892
July 1, 2004 to July 1, 2005	292,892	2,668	0.9	4,121	2,433	948	295,561
July 1, 2005 to July 1, 2006	295,561	2,802	0.9	4,178	2,418	1,006	298,363
July 1, 2006 to July 1, 2007	298,363	2,927	1.0	4,289	2,421	866	301,290
July 1, 2007 to July 1, 2008	301,290	2,769	0.9	4,329	2,448	889	304,060

Source: U.S. Census Bureau,
"Population, Population change and estimated components of population change: April 1, 2000 to July 1, 2008"
(released December 22, 2008).

Metadata / Header / Data / Metadata

Fig. 1. Sample of a table structure

The [1,3] proposed approaches for improving the quality of tables. Improving consists of transforming the physical structure of documents in a form which is closest to logical one and more suitable to further automated processing [6,7].

Processing and transformation of cells in a table header block is further discussed in [1]. The spatial analysis of block elements is used to estimate the possibility of empty cells merging with cells that contain information.

3 Contribution

This paper presents a heuristic algorithm for the physical structure recovery of spreadsheets cells for table headers. The algorithm is based on the alignments of text content within the cells and their visual borders. It heuristically evaluates the positions of the column boundaries by aligning the text content in the table body. The transformation of neighbouring cells in rows occurs only in the "header" (header part) of the table. The header position is evaluated by MIPS positions (search for minimum indexing points) algorithm [8].

Typically state-of-the-art methods for the cell structure recognition deal with low-level documents (bitmaps or print instructions). Unlike them, our approach intends to correct a high-level physical structure of spreadsheet tables. The proposed approach avoids errors (optical recognition for example) of cell structure recovery that inherent in methods oriented to the low-level representation of the source data. In this research, we only process the table header. The aim of the transformation is to create a correct header structure. It is expected that the proposed approach will improve the effectiveness of table understanding due to the processing of documents with the correct physical structure.

4 Table Header Analysis

4.1 Logical and Physical Layers of Tables

Tabular data are presented and stored in cells. Cells might have different properties (e.g. text, style) and might be merged with neighbouring cell or the merged cells might be split. Cells form a physical layer of a table. Automatic processing of a table involves operations with this layer.

Logical layer is a representation of the visual structure of a table. For example, by estimating the relative position of objects in the table, a person can define the structure of a header, rows and columns of the table. However, the physical structure might be different. Visually, one cell might be presented by several cells in a physical level. Differences between logical and physical structures might complicate the processes of table understanding and affect its quality.

4.2 Cells Classification

In this approach, we divide all cells of header section into two categories. Firstly, cells that do not contain any value, we refer to them as "empty" cells here. Cells that contain some value we call as "non-empty". An "empty" cell should be merged with a "non-empty" cell if the former is not surrounded by visual boundaries (internal or external) and does not go beyond the header section. In case when an "empty" cell is surrounded by visual boundaries, we do not change it. Secondly, we define "non-empty" cells as cells that contain data. "Empty" cells will be joined with these type of cells. Correct merging of cells allows creating a correct physical structure in the header section.

4.3 Dataset Header Cells Analyses

Let us consider a part of the table from the SAUSE dataset (Table 1). In this example, a part of the table that is required to be checked (table header) has a white background colour. Table head structure allows us to extract categories, labels, data, etc. [10].

Table 1. Sample of a table fragment

Census date	Resident population		
	Number	Increase over preceding census Number	Persent
1790 (August 2)	3,929,214	(X)	(X)
...	...	...	...

However, if we consider a physical structure of the table header it will be different from its logical structure, which is illustrated in the Tables 2 and 3 correspondingly. For instance, not all cells are merged or contain text in the Table 2. However, in most cases, it is possible to describe features of table head layout. Solid line denotes a visual border of a cell and the dash border line denotes a physical border of a cell in the Table 2 and the following tables. The physical borders do not always match with the logical structure as could be seen in the Tables 2 and 3.

Table 2. Sample of table fragment (physical structure)

Census date	Resident population		
	Number	Increase over preceding census	
		Number	Persent
1790 (August 2)	3,929,214	(X)	(X)
...	...	...	...

Table 3. Sample of table fragment (logical structure)

Census date	Resident population		
	Number	Increase over preceding census	
		Number	Persent
1790 (August 2)	3,929,214	(X)	(X)
...	...	...	...

5 Tables Physical Structure Recovery Algorithm

The goal of the proposed algorithm is to eliminate all "empty" cells in table header and transform physical structure according to visual representation. Elimination is achieved through merging one cell with other cells. Therefore, firstly we need to determine heuristical rules of cell merging. Visual analysis of random tables from SAUSE dataset led to a conclusion that the structure of all headers shares some common features.

5.1 Algorithm Hypotheses

The illustrative example of a table structure is presented in the Table 4. We introduce the following notation to describe merging operations. Denote by **merge(a,b)** an operation of merging the cell **"a"** with the cell **"b"**. The merging operation is not commutative. The results cell inherits the style of the cell **a**. If both of the cells contain data - the data type will cast to string and concatenate.

The following assumptions regarding the heading of the dataset's tables were made:

- We assume that it is possible transform a physical structure of the table header to logical, i.e. cells could be merged according to the visual representation of the table elements.

Table 4. Sample of the table header

	a	b	c	d	e	f
1	State and island areas	Department of Agriculture				
2					Educaton	
3		Food and nutrition service		(WIC)	No Child Left	Title Programs
4		Child nutrition	Food stamp			

- There should be no "empty" cells in the resulting physical structure. All these cells should be merged with neighbouring ones that have data. For example, operation **merge(a2, a1)** allows to escape of "empty" cell **a1**. Some tables might have empty cells which are completely surrounded by visual borders. In this case, the elements remain unchanged.
- Headers have a hierarchical structure. It means that the header might be divided into blocks by cells at the top level. All lower cells, in turn, form new blocks. Blocks of lower levels could not be wider (go beyond vertical boundaries) than upper-level blocks. The upper-level blocks are containers that contain all lower-level blocks (cells).
  For example, the top cell **b1** in the Table 4 determines vertical restrictions for lower rows in borders **b1:d1**. In turn, lower level cells **b3** determine vertical restrictions for lower rows in borders **b3:dc**.
- If two adjacent vertical cells have equal borders and there is no visual horizontal border between these cells - they should be merged. For example, **a1** = **merge(a2, a1)**; **a1** = **merge(a3, a1)**; **a1** = **merge(a1, a4)**. Finally, the merged cell **a1** = **a1:a4** will be created.
- The algorithm analyses cells in the table header, determine their positions, styles, data content and estimates the necessity of merging in vertical and/or horizontal directions. According to the hypothesis of the header elements hierarchy, the algorithm determines blocks of the header and analyses them. The blocks are processed sequentially starting from the top – left. Then the nested blocks are processed. Afterwards, the left block is processed, if one exists. The analysis is then carried out in the vertical surface from top to bottom. The process continues until the "non–empty" cell, the cell with a horizontal visual border, or the bottom border of the table header will be identified.

5.2 Algorithm Description

Based on the proposed assumptions, we consider a heuristic algorithm for transforming the physical structure of the table to its visual representation (logic layer). If the cell containing data is found, the previously "empty" cell(s) should be merged with it.

Consider, for example, the Table 4. Here, the cell **a2** should be merged with the **a1** forming the new cell **a1** = **merge(a2, a1)**. According to our assumption of block structure, we should analyse bottom cells at the possibility of merging.

Therefore, we will get a new cell **a1** = **merge(a1, a3)**, which has the address **a1:a3**. According to our assumption, we should analyse bottom cells at the possibility of joining. As a result, the new cell **a1** = **merge(a1, a4)** will be created and its address will be **a1:a4**. This cell has a visual border from the right side. This means that extending the cell to the right is impossible. The block, with address **b1:d4** will be considered next.

The right border of the block is defined by a border of the cell **d1**. The cell **b2** is empty. The lower **b3** cell is not "empty" therefore **b2** should be analysed to extend to the right side. Then, as the result of cell merging we get **b2** with address **b2:d2**. After the above transformations, b1 may be merged with b2. The address of the new cell will be **b2:d3**.

Transforming the physical structure to the logical one for the cell **c1** is conducted in the following way. Cell **e1** is "empty". Thus, firstly it needs to check lower cell **e2**. It is merged cell **e2:f2**. According to our assumption, the algorithm then checks the possibility of merging **c1** with cells on the right. As a result, the algorithm is formed new cell e1 with address **e1:f2**. The results of all transformations are shown in Table 5. The physical structure of the table is now the same as its logical representation.

Table 5. Results of table header physical structure transformation

<table>
<tr><th></th><th>a</th><th>b</th><th>c</th><th>d</th><th>e</th><th>f</th></tr>
<tr><td>1</td><td rowspan="4">State and island areas</td><td colspan="3" rowspan="2">Department of Agriculture</td><td colspan="2" rowspan="2">Education</td></tr>
<tr><td>2</td></tr>
<tr><td>3</td><td colspan="2">Food and nutrition service</td><td rowspan="2">(WIC)</td><td rowspan="2">No Child Left</td><td rowspan="2">Title Programs</td></tr>
<tr><td>4</td><td>Child nutrition</td><td>Feed stamp</td></tr>
</table>

If the algorithm could not verify any possibility of cells merging, the physical structure of the table will stay unchanged. Such cases are reflected in the results of the experiment.

6 Experimental Evaluation of the Algorithm

For the experimental evaluation of the proposed algorithm 200 random tables were extracted from the SAUS data set. The tabular data were gathered into one Excel workbook. All these tables were marked by the $START and $END tags. These tags were needed in order to determine the table in Excel worksheet for further extraction by TabbyXL [14]. The algorithm operates with the table objects in terms of TabbyXL object model. All described transformations are saved in the copy of the loaded workbook. After all operations, the copy of the workbook is saved to a separate file. This allows us to compare the results of the transformation logical structure with visual ones. The software tool for evaluating the proposed algorithm was created as a Java subprogram for TabbyXL.

The comparison was made by an expert, who estimated the accuracy of the transformation. The expert compared two tables (initial and modified) and made a conclusion. The following conclusions were made:

- *without (w/o) changes* – the physical structure of initial and modified tables is correct and equal;
- *modified correctly, fully* – the physical structure of the table was modified correctly, all cells are matched to their logical representation;
- *modified correctly, partially* – the physical structure of the table was modified correctly, but not all cells are matched to their logical representation;
- *incorrect modification* – the physical structure of the table was modified incorrectly, including with possible changes to its structure.

The results of the experimental evaluation are shown in the Table 6. They demonstrate good enough acceptability of heuristic methods for recovering a physical structure of spreadsheets header.

Table 6. Experimental results (in %)

W/o changes	Modified correctly		Modified incorrect
	Fully	Partially	
8.1	76.5	10.8	4.6

7 Conclusions and Future Work

Transformation of the physical structure table header to logical ones allows simplifying and improving table understanding processes.

We have demonstrated a heuristic method of correction the physical structure of the table headers to a logical once by utilising assumptions about the structure of table headers. The aim of our proposed approach is to achieve the same logical and physical representation of a table header. The results of such transformations may be used in the tasks of table understanding as the correct structure of the table allows reducing errors in table understanding tasks. The conducted experiment showed that the proposed heuristic algorithm achieved good results in terms of accuracy of the transformation. In the future work, we aim to improve the rate of fully correct transformation and decrease the rate of incorrect structure transformation. We are also planning to extend evaluation to other datasets.

References

1. Abraham, R., Erwig, M.: Header and unit inference for spreadsheets through spatial analyses. In: Proceedings of 2004 IEEE Symposium on Visual Languages and Human Centric Computing(VLHCC), pp. 165–172, September 2004. https://doi.org/10.1109/VLHCC.2004.29

2. Cafarella, M.J., Halevy, A., Wang, D.Z., Wu, E., Zhang, Y.: Webtables: exploring the power of tables on the web. Proc. VLDB Endow. **1**(1), 538–549 (2008). https://doi.org/10.14778/1453856.1453916
3. Cunha, J., Fernandes, J.P., Mendes, J., Saraiva, J.: Spreadsheet engineering. In: Central European Functional Programming School - 5th Summer School, CEFP 2013, Cluj-Napoca, Romania, pp. 246–299, 8–20 July 2013. https://doi.org/10.1007/978-3-319-15940-9_6
4. Eberius, J., Werner, C., Thiele, M., Braunschweig, K., Dannecker, L., Lehner, W.: Deexcelerator: a framework for extracting relational data from partially structured documents. In: Proceedings of the 22nd ACM International Conference on Information & Knowledge Management CIKM 2013, pp. 2477–2480. ACM, New York (2013). https://doi.org/10.1145/2505515.2508210
5. Embley, D.W., Hurst, M., Lopresti, D., Nagy, G.: Table-processing paradigms: a research survey. Int. J. Doc. Anal. Recogn. (IJDAR) **8**, 66–86 (2006). https://doi.org/10.1007/s10032-006-0017-x
6. Koci, E., Thiele, M., Romero, O., Lehner, W.: Table identification and reconstruction in spreadsheets. In: Dubois, E., Pohl, K. (eds.) Advanced Information Systems Engineering, pp. 527–541. Springer, Cham (2017)
7. Koci, E., Thiele, M., Romero, O., Lehner, W.: Cell classification for layout recognition in spreadsheets. In: 8th International Joint Conference, 9–11 November 2016, IC3K 2016, Porto, Portugal, pp. 78–100, January 2019. https://doi.org/10.1007/978-3-319-99701-8
8. Nagy, G., Seth, S.C.: Table headers: an entrance to the data mine. In: 23rd International Conference on Pattern Recognition, ICPR 2016, Cancún, Mexico, pp. 4065–4070, 4–8 December 2016. https://doi.org/10.1109/ICPR.2016.7900270
9. Panko, R.R.: Spreadsheet errors: What we know. what we think we can do. CoRR **abs/0802.3457** (2008)
10. Pasupat, P., Liang, P.: Compositional semantic parsing on semi-structured tables. In: Proceedings of the 53rd Annual Meeting of the Association for Computational Linguistics and the 7th International Joint Conference on Natural Language Processing (Volume 1: Long Papers), pp. 1470–1480. Association for Computational Linguistics (2015). https://doi.org/10.3115/v1/P15-1142
11. Rastan, R., Paik, H.Y., Shepherd, J., Ryu, S.H., Beheshti, A.: Texus: table extraction system for pdf documents. In: Wang, J., Cong, G., Chen, J., Qi, J. (eds.) Databases Theory and Applications, pp. 345–349. Springer, Cham (2018)
12. REASON, J.: Human error, pp. XV, 301, p. ill. 23 cm (1994). http://infoscience.epfl.ch/record/2249. bibliogr.: p. 258–290. Index
13. Shigarov, A., Altaev, A., Mikhailov, A., Paramonov, V., Cherkashin, E.: Tabbypdf: web-based system for pdf table extraction. In: Damaševičius, R., Vasiljevienė, G. (eds.) Information and Software Technologies, pp. 257–269. Springer, Cham (2018)
14. Shigarov, A.O., Mikhailov, A.A.: Rule-based spreadsheet data transformation from arbitrary to relational tables. Inf. Syst. **71**, 123–136 (2017). https://doi.org/10.1016/j.is.2017.08.004
15. Shigarov, A.O., Paramonov, V.V., Belykh, P.V., Bondarev, A.I.: Rule-based canonicalization of arbitrary tables in spreadsheets. In: Dregvaite, G., Damasevicius, R. (eds.) Information and Software Technologies, pp. 78–91. Springer, Cham (2016)

A Web-Based Support for the Management and Evaluation of Measurement Data from Stress-Strain and Continuous-Cooling-Transformation Experiments

Ronny Kramer(✉) and Gudula Rünger

Department of Computer Science, Chemnitz University of Technology, 09107 Chemnitz, Germany
{kramr,ruenger}@cs.tu-chemnitz.de

Abstract. Mechanical engineers are often faced with the challenge to keep track of the results of experiments done within their own institutions or other research groups. The lack of this knowledge may cause a potential waste of resources and time since experiments might be repeated if the existence of previous experiments is unknown or the data cannot be reused. The main problem with managing scientific data is the huge amount of experiments, the lack of public access to the results, and often missing meta-data which would be needed to compare the results. This article considers experiments and data resulting from experiments of Stress-Strain and Continuous-Cooling-Transformation experiments. It is shown how a web-driven data management process starting from the planning of experiments through to the evaluation of the measurement data could be supported by an appropriate work-flow to make sure that all meta-data is preserved to find and reuse the results.

Keywords: Data management · Data evaluation · Web-based · Material science

1 Introduction

Public databases providing knowledge for scientists in different fields exist for decades. An example for material science are crystallography databases which exist for over 70 Years [10]. These databases have been designed independently for the specific use cases and, thus, they are heterogeneous in the sense that they store their data in different formats or use different terminology, so that a combination of their datasets is not trivial. To accommodate this issue, new ontologies, such as MatOnto [6] have been created to support the storage of knowledge in a structured way using a defined terminology. The definition and acceptance of ontologies like MatOnto combined with new Internet Standards,

L. Borzemski et al. (Eds.): ISAT 2019, AISC 1050, pp. 150–159, 2020.
https://doi.org/10.1007/978-3-030-30440-9_15

such as the Resource Description Framework (RDF) [8] and the Web Ontology Language (OWL) [4], can be explored by search interfaces, such as MatSeek [7], which allow ontology-based searches in multiple databases. These inventions have improved the cross usability of material science databases as well as the overall search process. The information gained by combining public databases is often not enough for data science processes in material science, as they normally only provide results and do not include measurement data with extensive knowledge about the provenance. The measurement data in conjunction with the provenance is required to gain new information, e.g. by the detection of similarities or differences in the datasets, which would be needed to make predictions about material compositions.

The goal of this article is to propose an approach which improves the available measurement data by including enough provenance necessary for data science analysis while keeping the overhead for gathering the data as low as possible. A thorough digital archiving including the collection of as much provenance data as possible can become a time-consuming task so that there is a need for a sustainable support accomplishing this task implicitly while performing the current research experiment. The approach proposed in this article is to provide a software system that is used by a scientist during the research work and that is able to collect the provenance data automatically. The challenge for such a software system is to define a work-flow that is convincing enough for the scientists so that they adapt their habits and take the extra burden to use a new support system instead of their individual work-flow they are used to. This can be achieved by providing assistance or even automation for repeating and time-consuming tasks. In the context of material science, those tasks are: The planning of experiments, the organization, the scheduling of the testing machine usage, the preparation of specimens or the actual execution of experiments and their evaluation. The execution of all of these tasks can be assisted or even automated except the preparation of specimens and the actual execution of experiments, as they still require manual work to be performed.

The contribution of this article includes the design and implementation of a data management and evaluation web-based software platform for a real world application in mechanical engineering, which has been developed in cooperation with the application scientists, and a proof-of-concept for a data management and evaluation web-based software platform to be used in other applied sciences. The rest of this article describes related work in Sect. 2 and introduces the software system and work-flow concept in Sect. 3. Section 4 presents the application from material science and Sect. 5 provides concluding remarks.

2 Related Work

The book "Building the Data Warehouse" [2] from 1992 already presents some general patterns for the design of a system called data warehouse which is suitable to extract information from legacy systems and makes them usable for further analysis. The idea of data warehouses was applied to material databases by Li, for the usage in large data collections with the goal to identify similarities and handle differences [5]. The problem with the data warehouses as described in these books is that they are designed to be compatible to specific databases, which makes the adaption to new systems more complicated [7]. The new idea in this article is to take the idea of data warehouses and place them in the center of a software system and build a work-flow about data generation and computer aided evaluation around it. While the disadvantage of this approach is, that the usage of existing data is still complicated, the advantage is that newly collected data can be stored with the focus on data science. This focus not only improves the applicability of data but also increases the portability as the storage can be built to support established ontologies.

A computer aided evaluation is already established in material science by software systems typically provided by the manufactures of the computer aided test machines used to perform the experiment. Since the main functionality of these software systems is to control the corresponding machine, the provided data evaluation is optimized for the specific use case. As a result of this optimization the software system is able to perform almost any computation that the user of this machine might want to perform, but its capabilities are limited to the specific use case.

In the area of mechanical material testing, two companies that provide their own software solution are ZwickRoell with the software testXpert [15] and MetLogix with L3 [12]. Besides the Stress-Strain evaluation via tensile testing, they also support three-point flexural, peel, holding, creep and creep rupture testing. They do not support any evaluations for which the data required cannot be determined using their machines. Thus, a thermal analysis is not possible. In the area of thermal analysis, two companies which provide their own software solution are NETZSCH with Proteus [14], and LINSEIS thermo analytical software [13]. As specialized software solutions they are able to evaluate additional experiments to CCT, such as Continuous-Heating-Transformation (CHT) and Time-Temperature-Transformation (TTT). These software tools allow the export of the measurement data and results in various file formats, such as ASCII CSV or Microsoft Excel files. The usage of these evaluation software tools would be an applicable solution to reduce the complexity of a data management system. However, this is not viable for the experiments considered here, since parts of the provenance meta-data is lost during the post processing and the evaluation in external tools. Thus, none of the existing software tools are capable to provide a consistent software support as needed for the specific experiments considered in this article, which leads to the software solution.

3 Work-Flow for Implicit Meta-data Collection

Data collection is an essential part for scientific experiments, but is very time-consuming. To support the scientist, we propose an implicit meta-data collection which accompanies all phases of the experiment and collects data during the planning phase, execution phase and evaluation phase of an experiment. The implicit data collection is modelled as a work-flow containing all steps in the experiment, including manual work in the lab as well as software-supported steps. Figure 1 illustrates such a work-flow. To guide the scientist through the work-flow, an appropriate software system is to be provided. From the point of view of the scientist, the advantage of the software system is a support in the background during the planning, execution and evaluation phases of an experiment.

The work-flow proposed contains steps consisting of multiple actions. The steps in Fig. 1 are presented in two different colors, white and grey, depending on their connection to the software system. The steps presented in white boxes can be supported or even automatized by the software system, while a grey box denotes actions that belong to the experiment and require manual labor executed outside of the software system. However, as the actions executed during these manual labor steps are intended to follow the created plan, the provenance is still guaranteed. Any incident needs to be logged to keep the provenance consistent.

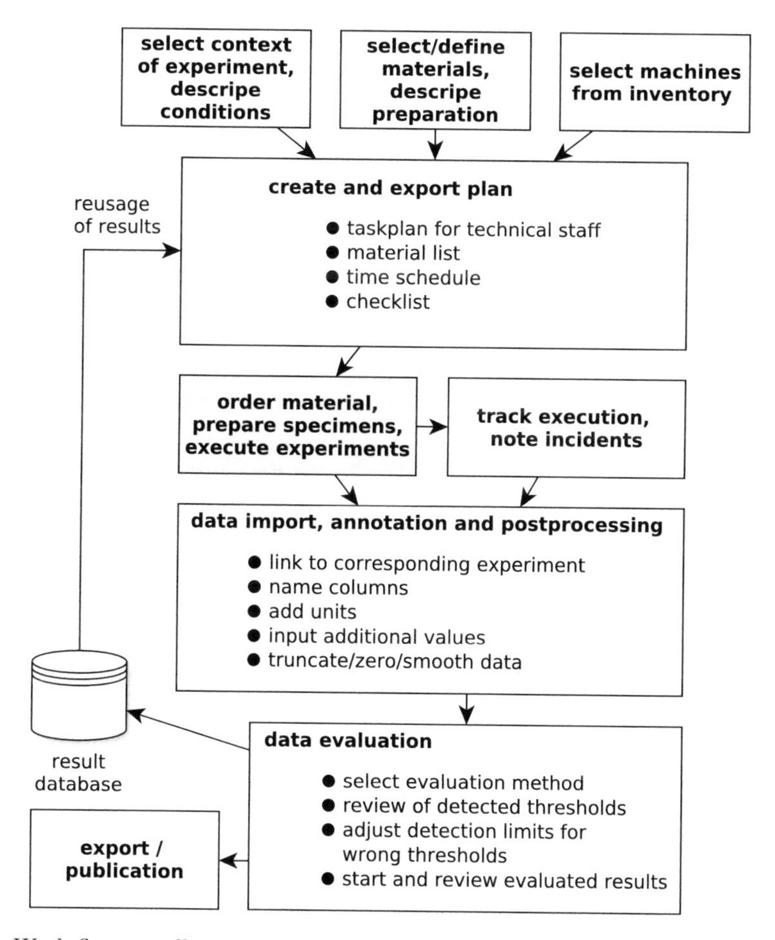

Fig. 1. Work-flow to collect meta-data from an experiment, starting with the planning of an experiment up to the evaluation of the experiment. A mechanism for the reuse of knowledge gained by previous experiments is included.

The **planning phase** of an experiment is based on four pillars:

1. The selection of the context for the experiment planned. This decision influences the templates that are going to be used for all other steps.
2. The preparation planning for the specimen based on materials and preparation steps.
3. The selection of machines to be used from a list of compatible machines based on pillar 1 and 2. This connects the final results to the available information about the machines.
4. The processing of the data, such as predictions or suggestions based on larger data sets and data analysis.

Based on those four pillars, the experimental plan is created containing a checklist, a task plan for the technical stuff, a material list and a time schedule.

As already stated, the **execution phase** is contained in the work-flow but only loosely connected to the software system. Thus, the interaction with the software system is limited to the logging of unexpected incidents.

The **evaluation phase** consists of two major parts, which are the preparation of the data and the final evaluation of the data. For the preparation, the raw measurement data gained from the experiments are imported, annotated and post processed so that data are in a form that is compatible to the software system. The annotation is done by linking the data sets to the experiment and by mapping the data columns to the contexts requirements. The physical units of the data columns are automatically assigned according to the mapping, but can be changed by the scientist. By a post processing of the imported data with methods, such as trimming, smoothing and zeroing, the compatibility to the software systems evaluation can be improved. For the final evaluation, the data is used to gain knowledge about the properties of the material, such as the stiffness, flexibility or the resistance to high temperatures. The material properties are presented in an appropriate form for exploration or exportation and are stored in the database for future data-mining. An example for the final evaluation based on Stress-Strain and Continuous-Cooling-Transformation experiments is given in the next section.

4 Stress-Strain and Continuous-Cooling-Transformation

The implicit collection of meta-data of the Stress-Strain, see Subsect. 4.1, and Continuous-Cooling-Transformation, see Subsect. 4.2, evaluation is captured in a work-flow as described in general in the previous section. To realize this work-flow, the software system provides templates for all steps and implements the specific equations required to calculate the material properties. As the evaluation of the experiments can be a very time consuming task, the automated evaluation is the main advantage of the software-system provided to the application scientist in material science. The automated evaluation is crucial to provide faster evaluations and visual support and, thus, supports to perform more experiments in shorter time frames supporting a faster knowledge acquisition.

4.1 Stress-Strain

The goal of Stress-Strain data is the calculation of material properties that, among others, describe how the specimen and therefore the materials or joins react when forces are applied [9]. The data is collected during tension testing where a force is applied with a given speed. A standardized specimen is strained until its failure while the change in length and the required force is being recorded [11].

The evaluation of Stress-Strain Data is defined in the DIN 6892-1 [11]. The base value for all calculations is the young modulus E in Pascal that represents the elastic value of the specimen and is calculated as $E = \frac{\sigma}{\epsilon}$ where σ is the stress and ϵ the proportional deformation. For metals and some other materials, the young modulus can be described by a law of physics, called Hooke's law. According to this law, the young modulus visualizes a linear slope at the beginning of the data region where $E = \frac{\sigma}{\epsilon} = const$ applies; this is called the Hookean slope [3]. Algorithm 1 is used to detect the young modulus and assumes that Hooke's law is valid for the specimen considered. In the ideal case, the Hookean slope is represented by a data range that starts with the first datum and ends at the beginning of a concave curve at the datum with maximum stress. However, real world data often shows a convex curve which changes into this concave curve. In such cases, an approximation is used to determine the inflection point which has the biggest slope [11]. As the measurement data may not be smoothed, the software system needs to cope with this problem described by calculating the average slope over a data window.

Algorithm 1. Detection of the young modulus

```
1  procedure findYoungModulus
2      Let C be a list of stress and proportional deformation
       pairs with m index of max stress and w as window size;
3      for (i=w,...,m-w) do
4          calculate average slope over C(i-w,...,i+w);
5      end for
6      find maximum of the calculated slopes;
7  end procedure
```

In all cases tested for this article, the automatic detection of the young modulus by the algorithm has been proven to be robust. Due to the error-prone nature of real world data, there is a chance that the algorithm makes a wrong detection. Thus, the scientist may visually review the final result diagrams and may decide whether the algorithm has found a plausible young modulus or an adjustment of detection parameters is required in some circumstances.

The results of Stress-Strain experiments consist of three different components; an example diagram generated by the software-system is given in Fig. 2.

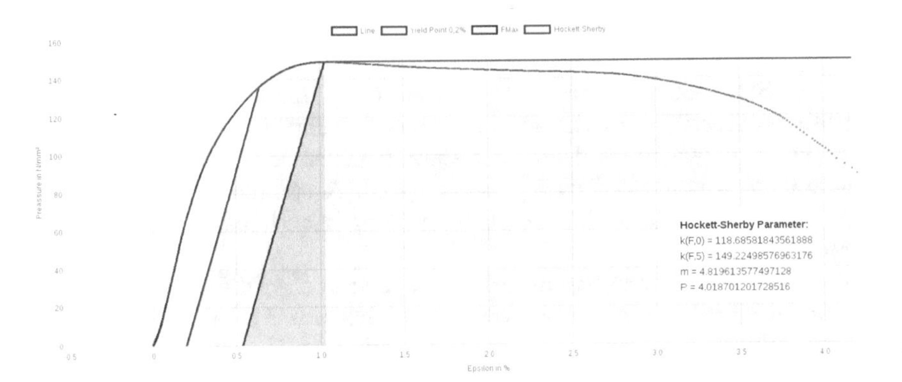

Fig. 2. Example diagram of the Stress-Strain evaluation generated by the software system.

The first component is a plot of the Stress-Strain curve to provide a visualization. The second component consists of the evaluated material properties, which are split up again into two groups. There are values that can be determined by searching local extremes on the curve such as the point at which the maximum force was applied. Other values require the construction of the Hookean slope in predefined points, such as the yield point at 0.2% proportional deformation, and a subsequent calculation of their intersection with the curve or the zero-crossing. The third component provides predictions, such as the Hockett-Sherby equation [1], which are calculated by curve fitting.

4.2 Continuous-Cooling-Transformation

The goal of the Continuous-Cooling-Transformation (CCT) is to determine the phase transformation temperatures during the cool-down process. Phase transformations are structural changes within the material considered. They are visualized as a change of the length of the specimen that is not related to the linear shrinking based on the expansion coefficient. The temperatures of the phase transformation represent an important part of CCT diagrams, which are required to judge the compatibility to a specific use case [9]. While the recognition of the phase transformation within a diagram is trivial for a human with experience in this domain, the adaptation of this experience in a software system is a challenge. In the following, it will be discussed how it is possible to predict the theoretical shrink process by expanding the linear sections close to the phase transition. The extraction of the transformation temperatures is achieved by comparing the theoretical shrink with the real shrink.

The Algorithms 2 and 3 used for the detection of the phase transformation, are based on the assumption that the expansion and shrinking process during heat-up and cool-down is only influenced by the expansion coefficient. Algorithm 2 works by searching all linear sections of T, where the temperature range for the slope calculation is at least $k\ Kelvin$. The curve data is expected

Algorithm 2. Find all linear sections with a temperature range of over k $Kelvin$

```
1  procedure findLinearSections
2      Let C with |C|=m be a list of length variation over
       temperature;
3      Let k be the minimal temperature range in Kelvin and w
       the window size;
4      Let T be an empty list;
5      Let trange be a tuple initialized with the temperature
       of the first element of C;
6
7      for (i=w,... ,m -w) do
8          calculate average slope over C(i-w,... , i+w);
9          calculate average temperature over C(i-w,... , i+w);
10         if (slope change > threshold or last window) then
11             if (trange > k) then
12                 Add trange to T;
13             end if
14             set low temperature of trange;
15         end if
16         set high temperature of trange;
17     end for
18 end procedure
```

Algorithm 3. Filter linear sections by largest curve coverage

```
1  procedure filterLinearSections
2      Let T with |T|=m be a list of linear section
       temperature tuples;
3      for T(i=1,.... ,m) do
4          filter adjacent pair (i, i+1) of linear sections
       with largest range between temperatures;
5      end for
6      transform pair into linear functions t_h and t_l;
7  end procedure
```

to consist of a heat-up and a cool-down stage, including a plateau p where the temperature has a peak for a short amount of time. By splitting the curve data C at the plateau p, Algorithm 2 can be used to find the linear sections for each stage independently. Algorithm 3 filters the linear sections close to the phase transformation by the largest curve coverage and converts them into the linear functions t_h and t_l. The percentage of transformation over the temperature t can be evaluated by calculating $\frac{|t_h(t)-C(t)|}{|t_h(t)-t_l(t)|}$ $\forall\, t \in Inputdata$. Figure 3 shows the evaluated results in form of a diagram created by the software system.

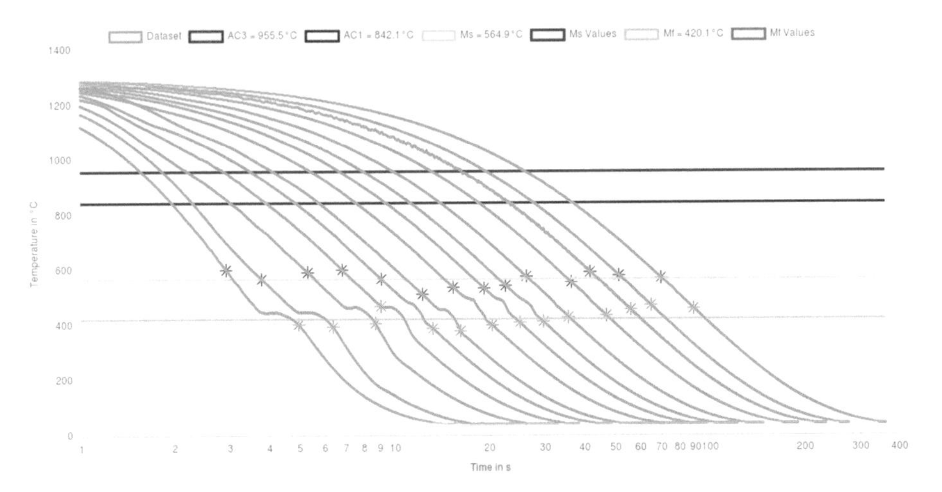

Fig. 3. Example diagram of the Continuous-Cooling-Transformation evaluation generated by the software system.

A user survey comparing the automated evaluation of these exemplary experiments to the traditional evaluation method using multiple disjoint software-tools showed a time saving potential. The scientists who have tested this software-system reported an individual time reduction from a couple of hours down to a few minutes per experiment, which is considered to be a success.

5 Conclusion

In this article, we have proposed a software system which is designed to support experiments in material sciences with the specific functionality to collect a large variety of experimental data and meta-data with the aim to provide the possibility for successful data-mining and data analysis processes in the future. The core of the software system proposed is a specific work-flow supporting this functionality. The goal is to provide a support of the scientists in material science which serves their needs by adding benefits that ease their experimental work evaluation. Our approach is to build a software system which makes it possible for the scientists to mainly concentrate on the experiments while the software and the data gathering is executed in the background and makes it possible to come back to the software system when needed or the next step is to be done. The software implementation is in the state of a proof-of-concept, which has successfully been applied to a specific experiment described in this article.

Acknowledgements. This work is supported by the Welding Engineering group of the TU Chemnitz and by the Federal Cluster of Excellence EXC 1075 *MERGE Technologies for Multifunctional Lightweight Structures* financed by the German Research Foundation (DFG).

References

1. Hockett, J.E., Sherby, O.D.: Large strain deformation of polycrystalline metals at low homologous temperatures. J. Mech. Phys. Solids **23**(2), 87–98 (1975). https://doi.org/10.1016/0022-5096(75)90018-6. ISSN: 0022-5096
2. Inmon, W.H.: Building the Data Warehouse. Wiley, New York (1992). ISBN: 0471569607
3. Ugural, A.C., Fenster, S.K.: Advanced Strength and Applied Elasticity, 4th edn. Prentice Hall, Upper Saddle River (2003)
4. Bechhofer, S., et al.: OWL Web Ontology Language Reference, February 2004
5. Li, Y.: Building the data warehouse for materials selection in mechanical design. Adv. Eng. Mater. **6**(1–2), 92–95 (2004). https://doi.org/10.1002/adem.200300522
6. Cheung, K., Drennan, J., Hunter, J.: Towards an ontology for data-driven discovery of new materials, January 2008
7. Cheung, K., Hunter, J., Drennan, J.: MatSeek: an ontology-based federated search interface for materials scientists. IEEE Intell. Syst. **24** (2009). https://doi.org/10.1109/MIS.2009.13
8. Wood, D., Lanthaler, M., Cyganiak, R.: RDF 1.1 Concepts and Abstract Syntax. W3C Recommendation. W3C, February 2014. http://www.w3.org/TR/2014/REC-rdf11-concepts-20140225/
9. Hahn, F.: Werkstofftechnik-Praktikum: Werkstoffe prüfen und verstehen; mit ... zahlreichen Tabellen. München: Fachbuchverl. Leipzig im Carl Hanser Verl., 260 S (2015). ISBN: 9783446432581
10. Bruno, I., et al.: Crystallography and databases. Data Sci. J. **16** (2017). https://doi.org/10.5334/dsj-2017-038
11. ISO. Metallic materials - Tensile testing - Part 1: Method of test at room temperature. Beuth Verlag, Berlin (2017)
12. L3 - Kraft Prüfsoftware für die Materialprüfung. http://www.metlogix.de/pf/l3. Accessed 09 May 2019
13. LINSEIS Thermoanalyse Software. https://www.linseis.com/produkte/dilatometer/l78-rita/#software. Accessed 09 May 2019
14. NETZSCH Proteus® Software für die Thermische Analyse. https://www.netzsch-thermal-analysis.com/de/produkteloesungen/software/proteus/. Accessed 09 May 2019
15. Prüfsoftware testXpert III. https://www.zwickroell.com/de-de/pruefsoftware/testxpert-iii. Accessed 09 May 2019

Software Toolkit for Visualization and Process Selection for Modular Scalable Manufacturing of 3D Micro-Devices

Steffen Scholz[1,2], Ahmed Elkaseer[1,3](✉), Mahmoud Salem[1,4], and Veit Hagenmeyer[1]

[1] Institute for Automation and Applied Informatics, Karlsruhe Institute of Technology, Karlsruhe, Germany
ahmed.elkaseer@kit.edu
[2] Karlsruhe Nano Micro Facility, Hermann-von-Helmholtz-Platz 1, Karlsruhe, Germany
[3] Faculty of Engineering, Port Said University, Port Said, Egypt
[4] Faculty of Engineering, Ain Shams University, Cairo, Egypt

Abstract. SMARTLAM is a European funded research project, intended to develop a modular manufacturing platform and capability database, for three-dimensional integration/manufacturing of component parts constructed from laminated polymer films. To enable communication/interaction between SMARTLAM integrated manufacturing modules and the capability database, a software toolkit has to be developed, which must comply with the modularity concept of the SMARTLAM manufacturing system, and with the capability to integrate changes in the manufacturing platform, e.g. adding a new manufacturing module. This paper presents the implementation of the SMARTLAM software toolkit, i.e. the software and graphical user interface designed and developed to conform to these requirements, which allows for automated process chain selection while enabling the operator to load and modify the selected process chains and to alter parameters' values. The developed software toolkit was tested and it was found able to compile and execute complete manufacturing recipes with the given equipment and it was directly linked to the SMARTLAM database. In addition to its ability to exchange data between different manufacturing modules, it gives the users the opportunity to modify the manufacturing processes in real time. The main result of the overall software development is a fully functional and a commercially applicable solution for the design, planning, configuration and coordination of a modular manufacturing platform for micro-devices with minimum cost/time of the identified process chain. As a validation example, a flexible lighting device demonstrator (DLED) was manufactured using the SMARTLAM platform, and the software toolkit was successfully utilized to visualize and select the appropriate production process chain.

Keywords: Software toolkit · Socket programming · Modular manufacturing · Process visualization · Process selection

L. Borzemski et al. (Eds.): ISAT 2019, AISC 1050, pp. 160–172, 2020.
https://doi.org/10.1007/978-3-030-30440-9_16

1 Introduction

Current trends in the manufacturing of micro components indicate increasing demand for customized parts [1]. The concept of a flexible production cell, able to combine multiple manufacturing techniques, materials and software solutions provides a starting point for SMEs' production of customized products [2]. Especially, the ability to exchange single modules to fabricate different products in close succession, or even in parallel, without having to adapt sub-systems or peripheral devices gives the opportunity to promptly react to changing demands, while limiting the cost of tooling, that may be required for other production alternatives [3]. Whilst there are few examples of truly high volume 3D micro-product manufacturing, there is a growing demand for fabrication facilities that can cope with variable volumes and be cost effective for small batch sizes as well as high volume manufacture. This requires the development of flexible and scalable manufacturing approaches applicable in the micro domain [1, 4].

Presently, most of the applications in the multi material micro technology area are addressing small and medium batch size manufacturing. There is also a high cost associated with establishing production facilities for medium volume manufacturing which effectively hinders small companies in bringing new products to the market (production cost are too high to allow for reimbursement) [2, 5, 6]. Thus, one can argue that there is a real need to address technological challenges for cost-effective and high throughput manufacturing of functional micro-products with variable batch sizes. Therefore, a new generation of fabrication methods and manufacturing systems are required to address the industrial needs for highly productive, reliable, innovative and efficient processes in 3D micro manufacturing [7–9]. In this context, SMARTLAM, a European funded research project intends to create a new concept in the production of functional micro-devices out of laminated polymer films, based on a flexible, scalable and modular manufacturing platform. The SMARTLAM platform encompasses several manufacturing capabilities (modules), i.e. laser structuring, aerosol-jet printing, laser joining and stacking, positioning of pre-manufactured components and inspection (see Fig. 1).

Fig. 1. SMARTLAM manufacturing modules.

One challenging task in modular manufacturing approaches is to develop a software solution to realize visualization and process selection and optimization for the manufacturing process chain [10–14]. There have been some attempts to develop software solutions for modular manufacturing. However, to the best of the authors' knowledge, the large body of the proposed software solutions to address the communication between the manufacturing modules but with restricted capability to handle potential upgrades in the manufacturing/assembly techniques and/or with limited ability to visualize and optimize the process chain [14]. Nevertheless, in other research studies that addressed process chain optimization using decision support system, the developed software solution does not allow for manual real-time amendment of the identified process chain that generated automatically [15]. Looking at the reviewed literature, it is possible to conclude that for successful implementation of modular manufacturing platform, it is necessary to implement a software solution (toolkit) to enable the communication between (sub-systems) fabrication platforms, and to allow for the visualization and selection of optimal process chain for the manufacturing of 3D micro-devices.

The aim for the research presented here, is to develop a software toolkit that facilitates the visualization and selection of process chains for the manufacture of functional micro-devices starting from the conceptual design stage to production of the final product. This software toolkit will include different software modules/sub-modules and enable communication between different manufacturing modules within the developed platform via a communication layer based on socket programming.

Following this introduction, the reminder of the paper is organized as follows. First, the proposed software solution entailing module architecture and communication scenario, process selection algorithm and software implementation is detailed. Then, the results are presented and discussed. Finally, conclusions are drawn based on the findings of this research endeavor followed by giving some future work.

2 Proposed Software Solution

2.1 Module Architecture and Communication Scenario

The software toolkit, developed in this research, is equipped with data and communication interfaces, which are depicted on the left and right sides of the image (see Fig. 2). Data that is communicated via the interfaces needs to be stored internally using data models. Colored rectangles refer to internal functional software component modules. The configuration tool uses a Process Chain Loader in order to import a selected process chain via a HTTP request or by loading it from an XML encoded file. Internally, the process chain data is stored and encoded based on a representative data model. By using the Process Chain Editor, which is part of the GUI, the process chain data can be modified. After the process chain has been edited, a recipe generator is used to perform a model-to-model transformation. The process chain is transformed into an executable manufacturing recipe, which is enriched with data from the material and processes database as well as from the component module models, via HTTP and TCP/IP. By triggering the recipe execution, the generated recipe can be transferred to the SMARTLAM main controller system for manufacturing execution. After that the main controller sends the recipe execution to desired sub-system to execute the sub-task of manufacturing process based on socket programming via TCP/IP protocol. During recipe execution, status information from process modules as well as the status of recipe execution is received by the configuration tool using the designated interfaces. All states are visualized in the GUI. The execution results are returned and can also be visualized. After execution has finished, an updated process chain can be stored in the SMARTLAM data base using the HTTP based API. The process of generating a control recipe, based on a selected process chain.

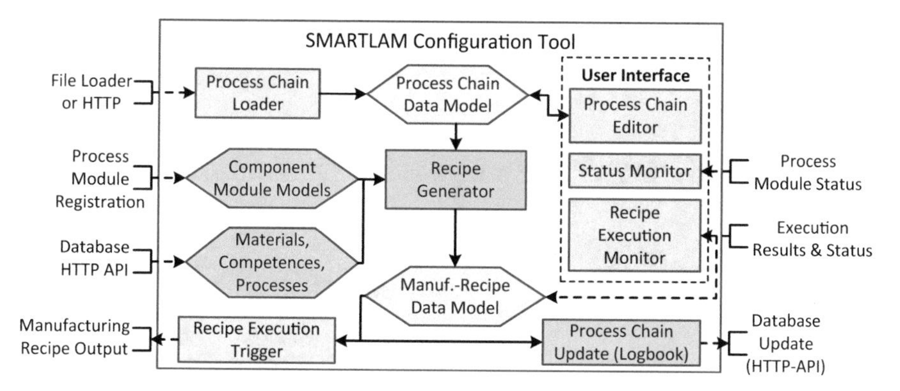

Fig. 2. Schematic of the configuration/selection tool architecture.

In order to establish a communication link, a worker requires to configure at least two endpoints. One or more endpoints are used to receive messages from other workers. Moreover, the worker needs at least one endpoint in order to send data (see Fig. 3).

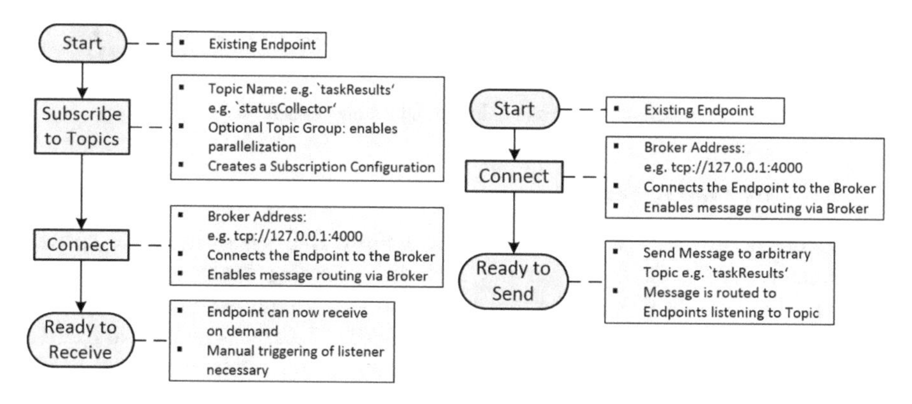

Fig. 3. Flowcharts for endpoint configuration, left chart: configuration as receiver and right chart: configuration as sender.

2.2 Process Selection Algorithm

The process selection algorithm works to reject the invalid process based on knowledgeable database. All black blocks are rejected process (see Fig. 4). Then the algorithm selected the optimum process chain in order to minimize the cost. In our case studies total number of process chains is 32 and the valid process chains is 6 process. After that the algorithm select the optimal one with the minimum cost form the process chain (see Fig. 5). The algorithm calculated the cost and time based on the calculation of the product size/features/amount/machining rate. These data was aggregated form many sub-modules via the communication approach and stored in the database. The developed software visualized all these processes and enables human adaption/correction using the developed GUI.

Fig. 4. Process chain validation and rejection.

2.3 Software Implementation

The proposed software was developed using Eclipse IDE, an open-source software development framework based on the Java programming language. The Eclipse Modelling Framework (EMF) was utilized to provide simplified methodologies for defining the structured data models. The communication layer was developed based on zero message queuing (ZQM) [16], a high performance, asynchronous messaging library used in the fields of distributed computing/concurrency frameworks. Communication endpoints are based on common sockets, supporting protocols such as inter-process, in-process, TPC and multicasts. Furthermore, Google Protocol Buffers were used to define the data format in a text-based and human readable notation. Selection

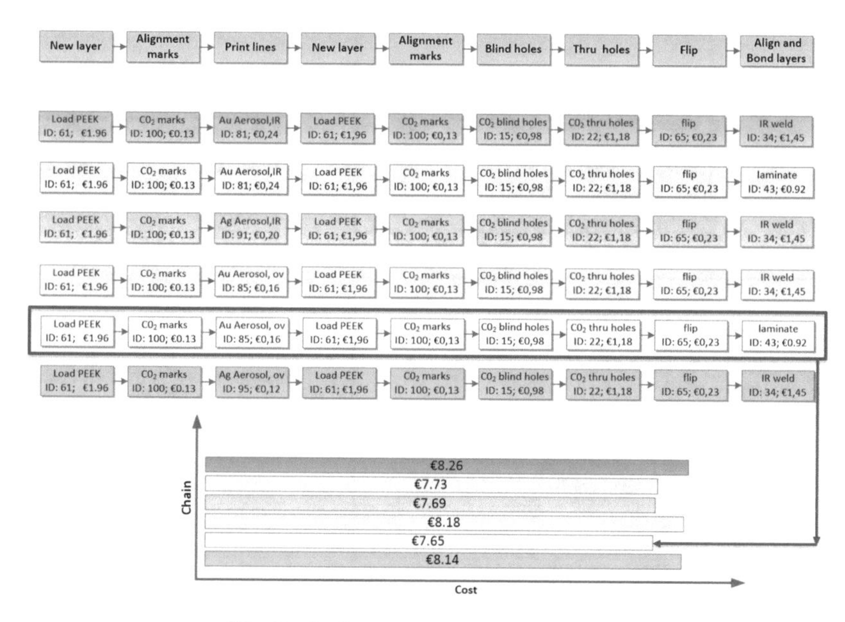

Fig. 5. Optimum process chain selection.

algorithms were executed using MATLAB package, which are used to generate code to integrate with the software module.

The ProcessModule block (see Fig. 6) is the central part of the component module meta-model. It represents a SMARTLAM process module and requires a number of process inputs and outputs. The meta-model package is used to describe process inputs and outputs. Three basic functionalities have been defined.

- startup() – Executed upon module startup. Operational conditions are checked e.g. required amounts of operation supplies.
- initialize() – The ProcessModule and its supporting equipment is initialized and set into the ready state, waiting for a task.
- shutdown() – The ProcessModule finishes its current operation. Afterwards the module and its supporting equipment is led to a safe and defined system state.

In a Manufacturing Environment, ProcessModules may include an arbitrary number of ModuleComponents, also referred to as Supporting Equipment, which internally process received process inputs, and also produce process outputs. Certain MaintenanceTasks can be defined and scheduled for a ProcessModule at given due dates. Additionally, the maintenance workers' qualifications can be assessed for the specific task. As ProcessModules, which are newly integrated into the SMARTLAM system, need to be integrated into the materials and process database, the block Module Capability is defined to specify the module's capabilities including its input, output, and capability parameters. Every Process Module requires PhysicalSpace which is represented through CAD data that describes the measurements and an estimated size of the volume. Finally, a collection of MechanicalInterfaces is needed which describe how the module is mechanically attached in the SMARTLAM workspace.

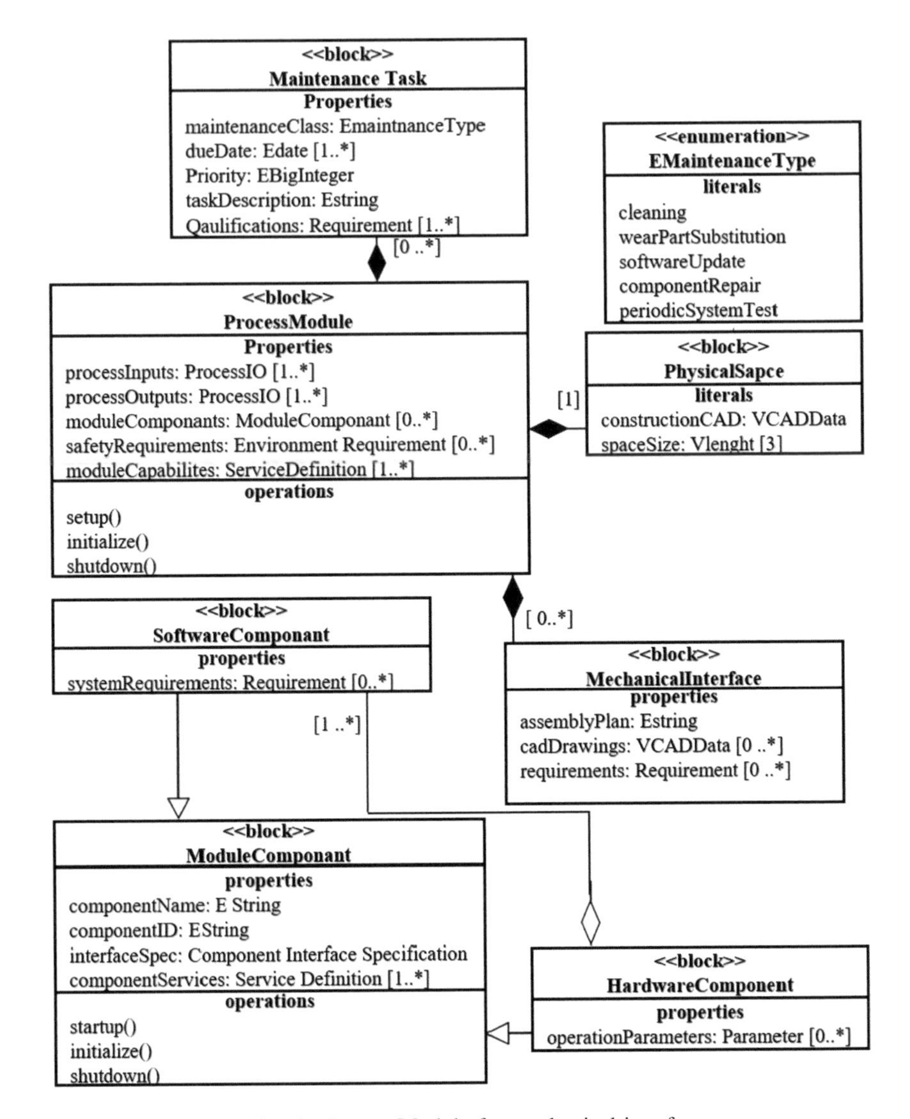

Fig. 6. The ProcessModule for mechanical interfaces.

3 Results and Discussion

The software toolkit is equipped with a GUI, which supports the operator in configuring/adapting the SMARTLAM manufacturing process chains. The process chain editor is the main view of the GUI. Once a process chain is imported into the configuration tool, either by loading from an XML file or by accessing the resource via the sREST interface, the process steps are depicted in order of their execution sequence. Figure 7 shows the Process Chain Editor with a process chain loaded. The chain steps are identified through their names which are given by the related manufacturing competence.

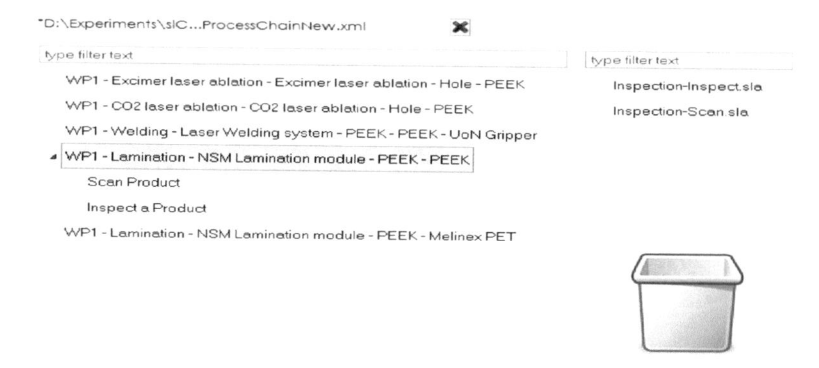

Fig. 7. Process chain editor.

The Parameter Editor is tightly connected to the Process Chain Editor, as it visualizes the parameters of the currently selected process chain step. Figure 8 shows the outline of the Parameter Editor. Process chain parameters are divided into Inputs, Results and Constraints. Inputs refer to technical parameters of the relevant manufacturing competences, which serve as an input to the manufacturing step. Results refer to technical parameters of the relevant manufacturing competence, which are returned after the execution of the step. These include values measured during the processing of the manufacturing step. Constraints refer to competence parameters of the database, denoting parameters that are related to the competence and do not directly affect the manufacturing process. Also, to visualizing the parameter values, new parameters can be performed. The software toolkit supports the same data types as used in executable manufacturing recipes. These also include array values, which can be set element-wise. This is necessary if the parameterization of a subsequent process chain step depends on parameter values resulting from the preceding operation. A reference is encoded using a string-value, consisting of the following parts: (Chain Step ID, Parameter Type and Parameter Name). (see Fig. 9).

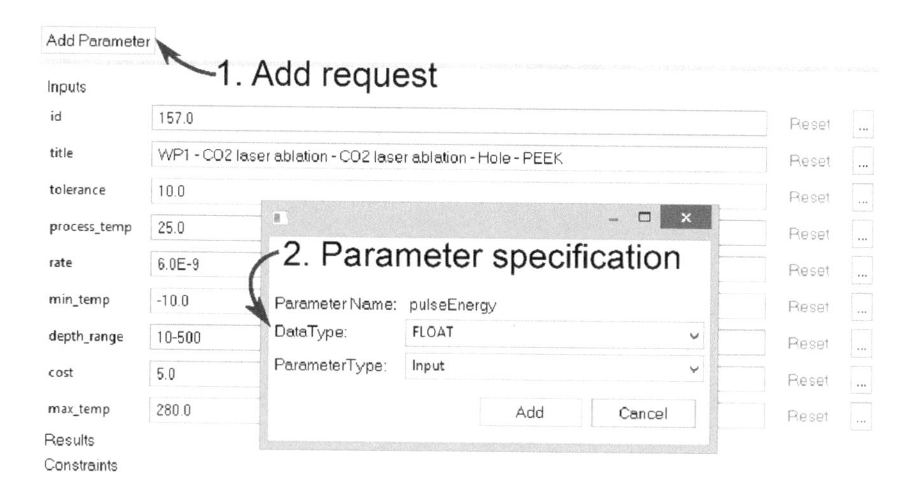

Fig. 8. Adding a parameter to a given process chain step.

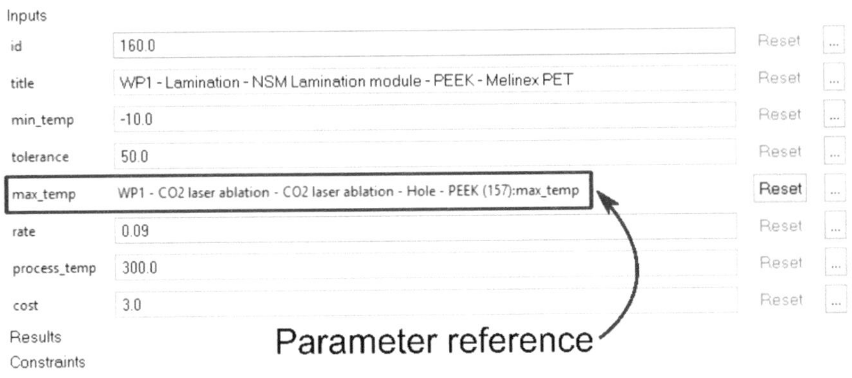

Inputs

id	160.0	Reset	...
title	WP1 - Lamination - NSM Lamination module - PEEK - Melinex PET	Reset	...
min_temp	-10.0	Reset	...
tolerance	50.0	Reset	...
max_temp	WP1 - CO2 laser ablation - CO2 laser ablation - Hole - PEEK (157):max_temp	Reset	...
rate	0.09	Reset	...
process_temp	300.0	Reset	...
cost	3.0	Reset	...

Results
Constraints
Parameter reference

Fig. 9. Parameter reference to a value of a previously executed process chain step.

In the given case, the value of the parameter max_temp is referenced from a previously scheduled process chain step. The configuration tool's GUI only permits reference to parameters of previously executed steps relevant to the currently considered chain step (see Fig. 9). The visualization of the current state of the recipe execution process is a main requirement. Using this information, the operator can track the manufacturing process and is informed about the status of each step. The visualization function is implemented as a part of the Process Chain Editor. In Fig. 10, the process chain steps that were executed successfully are marked with a green flag. In the case of an error a red flag is painted next to the chain step's name. Chain steps currently being processed are highlighted with green background colour and are marked with a "processing" symbol (gear wheels). In the present example, two process modules – a Lamination module and an Inspection module – are active within the SMARTLAM system. The current state of these modules and of the recipe execution system (MainControl) is depicted in the Status Monitor view. As the process chain step "Laminate" is currently active, only the Lamination module is busy. The recipe execution system stays busy during the entire execution of the manufacturing process.

*D:\Experiments\slC...pectAndLaminate.sla
type filter text
Inspect Paths
Laminate

Component ID	Component Name	Status
0	MainControl	EP_BUSY
0	InspectorA	EP_IDLE
0	LaminatorA	EP_BUSY

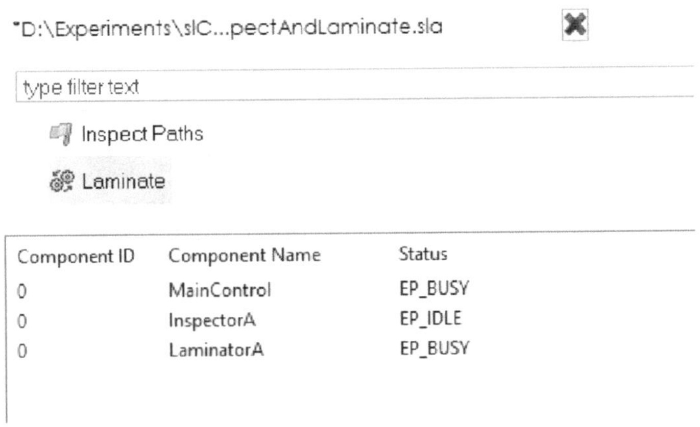

Fig. 10. Execution visualization for a process chain.

After all modules and sub-systems achieved their final stage of completion (development and build-up), a stand-alone demonstration manufacturing cell was set up. The SMARTLAM manufacturing cell consisted of all six modules for the production demonstrator together with the main control system and a clean room housing for appropriate manufacturing conditions. Figure 11 presents the user interface for identifying the optimal process chain based on the chosen user criteria. All relevant data has to be sent over the communication system in XML format via TCP/IP protocol, including a short description of the configuration step, which prepares the hybrid platform for manufacturing. If the user needs to review this information, the relevant data can be also exported in the form of an XML file, or simple text file format. These outputs can be tailored to supply the desired level of detail for each process step. These results are evaluated in terms of manufacturing time and cost and thus to select the optimal process chain to be executed, which in turn increases the efficacy of manufacturing processes.

Fig. 11. Process chain visualization and selection.

The main result of the overall software development is a fully functional and commercially applicable solution for the design, planning, configuration and coordination of an autonomous manufacturing cell for nano- and micro-devices, see Fig. 12. A flexible lighting device demonstrator (DLED) was successfully produced (see Fig. 13).

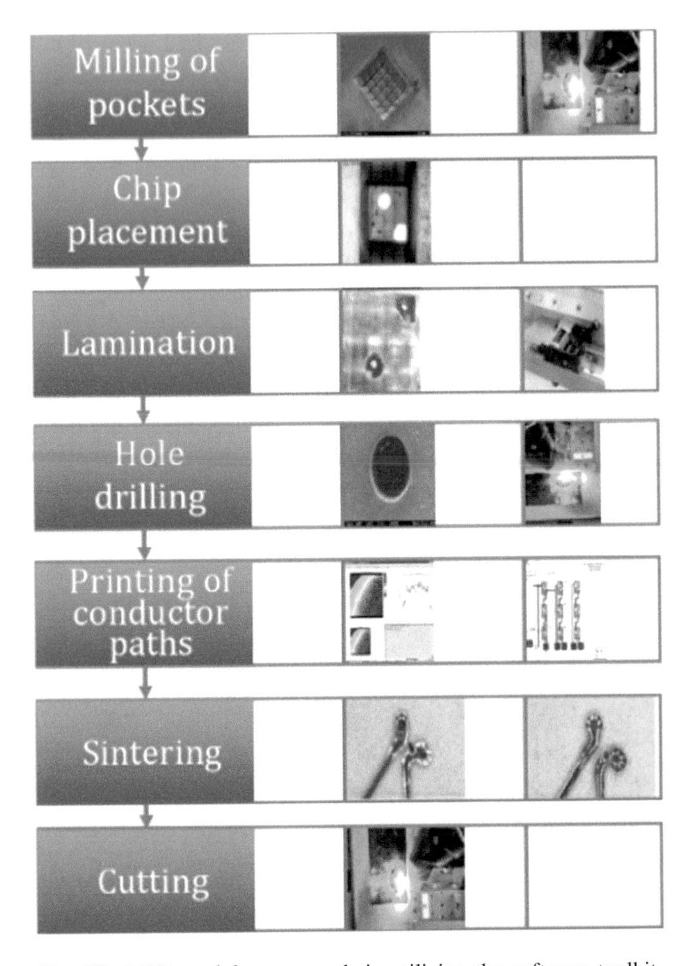

Fig. 12. LED module process chain utilizing the software toolkit.

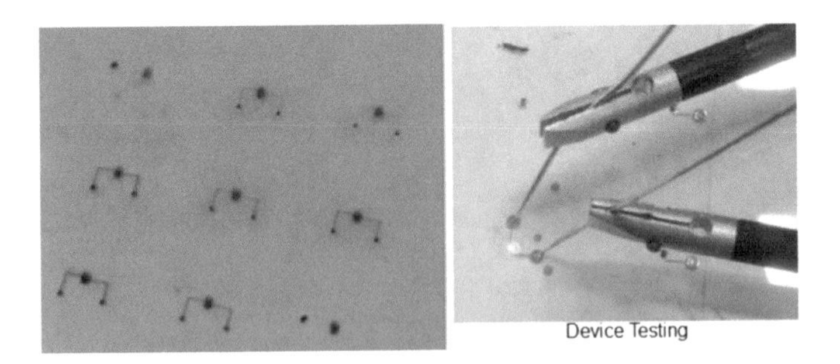

Fig. 13. LED module demonstration.

4 Conclusion

The SMARTLAM concept combines state-of-the-art fabrication techniques for micro manufacturing, and adds an overall control system, making it possible to simply exchange modules to match different products and product requirements. By basing the approach on already available machines and techniques, it is possible to quickly transfer the knowledge obtained during the prototyping phase to actual production by using the same technology and process constraints. This allows for a faster assessment and reduces the effort required for process qualification. In addition, the modular approach in combination with the design tools enables engineers and designers to quickly assess different approaches to manufacturing their products. In this research a software toolkit (configuration/selection tool) to facilitate visualization and selection of process chains for the manufacture of functional micro-devices starting from the conceptual design stage to production of the final product, was developed. The database and configuration/selection tool help to assess different routes and their suitability for cost-effective and profitable production. The results show that the SMARTLAM approach has the power to contribute to easier and faster commercialization of new ideas, especially for multi-material micro components. SMEs can obtain a design solution that will help overcome the so-called "valley of death" funding gap, with the capacity to avoid costly and intensive development, ineffective processing routes and design flaws. The main result of the overall software development is a fully functional and commercially applicable solution for the design, planning, configuration and coordination of an autonomous manufacturing cell for nano- and micro-devices.

In future work, and based on the lessons learnt from this project, scalable modular manufacturing for 3D micro-devices based on hybrid additive and subtractive techniques will be implemented. Also, IoT-based communication technique will be utilized to manage the interaction between the manufacturing modules using distributed algorithm and thus minimize the data transfer dependency.

Acknowledgments. This work was carried out in the frame of the project SMARTLAM funded by the European commission under FP7 program Grant no. 314580. The support by the Karlsruhe Nano Micro Facility (KNMF-LMP) a Helmholtz research infrastructure at KIT, is gratefully acknowledged.

References

1. Scholz, S., Mueller, T., Plasch, M., Limbeck, H., Adamietz, R., Iseringhausen, T., Kimmig, D., Dickerhof, M., Woegerer, C.: A modular flexible scalable and reconfigurable system for manufacturing of microsystems based on additive manufacturing and e-printing. Robot. Comput.-Integr. Manuf. **40**, 14–23 (2016)
2. Klewitz, J., Hansen, E.G.: Sustainability-oriented innovation of SMEs: a systematic review. J. Clean. Prod. **65**(15), 57–75 (2014)
3. Qin, Y., Brockett, A., Ma, Y., Razali, A., Zhao, J., Harrison, C., Pan, W., Dai, X., Loziak, D.: Micro-manufacturing: research, technology outcomes and development issues. Int. J. Adv. Manuf. Tech. **47**, 821–837 (2010)

4. Koren, Y., Gu, X., Guo, W.: Reconfigurable manufacturing systems: principles, design, and future trends. Front. Mech. Eng. **13**(2), 121–136 (2018)
5. Kumari, S., Singh, A., Mishra, N., Garza-Reyes, J.A.: A multi-agent architecture for outsourcing SMEs manufacturing supply chain. Robot. Comput.-Integr. Manuf. **36**, 36–44 (2015)
6. Djatna, T., Luthfiyanti, R.: An analysis and design of responsive supply chain for pineapple multi products SME based on digital business ecosystem (DBE). Proc. Manuf. **4**, 155–162 (2015)
7. Chu, W.S., Kim, C.S., Lee, H.T., Choi, J.O., Park, J.I., Song, J.H., Jang, K.H., Ahn, S.H.: Hybrid manufacturing in micro/nano scale: a review. Int. J. Precis. Eng. Manuf. Green Technol. **1**(1), 75–92 (2014)
8. Qin, Y.: Micro-manufacturing research: drivers and latest developments (keynote paper). In: 23rd International Conference on Computer-Aided Production Engineering (2015)
9. Scholz, S., Elkaseer, A., Müller, T., Gengenbach, U., Hagenmeyer, V.: Smart modular reconfigurable fully-digital manufacturing system with a knowledge-based framework: example of a fabrication of microfluidic chips. In: 2018 IEEE 14th International Conference on Automation Science and Engineering (CASE), Munich, pp. 1012–1017 (2018)
10. Wang, S., Wan, J., Li, D., Zhang, C.: Implementing smart factory of industrie 4.0: an outlook. Int. J. Distrib. Sens. Netw. **12**, 3159805 (2016)
11. Afazov, S.M.: Modelling and simulation of manufacturing process chains. CIRP J. Manuf. Sci. Technol. **6**(1), 70–77 (2013)
12. Ogorodnyk, O., Lyngstad, O.V., Larsen, M., Martinsen, K.: Application of feature selection methods for defining critical parameters in thermoplastics injection molding. Proc. CIRP **81**, 110–114 (2019)
13. Müller, T., Schmidt, A., Elkaseer, A., Hagenmeyer, V., Scholz, S.: A knowledge-based decision support system for micro and nano manufacturing process chains. In: 44th Euromicro Conference on Software Engineering and Advanced Applications, pp. 314–320 (2018)
14. AlGeddawy, T.: A new model of modular automation programming in changeable manufacturing systems. Proc. Manuf. **11**, 198–206 (2017). FAIM 2017
15. Hoffmann, M., et al.: Optimized factory planning and process chain formation using virtual production intelligence. In: Enabling Manufacturing Competitiveness and Economic Sustainability, pp. 153–158. Springer (2014)
16. ZeroMQ (ZMQ). http://zeromq.org/. Accessed 15 July 2019

Security in IoT and Web Systems

Verification of IoT Devices by Means of a Shared Secret

Tomasz Krokosz(✉) and Jarogniew Rykowski(✉)

Department of Information Technology,
Poznań University of Economics and Business, Poznań, Poland
{krokosz, rykowski}@kti.ue.poznan.pl

Abstract. The paper presents a new proposal to solve the problem of confirming the identity of devices in urban Internet of Things environment. The approach is dedicated to Bluetooth Low Energy devices and broadcast transmission. The main goal of the approach is to obtain a suitable level of security and trust with reasonable energy (resources) usage and minimal cost of cryptographic functions. In the case direct interaction with an IoT device is not possible, all the traditional approaches for the verification of trust fail. Moreover, due to the limited resources of a typical IoT device (memory, CPU, battery life), implementation of traditional verification mechanisms is not feasible. In our approach, the IoT devices are detected and inspected by a dedicated application (executed in a smartphone) prepared by a trusted third party (usually city administrator). The system acts as a base for a trusted, generic information system of the city area. The implementation of trust is based on a modified shared-secret algorithm and an exchange of some parameters at the installation phase for each device. Later on, each device is operating autonomously and off-line, transmitting the information only in broadcast mode. The application, while detecting a signal from a device, is able to assess the level of trust towards the device and the information received. The solution assumes backward compatibility with devices and applications which do not require verification of the trust.

Keywords: Device identity · Proof of identity · Secret sharing · Identity of devices on the Internet of Things

1 Introduction

The Internet of Things (IoT) solutions that are currently proposed by service providers in the scope of "intelligent cities", more and more are applied for everyday usage, following success of mobile telephony and smartphones [1]. Different kinds of sensors, distributed across a city area, interact with humans and their personal devices to provide all the information they need at the moment and at the place. As a result, the whole urban ecosystem is monitored by IoT devices (e.g., RFIDs, beacons, natural-environment measurement units, etc.), which are either a new element of the city (such as pollution-level monitoring station), or an integrated part of an existing object, e.g., a tram stop with an interactive time-table. As for the latter, a client is able to find a real-time information on tram/bus arrivals, the degree of the congestion of the next

L. Borzemski et al. (Eds.): ISAT 2019, AISC 1050, pp. 175–186, 2020.
https://doi.org/10.1007/978-3-030-30440-9_17

vehicle, detailed routing and suggested exchange points, and many more. Collected data from all sensors may be further analyzed off-line [2] and, for example, in the reference to the tram time-tables mentioned above, if the congestion at certain time/place is too high, a decision may be taken to add another vehicle to the timetable.

Communication between a client and a device installed within the city area (e.g., a geolocation beacon) is assured by exchanging some data packets. However, during the process of data exchange, for most of nowadays solutions, no confirmation is possible of the sensor identity and the level of trust. This task requires (1) direct interaction of the IoT device and client personal device at-the-place and in a very short period of time, (2) complicated cryptographic procedures to be undertaken on both interacting devices, and (3) independent verification from a third-party server, at least to confirm the digital signatures and public keys. For some devices and applications, in particular, those requiring a payment, proper verification of the communicating parties is a must. Unfortunately, due to very limited resources of IoT devices, the classical cryptographic procedures for e.g., the digital signature cannot be applied. As a result, a client is not able to confirm the device he is currently interacting with is really operated by e.g., a city. Moreover, as most of the smallest IoT devices use only broadcast mode for the communication, direct application of the classical public-key-infrastructure (PKI) solutions is questionable. Although not widely reported, the danger for abusing the clients with the substantiated beacons or record-and-play attacks seems to be very serious.

The aim of the paper is to propose a new way to solve the problem of the lack of an effective method of device verification in the "urban" Internet of Things environment. We assume that the city has placed some IoT devices across the certain area, access to which is carried out with the use of a dedicated application, also prepared and managed by the city. Access to city information does not require registration, and the use of devices is not preceded by the login process. We take into account the limitations of devices/sensors and the inability to perform complex cryptographic algorithms. Enforced by the requirements for energy saving and thus the use of economical solutions such as BLE [3], we found further restrictions on the final solution, relating to, among others, the maximum size of transmitted data packets (e.g., up to 47 bytes in the MCh/iBeacon standard). In the text, we offer a solution that, despite the above restrictions, allows to assess the degree of confidence in each accidentally encountered device. This solution is dedicated to Bluetooth Low Energy technology [4] and broadcasts the transmission in the BLE Marketing Channel (MCh). The proposal is based on a modified trivial-secret-algorithm to confirm the identity of devices.

The further organization of the paper is as follows. The second chapter describes how to solve the problem of authentication of the transmission using a trivial secret sharing algorithm. The schematic diagram of the algorithm has been supplemented with an appropriate description. The third chapter contains a description of the case of using the proposed implementation of the algorithm. The fourth chapter presents a comparison of the proposed solution with some existing systems. The summary of the paper is given in the last section. It contains not only a summary of the features of the proposed method but also points out some alternatives and further development paths for the solution presented.

2 Secret Sharing

The scheme of sharing (a division) a secret is based on an efficient implementation of a cryptographic method that allows dividing certain information into smaller (of certain length) parts. These parts are called shares and are sent to target clients (shareholders). A single shareholder, possessing one part of the secret, is not able to reproduce the whole secret. Such reproduction is possible only by means of cooperation of certain (defined by a parameter called a threshold) number of shareholders.

Basic algorithm was designed by Adi Shamir and announced in 1979. It is based on Lagrange interpolation polynomials, defined over finite fields [5]. There are different variants of the mentioned algorithm, but in relation to our problem, the simplest version seems to be sufficient, so-called a trivial secret sharing algorithm.

2.1 Sharing Secrets via Interaction

In the case of two-way (mutual) communication, devices can simultaneously send and receive data, which creates the possibility of, among others, parameter negotiation. The beginning of the identity-confirmation process must include a method to agree on the session key. Due to the fact that the Authentication Center (AC, a device performing the role of a trusted third-party) has unlimited memory, it may be a value of any length of bytes. This observation gives the possibility to use Diffie-Hellman key agreement algorithm, with parameter p being a large prime number, as well as a parameter g, called a generator, both resulting as a single-time installation device procedure, i.e., it does not increase energy consumption at run-time. These parameters may be made public, allowing, as a next step, an implementation of the Diffie-Hellman algorithm [6] (Table 1). Devices with limited resources (including memory) to their purposes may use flash memory, which is the source of all variables (e.g., for cryptographic methods). The operation of installing a device and storing data in its flash memory is a one-time procedure.

Table 1. Diffie-Hellman key agreement algorithm

Authentication Center (AC)	Open (unencrypted) communication channel	Client (C)
	Public p module, and generator g	
Selecting a random private number $\boldsymbol{a}$ A = $g^a \bmod p$ (public value)		Selecting a random private number $\boldsymbol{b}$ B = $g^b \bmod p$ (public value)
	AC: Client → A Client: AC → B	
K = $B^a \bmod p$ (key)		K = $A^b \bmod p$ (key)

From that moment, the client and AC have an agreed session key. In the next step, the client generates GUID (Globally Unique Identifier), which is created, among others, based on its creation time and pseudo-random numbers. It has a form of a 32-byte-long string. The previously agreed key and the created GUID are the parameters of the hash function, i.e., HMAC(K, GUID) [7], which output is sent and saved in the AC. As a response, the client receives from AC the value of the generated secret returned by the HMAC function with the previously agreed key K. AC also responds to the request made by the IoT device. AC chooses a group of devices (according to, for example, their type or location) and instructs them to generate their numbers and send them back. As a result, AC stores the numbers generated by other devices and is able to send them back (as hash values generated with the key K) to the customer at request. Then, each device to which the request is addressed sends its secret number to the device which initialized the process (asked by the client). After receiving the data, it sums them up and sends the result to the client who performs the necessary arithmetic operations [8]. At the end of the process, the client compares a value obtained from the device with a value obtained earlier from AC. The client-side value is created as a hash-function result over the value obtained from the device.

The validity period of the secret (time window) is a configurable value. It may be expressed in seconds or minutes, during which the key K may be used for binding.

3 Description of the Proposed Solution

Currently, most market proposals applied to battery-powered solutions. The functionality provided by battery-supplied devices should be effective for the uninterrupted and longest working time without the need to replace the battery. Therefore, the overriding requirement in relation to low-energy wireless sensor networks is the use of a dedicated technology (protocol), such as Bluetooth Low Energy, to save the energy at the maximum extent. In our solution, we use the mentioned technology as a base to implement the confirmation of device identity.

The presented solution is a proposal for a new pattern of the credibility of broadcast transmission in the Internet of Things, in particular in ad-hoc applications and for the purposes of incidental interaction. Broadcast transmission in IoT is most often implemented for the devices with very small resources (CPU, memory, energy), lacking the possibility of two-way transmission. Consequently, classical methods of authentication, involving interaction to exchange encrypted messages, are not available. In addition, classic encryption procedures are extremely expensive, which shortens the device's operating time from years to often months or even weeks. Another widely applied solution is to confirm the credibility of e.g., an address (MAC, IMEI) of a device in an external database – this, however, requires on-line network transmission and seriously limits the privacy. Therefore, we propose to introduce an additional, encrypted transmission, which allows assessing the degree of reliability of data transmitted cyclically in an unencrypted manner. The encryption scheme uses relatively simple operations, such as hash functions, using the strategy of "zero knowledge" and a shared secret.

Recently more and more popular BLE standard is a solution for at least two problems characterizing ad-hoc interaction. The standard eliminates a need for a continuous pairing of already connected devices and limits the rate of exhaustion of the device's battery. The main advantage of BLE (specification 4.0 was published in June 2010, the latest version comes from the turn of 2018 and 2019 and bears the mark 5.0), is to cancel the approach of constantly maintaining the stream of information, in favor of sending smaller portions of data at the moment when it is needed, in the scope of Marketing Channel [4]. After sending or receiving data, each device switches to "sleep" mode until the next communication. For example, a BBMagic device powered by a 220 mAh CR2032 battery can operate efficiently for hundreds of thousands of hours (a few or even a dozen years), "waking up" once every several-dozen seconds, while a typical iBeacon broadcasts its identity for four-five years.

3.1 Installation of Devices

In order to be able to apply the BLE protocol and MCh transmission to solve the problem of trust and identification of broadcasting devices, a number of assumptions should be introduced as to the final architecture of the system. Firstly, we propose to use a homogeneous environment for the purposes of downloading and visualizing data, including sensors (in particular location beacons), BLE, and a dedicated smartphone application for their detection and interpretation. We divide BLE devices into specific groups, called types - each type is equally identified by the application regardless of the place and time of communication. For each type, we set a number of parameters that are programmed during deploying the device within a city/region. These parameters determine the autonomous mode of operation of the device and are used throughout the full time of operation of the device (that is until it is disassembled or reprogrammed into another type). This assumption allows for the operation of devices in the broadcast mode (without confirmation) and fully automatic, but also repetitive and programmable. Data stored in the non-volatile memory of the device during its installation in particular include:

(1) private key – used as a parameter of the hash function,
(2) group identifier – a string of characters that uniquely identifies a given group of devices,
(3) range of secret numbers – it is a set interval, from which one part of a secret will be drawn, required to determine the identity of the device. Its size may be, for example, 16 bits (2 bytes), which creates a set of 65535 possible parts of secret.

In addition, each device must implement an algorithm for generating the hash values and an algorithm for generating the time stamp. Each data packet, sent by e.g., a location beacon contains the device identifier and its hash - these are permanent data. Except that, a device sends a secret (randomly drawn for each packet sent), associated with a time stamp and being a result of the execution of the hash function.

3.2 Online Access to Devices

The solution presented in Sect. 2.1 was focused on mutual communication. A client reported the need to obtain data, confirm the identity of the device and, in order to implement the challenge, the devices exchanged some data. This section presents the approach using Bluetooth Low Energy, in which the device (beacons) works as a lighthouse, i.e., it sends a portion of data every defined period of time. The operation of the bi-directional exchange of data with the device it is not possible because it uses only the broadcast mode. Therefore, in order to solve the problem of identity confirmation, there is a need to add certain assumptions and improvements.

Communication, or actually receiving data from a device, starts when a client is within the range of its operation area (i.e., a radio-signal range, usually a few meters for a typical beacon). When this happens, a person with a dedicated application is able to read the broadcast message. However, as already mentioned above, it is not possible to send back any message to the device. In addition, the devices do not have an Internet connection, therefore all the data for broadcasting must be programmed prior to the installation of the device at the target location.

The client application, operating at client's smartphone and obtained from the city (the same that dislocated the devices across the area), is given by the information on all types of devices available in the system, and their corresponding public keys. Due to this fact, after receiving a specific message from a device of a given type, the application will be able to independently identify and confirm the data source. For this purpose, using a HMAC for a given parameter, the application independently creates a device group hash value to be compared with the value just received from the device. The application can download the verification information data for both existing and new types of devices cyclically, in a similar way as antivirus systems update their local code and virus signature databases. Thus, the set of possible device types and their corresponding validity data are to be updated as needed.

The application, in order to confirm the identity of the device (e.g., a location beacon), must send a request to obtain the second part of the secret from the module called the Authentication Center. For this purpose, it sends a confirmed group device identifier received from the device. It is also (once for the duration of the session) obliged to generate a GUID which will act as the session identifier. AC receives the group ID, its hash, and the GUID also along with the hash. The last sent package contains a time stamp with a hash for generated timestamp value. If AC confirms data, it checks whether there is a record in the database for the received GUID value and device group. If such information exists and the session has not been finished, it sends to the client a secret and its hash, which the client confirms itself. However, if the record does not exist, or exists but the session ended, the server generates the next part of the secret (S_n) and the main value of secret S, and creates their hash values, to be sent to the client. The reason for querying the database, carried out by AC, is to check whether a secret has been generated for the client with the given GUID and if there is an active connection session. If a record exists and the session is active, then the process of generating the second part of the secret is skipped, as the previous secret is still valid. If the session is not active, or there is no record for the given GUID, then a full-secret-generation procedure should be performed. Regardless of the result of the query, the client, after receiving the data from AC, independently calculates the secret and compares it with the hash value received from the AC. If they are equal, then the identity of a device (a beacon) is confirmed.

It should be noted that client privacy is preserved, even if the server is contacted each time a device is to be verified. This is due to the fact the identity is related to the group of devices of the same type rather than a single device. Devices from the same group located in different parts of the area/city share the same type identifier, which makes it impossible to link the interaction with a certain place, i.e., to track actual clients' locations.

3.3 Approach with the Use of BLE Marketing Channel

The consequence of unidirectional communication of the BLE device is the earlier necessity of assigning the private key to it. For this purpose, we use the non-volatile memory (flash) of each device to save there above-mentioned parameter. We assume that a group of devices of the same type shares one common key and the same identifier. This means that all deployed devices (e.g., beacons) are assigned to one of the defined categories of devices and for all of them there is a common (for a category) private key and a common parameter that acts as the identifier of the device group. This also means that it is not possible to determine in which broadcast-transmission region a client is at the moment, because the devices of the same type in several different parts of the city show the same identifier. We do not see the need for such a distinction and verification of each device/place, in addition, such functionality can easily be provided by other means, e.g., by tracking a MAC address of the device. Due to this fact, in client application, there is no need to save individual keys for each device because there is a connection between the private key and the device type (category). Devices, by broadcasting messages, send an unencrypted (explicit) value of the part of the secret (randomly and linked to the timestamp), which is supplemented by the output of the hash function with the key for parameters: secret and earlier defined key. At the same time, each device sends the identifier of its group and result of hash function with the same key for that identifier and time stamp with its hash. The data transmitted by a device are shown in Fig. 1.

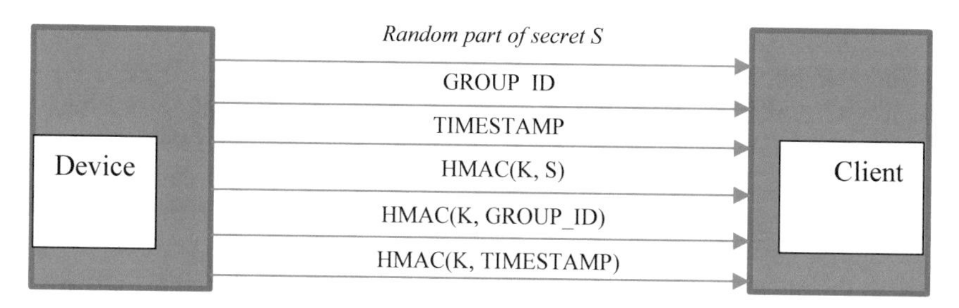

Fig. 1. Broadcasted information received by a client application from a device

The first task performed by the client application is a calculation of a hash function values with the key for the given parameter GROUP_ID. If the result of the comparison proofs the equity of parameters, then the client sends the hash value GROUP_ID to AC, a result of the hash function with the key for the GUID generated for the duration of the session and a timestamp (and its hash), which determines the session. In this way, this transmission begins the process of confirming the shared secret. AC, after receiving the request, performs a query to the database in order to determine whether for a pair of parameters: device group and client (GUID) there is a record. If AC finds record for the transmitted parameter pair, it checks a state of the session (whether from the received and confirmed timestamp did not pass enough time to finish it) and, depending on its status, performs the process of generating a secret or sends the saved value of the secret and its hash to the client. If AC, however, does not find a record or confirms the end of the session for the given parameters, then it draws the secret S. Using hash function with the key, it generates its hash, and additionally generates the second part of the secret, which along with its hash value it sends to the client. The client determines whether the second secret value comes from AC (by comparing the received and self-generated hash values). Then it creates a hash value for the calculated result and compares it with the one that was obtained from AC. Above-described communication scheme between the client and AC is presented in Fig. 2.

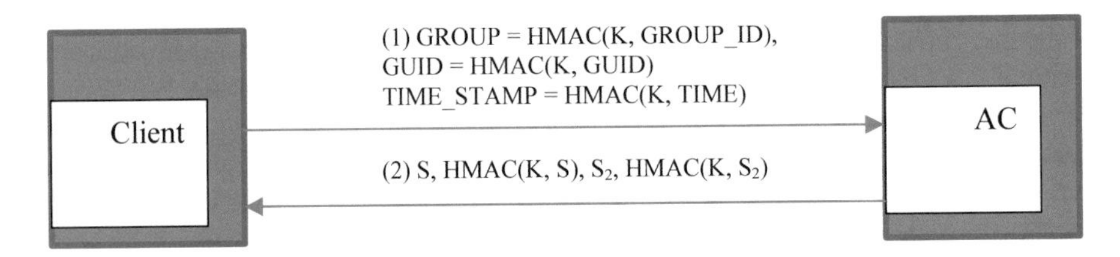

Fig. 2. Parameters reconciliation between a client and AC

For example, the client sends to AC the result of the hash function with the key for the parameters K and GROUP_ID, hash of session identifier and time stamp with hash (1). For K equal to 1234567890, and GROUP_ID d2c8440c-490a-4393-9369-e827b20f631d, transmitted value GROUP is 1451aee6fc216b27bcf39df8 858ecd535b208b2e. With the same key, the client creates a hash for the GUID parameter, which for the value 9c6557a4-0d8d-44bc-b78c-882ac929eac7 is equal to 8e81b5f49b38a09525e5f9a4d02f29601b8c558e. AC checks whether for the sent GROUP and GUID values, there exists a hash in the database and if there is a secret S associated with it. If so, then (if the timestamp generated by the device, saved and confirmed by comparison with the current one, has not elapsed sufficiently), AC selects the secret and its hash with key K and sends it back to the client. However, if the answer to the query is an empty set, then AC generates a new secret S (e.g., S = 42) and with key K passes as parameters of the hash function with the key K (HMAC (K, S) = a2bcd735950ffdc320e024a3a3dd5d19bf890b4c), which together with secret S are being sent to client. AC also generates the second part of the secret, which is sent to the client in the plaintext and hash thereof part with key K (2). The data sent are also saved in the database. After receiving the data, the client does not know the secret. It must calculate the secret and compare it with the received one, therefore for this purpose it calculates the value of secret S using other parts of the main secret (S_n) and performs the necessary arithmetic operations. Suppose that at the time of the broadcast transmission, a device sent a part of secret (which the client confirmed by creating a hash value with key K) equal to 15. AC has sent its second part, e.g. 23, along with hash and the value and hash of secret S. At this moment, the client calculates the secret himself:

$$R = (S - 15 - 23) \bmod 100 = (42 - 38) \bmod 100 = 4 \quad (1)$$

$$(15 + 23 + R) \bmod 100 = (42) \quad (2)$$

At this point, the client using HMAC function for the K key and the independently calculated result compares the secret hash obtained from AC with the hash of the secret which is calculated. If the result of the conditional statement is the equality of the parameters, the client has confirmed values and information that the session is active. In another area where exist device from the same group, if the session is active, key negotiation will not be necessary.

The timestamp generated by device stands for (in case of further correct operation of the identity confirmation operation) the beginning of the session, which will become active after all values are saved in AC. Note that the set of device messages is different for each transmission, which prevents the use of a record-and-play attack.

Regarding ensuring the security of used keys, a security policy may be adopted, with the keys updated at certain, designated moments of time (e.g., automatically after a specified period).

In order to use the BLE Marketing Channel technology, it was necessary to take into account its requirements, among these, the maximum size of the data packet is the most important parameter. In the frame of broadcasting of 47 bytes of data at once[1], it was necessary to fit all data in relation to each pair of parameters. Each device broadcasts six parameters: one part of the secret (with a hash), a device identifier (with a hash) and a timestamp (with a hash). The hash function accepts the private key as the second parameter, thanks to which the client is able to independently confirm the origin of the data by creating a hash value. Above-mentioned six parameters may be successfully grouped into three sets, two parameters in a pre-ordered order and then broadcasted as a single MCh frame, which means that one identity confirmation needs three transmissions. Each of them shows a sequential number in the header, and the client would expect to get all three (not necessarily in the order of sending, which means that it could pass the queue if it received data from the second transmission at the beginning).

The number of broadcasts, depending on the implementation, may be reduced to two or even one transmission. To do this, it is not necessary to create a separate hash value for each of the three parameters: SECRET_S, GROUP_ID, and TIMESTAMP,

[1] This is the limitation of the standard BLE scanner available for most smartphones, not the BLE and MCh standards [9]. We are now working on the implementation of our own BLE scanner, which will allow us to bypass this limitation and transmit all the data described in the text in one frame.

but only to concatenate them and pass as a single parameter of the hash function with the key. In this way, we send one hash value for all parameters, which we also add to the data package. Then the client, to confirm the data, concatenates the explicit values and creates a hash value using the key designated for the current group of devices. If, as a result of the comparison, the conditional statement confirms the equality, then the client accepts the obtained parameters as binding. Calling hash method would be as follows: HMAK(K, SECRET_S + GROUP_ID + TIMESTAMP)[2]. For the selected hash function, which takes 20 bytes, the remaining 17 bytes will be enough to pass the parameters that create the hash value. Placement of the mentioned parameters (in one transmission) is presented in Fig. 3. The broadcasting parameters are located in the part of the data packet in the frame labeled "Data", whose size is less than or equal to 37 B.

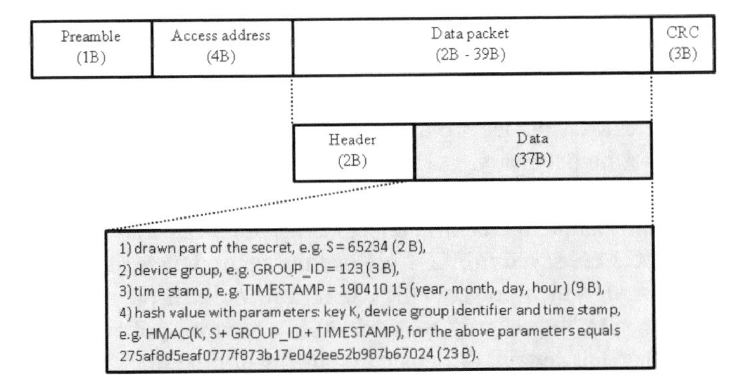

Fig. 3. Format of a data packet in BLE MCh

4 Final Conclusions

The purpose of the article was to present a way to confirm the identity of devices in the urban Internet of Things. To this goal we applied the sharing-secret algorithm. Use of BLE technology and broadcast-only messaging required an adaptation of this algorithm to bypass several limitations, e.g., in relation to the size of packets sent. Grouping transmitting devices into types, and assigning cryptographic keys to a type rather than a single device made it impossible to track the device and the clients, still assuring the high level of trust. Due to the fact the cryptographic parameters are fully initialized at the installation phase, each IoT device operates off-line, with no need for the network transmission to/from authorization center. The proposed solution is very effective, which makes it suitable especially for hardware-restricted environment and battery-operated mode.

In parallel, we conserve the traditional (non-trusted) way of accessing the IoT devices for those who do not require improved trust and security. Note that the way of interpreting the information from the devices depends on end-client application, so this is an individual choice of each client.

[2] "+" (plus) symbol means classical concatenation of string-based values.

To our best knowledge, there is no similar proposal of IoT-device verification working in broadcast mode and not based on classical solutions such as PKI. We developed previously such a solution (based on classical PKI-based cryptography and public/private keys for a digital signature) and we found the energy usage enlarged by a factor of hundreds (shortening battery life for a typical iBeacon from years to weeks) [9]. Thus, even if fully developed and patented, we found our previous solution, being a representative for any "classical" proposals, completely impractical.

Apart some very new proposals of using the shared secret [10], there is a new proposal from Estimote towards the verification of iBeacon devices [11, 12]. However, we were not able to find any details of this solution except for some marketing information and the technical discussions of the Estimote developers [13]. We see some mechanisms and techniques being quite similar as for the final result, however, due to the lack of information, we are not able to compare our solution with the one from Estimote. Note also that the Estimote approach is dedicated to individual devices rather than device types, which enables tracking of clients' activity by the authorization center (managed by Estimote). Our solution, on the contrary, is based on device types, which makes it impossible to track individual device usage.

References

1. Zanella, A., Bui, N., Castellani, A., Vangelista, L., Zorzi, M.: Internet of things for smart cities. IEEE Internet Things J. **1**(1), 22–32 (2014). Article number 6740844
2. Sun, Y., Song, H., Jara, A.J., Bie, R.: Internet of things and big data analytics for smart and connected communities. IEEE Access **4**, 766–773 (2016). Article number 7406686
3. BLE documentation. https://www.bluetooth.org/docman/handlers/downloaddoc.ashx?doc_id=229737
4. BLE. http://microchipdeveloper.com/wireless:ble-link-layer-channels
5. Yang, C.-C., Chang, T.-Y., Hwang, M.-S.: A (t, n) multi-secret sharing scheme. Appl. Math. Comput. **151**(2), 483–490 (2004)
6. Stulman, A.: Spraying techniques for securing key exchange in large ad-hoc networks. In: 11th ACM Symposium on QoS and Security for Wireless and Mobile Networks, Q2SWinet 2015, Mexico (2015)
7. Du, W., Han, Y.S., Deng, J., Varshney, P.K.: A pairwise key pre-distribution scheme for wireless sensor networks. In: Proceedings of the ACM Conference on Computer and Communications Security, pp. 42–51 (2003)
8. Bellare, M., Canetti, R., Krawczyk, H.: Keying hash functions for message authentication. In: Lecture Notes in Computer Science (Including Subseries Lecture Notes in Artificial Intelligence and Lecture Notes in Bioinformatics), vol. 1109, pp. 1–15 (1996)
9. Rykowski, J.: Multi-dimensional identification of things, places and humans. In: Proceedings of The 8th Annual IEEE International Conference on RFID Technology and Applications, IEEE RFID-TA 2017, Warsaw, 20–22 September 2017, pp. 152–157 (2017)
10. Ali, Z., Imran, M., McClean, S., Khan, N., Shoaib, M.: Protection of records and data authentication based on secret shares and watermarking. Future Gener. Comput. Syst. **98**, 331–341 (2019)

11. How does Secure UUID work? Estimote Community. https://community.estimote.com/hc/en-us/articles/201371053-How-does-Secure-UUID-work
12. https://blog.estimote.com/post/172115262320/presence-verification-and-security-is-more-refined
13. Secure UUID/I don't understand how it works, Estimote programmer forum. https://forums.estimote.com/t/secure-uuid-i-dont-understand-how-it-works/3769

A Secure IoT Firmware Update Framework Based on MQTT Protocol

Nai-Wei Lo(✉) and Sheng-Hsiang Hsu

National Taiwan University of Science and Technology, Taipei, Taiwan
nwlo@cs.ntust.edu.tw

Abstract. Recently massive Internet of Things have been deployed around the world. With data collected from sensors and functionalities provided by micro-controller based devices, new applications have emerged through big data analytics and autonomous real-time system responses. To support quality of service for deployed IoT devices, firmware update is a necessary task for IoT vendors. However, malicious attackers have been penetrated traditional firmware update processes and mechanisms to compromise deployed IoT devices, and launch destructive attacks through these controlled devices. In this paper, a secure IoT firmware update framework based on MQTT protocol is proposed. We picture a general firmware update model with IoT devices, gateway devices, firmware distribution broker servers, and firmware deployment servers of IoT vendors. Based on this model, a secure firmware update mechanism is developed to help IoT devices authenticate the source of received firmware and verify the integrity of the received firmware. MQTT protocol is adopted in the proposed framework to efficiently distribute new versions of firmware for IoT vendors. Cryptologic primitives such as Elliptic Curve based Diffie-Hellman key exchange and key-hashed message authentication code are used to secure the proposed process and corresponding protocols. Security analysis is conducted to evaluate security strength of the proposed framework.

Keywords: MQTT · IoT · Firmware update · ECDH

1 Introduction

As massively deployed Internet of Things (IoT) and corresponding management systems around the world have constructed various practical application scenarios such as smart home [1], smart city [2], and intelligent factory [3], information security risk of these IoT-based systems have increased because of large amount of sensed data collection and high physical accessibility of deployed IoT devices by malicious attackers. Therefore, how to secure IoT systems in various aspects has become a hot research topic in recent years.

Among different IoT security issues, firmware update for IoT devices has grabbed more attention since multiple devices-breaching instances occurred such as IP cameras were compromised and controlled by adversaries through wireless network connections. One of the reasons that malicious attackers can compromise IoT devices easily is non-effective device authentication mechanism. The other important factor is security

L. Borzemski et al. (Eds.): ISAT 2019, AISC 1050, pp. 187–198, 2020.
https://doi.org/10.1007/978-3-030-30440-9_18

vulnerability of IoT devices caused by out-of-dated firmware. Hence, in this study we focus on designing a secure IoT firmware update framework to deliver and install new version of firmware over-the-air to targeted IoT devices in a short period of time, and authenticate these targeted IoT devices at the same time.

In order to distribute new version of firmware through Internet as soon as possible, MQTT protocol [4] supporting message publishing/subscribed mechanism is an excellent tool to be adopted in our solution. In addition, Elliptic Curve based Diffie-Hellman key exchange and key-hashed message authentication code are adopted to design an effective device authentication mechanism for the proposed IoT firmware update framework.

The remainder of this paper is structured as follows. Section 2 discusses related work. The proposed IoT firmware update framework is depicted in Sect. 3. Section 4 addresses security analysis for the derived protocols used in the proposed framework. Finally, conclusion is given in Sect. 5.

2 Related Work

Thota and Kim in [5] pointed out MQTT protocol is more suitable for one-to-many communication scenario in comparison with CoAP protocol [6]. As IoT firmware update scenarios usually involve with one IoT vendor server to broadcast updated firmware to massive IoT devices simultaneously within a short period of time, MQTT protocol seems to be a good candidate to support good QoS for IoT firmware update framework. However, MQTT protocol does not support security features required for IoT firmware update framework. Therefore, security mechanisms need to be introduced to enhance the security of MQTT protocol adopted for IoT firmware update framework.

Authentication for Device-to-Device (D2D) communication is general mechanism used to verify both communicating parties are the genuine ones. To construct a secure IoT firmware update framework, it is necessary for the server of an IoT device vendor to identify itself to targeted IoT devices made by this vendor and get trust from those IoT devices to patch new firmware downloaded from the firmware update server. Ranjan and Hussain in [7] introduced a terminal authentication protocol to authenticate communicating devices at both sides. This protocol utilizes both digital signature and blind signature techniques to defend against adversaries launching impersonation attacks. Lavanya and Natarajan in [8] proposed a certificate-less Internet key exchange protocol, which adopts elliptic curve digital signature algorithm (ECDSA) to reduce the key length and the energy consumption during protocol execution. Bamasag and Youcef-Toumi in [9] presented a lightweight D2D continuous authentication scheme for IoT devices. In [9], tokens are used to construct a secret sharing mechanism to speed up the process of device identification and authentication. In recent years, multiple D2D authentication frameworks have been proposed. Butun et al. in [10] proposed a CMULA framework. In this framework, cloud service provider plays the role of certificate authority and is responsible for generating public and private key pairs for targeted IoT devices/wearable devices. Message authentication code is used between gateways and IoT devices in CMULA framework to evaluate device identities and preserve data integrity of transmitted messages. Hernandez-Ramos et al. in [11] presented a lightweight authentication and authorization framework for smart objects. This framework utilizes a new protocol called SEAPOL to authenticate smart objects over LAN with less memory consumption on smart objects and less network bandwidth usage during message transmission. Kumar et al. in [12] designed a lightweight session key establishment scheme for smart home environments. This scheme uses silicon ID, which is a unique identity associated with individual smart device, to mutually authenticate two communicating devices with the adoption of AES encryption algorithm, message authentication code and HMAC.

In terms of IoT firmware update issue, Hassan et al. in [13] categorized firmware update into three different scenarios: a skilled person is required, general people can do the task and updating task is automatically performed without human intervention. They also identified two general firmware update mechanisms: pulling and pushing. Pulling mechanisms are generally invoked by IoT device owners from time to time and pushing mechanisms are automatically triggered by the firmware update server of an IoT device vendor. Chandra et al. in [14] proposed an Over-the-Air firmware update scheme based on lightweight mesh network protocol. Nilsson et al. in [15] introduced a firmware update framework for electronic control units (ECUs) in vehicles to perform self-verification on the newly installed firmware patch over the air.

3 The Proposed IoT Firmware Update Framework

In this section a secure IoT firmware update framework utilizing MQTT protocol is proposed. The design goal of the proposed IoT firmware update framework is to construct a secure firmware update dispatching flow based on practical IoT environments by utilizing currently available technologies such as MQTT protocol and ECDH key agreement protocol. In current IoT environments, many IoT devices do not have IP addresses assigned with them. In addition, many of them do not have Internet access capability and usually IoT devices connect to a gateway device through short distance wireless technologies such as Bluetooth and Zigbee to save energy consumption on communication. In general, a gateway device has more computing resource and possesses Internet access capability.

Figure 1 shows the system architecture diagram of the proposed IoT firmware update framework. There are six roles in the proposed framework including firmware patch server, firmware broker server, gateway device, IoT device, broker discovery service and trusted third party key generation center. A firmware patch server belongs to individual IoT devices manufacturer, which is responsible for dispatching new versions of firmware to multiple firmware broker servers or gateway devices directly through Internet. A firmware broker server is owned by individual firmware broker company, which receives and stores new versions of firmware generated from different IoT devices manufacturers, and later distributes these new versions of firmware to gateways connecting to multiple IoT devices through Internet.

A gateway device has Internet access capability and generally connects to multiple IoT devices or sensors using various wireless technologies such as Bluetooth, Zigbee, and WiFi. An IoT device usually equips with one or more sensors and a microcontroller to collect sensed data and send those data to corresponding gateway for further

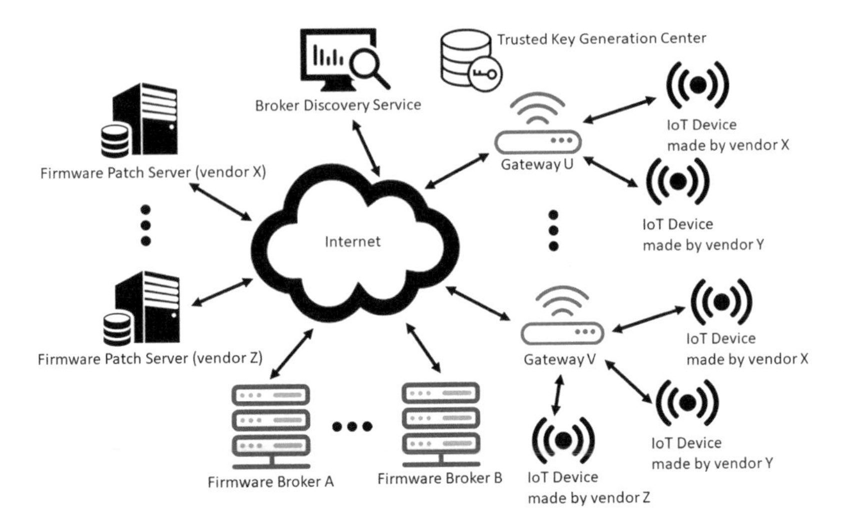

Fig. 1. System architecture diagram of the proposed IoT firmware update framework.

processing or analysis. The broker discovery service server is in charge of query service to support individual firmware patch server to find targeted firmware broker servers and their connection information such as the URL and associated port number. On the other hand, the broker discovery service also supports gateways to find appropriate firmware broker servers to acquire corresponding up-to-date firmware for these IoT devices associated with them. The trusted third party key generation center generates PKI-based key pairs for firmware patch servers, firmware broker servers and gateways. All private keys generated by the key generation center are delivered to corresponding servers or devices through secure channel. For each server or gateway, certificate associated with the assigned public key and the corresponding private key are installed before the servers launching online or gateways shipping out of factories.

In practical IoT environments, there exists a lot of IoT devices manufacturers. For each device manufacturer, various IoT devices with different models are massively sold and distributed all over the world. When there is a need to dispatch new versions of firmware to corresponding deployed IoT devices, it becomes a huge burden for individual device manufacturer. Therefore, third party firmware brokers might play an important role here since they can relief the heavy burden of firmware patch servers at the device manufacturer side and help end users, who usually manage their IoT devices through multiple gateway devices, to quickly find near-by firmware broker sites to download necessary firmware patches for their IoT devices. To help firmware patch servers and gateways to find appropriate firmware brokers based on different criteria, broker discovery service is introduced. Moreover, a trusted third party key generation center is required to generate long term public-private key pairs and corresponding certificates for firmware patch servers, firmware brokers and gateway devices to implement secure firmware update process. The general firmware update flow for the proposed framework is categorized into five phases as shown in Fig. 2. To simplify the diagram, MQTT servers (or broker) are not shown in Fig. 2; MQTT protocol icons are drawn in Fig. 2 instead. We also omit the well-known details of MQTT publish-subscribe operations flow when describing the proposed protocol flow. The first and second phases utilize MQTT protocol to mutually authenticate both communicating parties and establish the session key between a firmware patch server and a firmware broker server, and then publish the new firmware from the firmware patch server to the firmware broker server along with firmware integrity verification and firmware source confirmation. The third and fourth phases utilize MQTT protocol to mutually authenticate both communicating parties and establish the session key between a firmware broker server and a gateway, and then publish the new firmware from the firmware broker server to the gateway along with firmware integrity verification and firmware source confirmation. The fifth phase mutually authenticates both communicating parties and then complete the firmware update process by dispatching the new version of firmware from the gateway to corresponding IoT device and installing the patch onto the device.

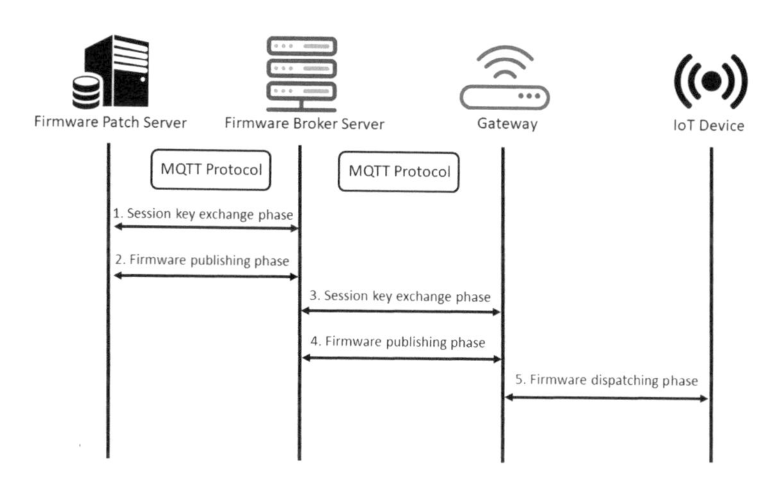

Fig. 2. The five phases of firmware update process in the proposed framework.

There are several assumptions for the proposed framework. All new versions of firmware are freely distributed to sold IoT devices without any surcharge. As each firmware is packaged in binary format and openly downloadable through Internet, there is no need to encrypt individual firmware. However, firmware integrity has to be verified and asserted before a received (or downloaded) firmware been installed to corresponding IoT device. All IoT devices have pre-installed one secret value and one secret key into their hardware security modules and they have stored the identities (ID_{MF}) of their manufacturers. We assume firmware patch servers and firmware broker servers will not be compromised by adversaries. The MQTT topics published by individual firmware patch server, which is associated with a particular IoT devices

Table 1. Notations for the proposed firmware update protocols.

Symbol	Definition
ID_{MF}, ID_{BK}, ID_{GW}, ID_{SN}	Identities for firmware patch server, firmware broker server, gateway device and IoT device, respectively
MD_{SN}	The device model string of an IoT device SN, which is stored in the ROM memory of this IoT device
TMD_{SN}	The device model string of an IoT device SN transmitted through network connection
FW	New version of firmware to be distributed from a firmware patch server to firmware brokers or gateways
V_{FW}	The newest version number of a firmware FW, which is ready to be dispatched by a firmware patch server
CV_{FW}	The current version number of a running firmware FW on an IoT device
SC_{SN}	The pre-installed secret value in an IoT device
SK_{SN}	The pre-installed secret key in an IoT device
TK	The token generates from a firmware patch server, which will be used by a corresponding IoT device to verify whether the associated firmware is actually from a legal firmware patch server
$Topic_{MF}$	A MQTT topic published by a firmware patch server *MF*
PK_{TTP}	The master public key owned by the trusted key generation center *TTP*
s_{MF}, s_{BK}, s_{GW}	The long term private keys for firmware patch server, firmware broker server and gateway device, respectively
PK_{MF}, PK_{BK}, PK_{GW}	The long term public keys for firmware patch server, firmware broker server and gateway device, respectively
$Cert_{MF}$, $Cert_{BK}$, $Cert_{GW}$	The certificates for firmware patch server, firmware broker server and gateway device, respectively
G	The base point of the elliptic curve cryptosystem, where $G = (x, y)$ in $E_q(a, b)$. a and b are the parameters of the selected elliptic curve. q is a selected prime number
d_{MF}, d_{BK}, d_{GW}	The short term ECC-based private keys for firmware patch server, firmware broker server and gateway device, respectively. Notice that $d_{MF} < n$, $d_{BK} < n$ and $d_{GW} < n$ where n is the order of G and indicates the smallest positive number such that $nG = 0$ (the point at infinity)
Q_{MF}, Q_{BK}, Q_{GW}	The short term ECC-based public keys for firmware patch server, firmware broker server and gateway device, respectively. Notice that $Q_{MF} = d_{MF}G$, $Q_{BK} = d_{BK}G$ and $Q_{GW} = d_{GW}G$
$S_{MF\text{-}BK}$, $S_{BK\text{-}GW}$	Session keys derived through ECDH key agreement protocol for communication between one *MF* and one *BK*, and communication between one *BK* and one *GW*, respectively
$Sign_{key}(\bullet)$	Signature operation, where the value of *key* is in $\{s_{MF}, s_{BK}, s_{GW}\}$
M_1, M_2, M_3, M_4	Digital signature values
$HMAC_S(\bullet)$	Keyed-hash message authentication code operation, where the value of key S is in $\{S_{MF\text{-}BK}, S_{BK\text{-}GW}\}$
$Verf_{key}(\bullet)$	Verification operation for certificate or digital signature, where the value of *key* is in $\{PK_{TTP}, PK_{MF}, PK_{BK}, PK_{GW}\}$

manufacturer, are unique and distinctive among all MQTT topics published in the proposed framework. All MQTT messages adopt QoS level 1 (at least once) for delivery guarantee. The clean session flag of variable header of every CONNECT control packet is set to zero. The message publishers, i.e., firmware patch servers, always set the RETAIN flag of fixed header of every PUBLISH control packet to one. There are two time-interval parameters T_R and T_I for a gateway manager (or owner) to configure: T_R indicates the time interval for a gateway to re-connect the same firmware broker server after this gateway actively disconnects itself with a firmware broker, and T_I indicates the time interval for a gateway to maintain a connection with a firmware broker without receiving any new message before breaking this connection actively.

Notations used for the proposed firmware update protocols are shown in Table 1. To support mutual authentication between two communicating parties while MQTT protocol is adopted, we utilize the two reserved control packet types of MQTT specification (type 0 and type 15) to define two new control packets called PUSH and RESPONSE, respectively. We assume these two control packets can associate with payloads, i.e., messages or parameters. The proposed MQTT-based firmware update protocols among firmware patch server, firmware broker server and gateway will use the newly defined control packets and their payloads to authenticate two communicating parties and construct a session key for these two parties. All gateways should establish connections to one or more firmware broker servers in order to subscribe firmware updating MQTT topics related to corresponding sensor devices in their management domains from those brokers. Figure 3 shows the general topics subscribing flow for gateways based on MQTT protocol.

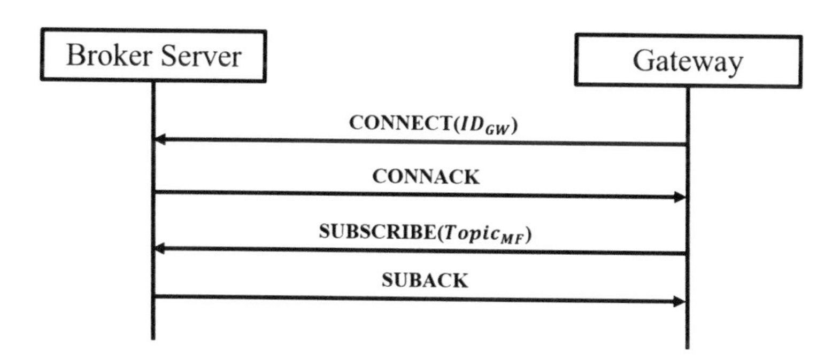

Fig. 3. The general topics subscribing flow for gateways based on MQTT protocol.

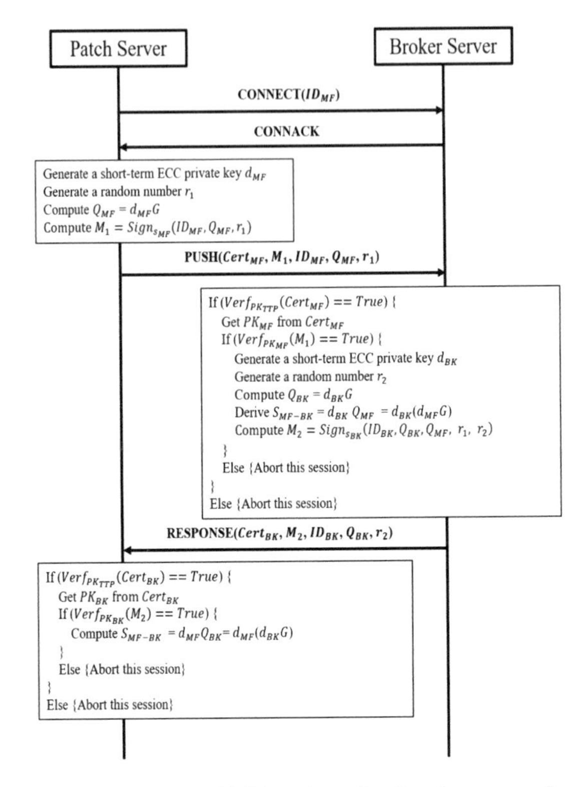

Fig. 4. The protocol to construct an ECC-based session key between a firmware patch server and a firmware broker server through MQTT scheme.

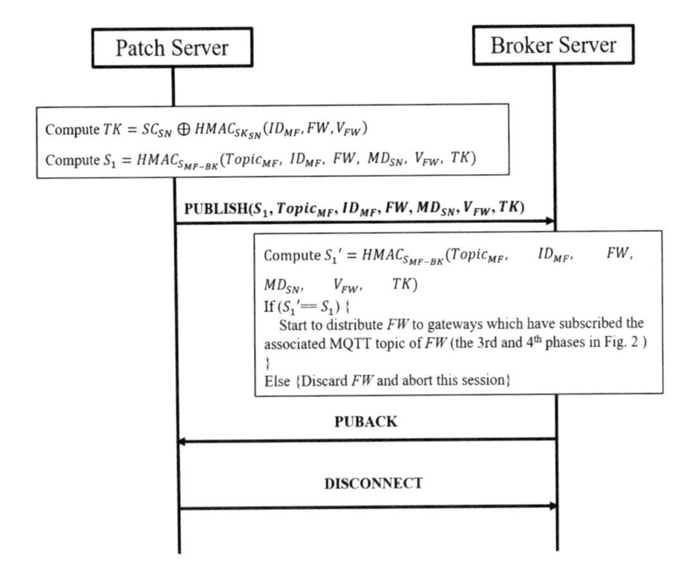

Fig. 5. The protocol to publish new released firmware *FW* from a firmware patch server to a firmware broker server through MQTT scheme.

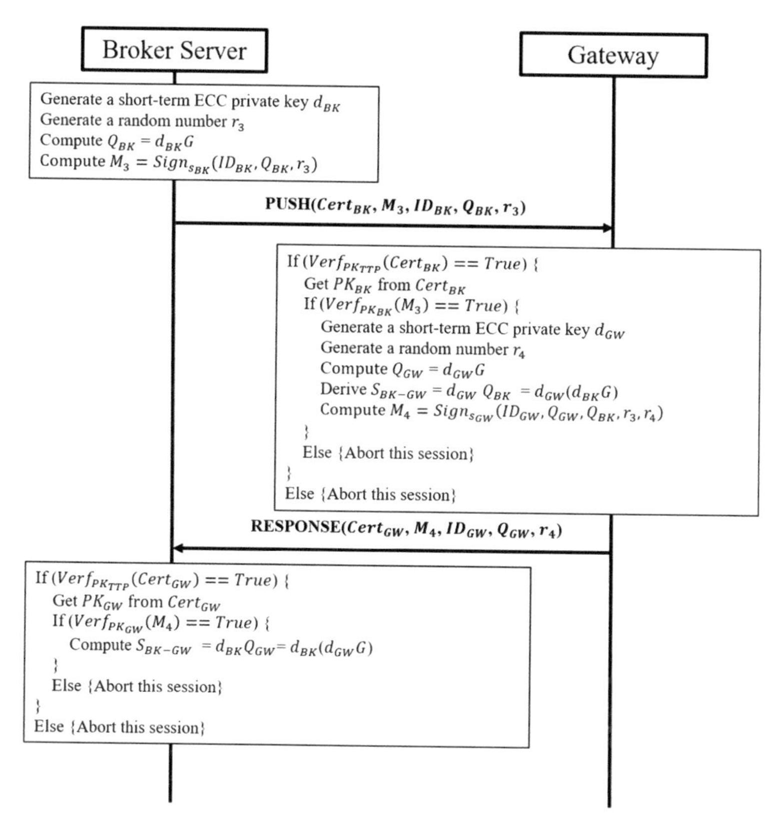

Fig. 6. The protocol to construct an ECC-based session key between a firmware broker server and a gateway through MQTT scheme.

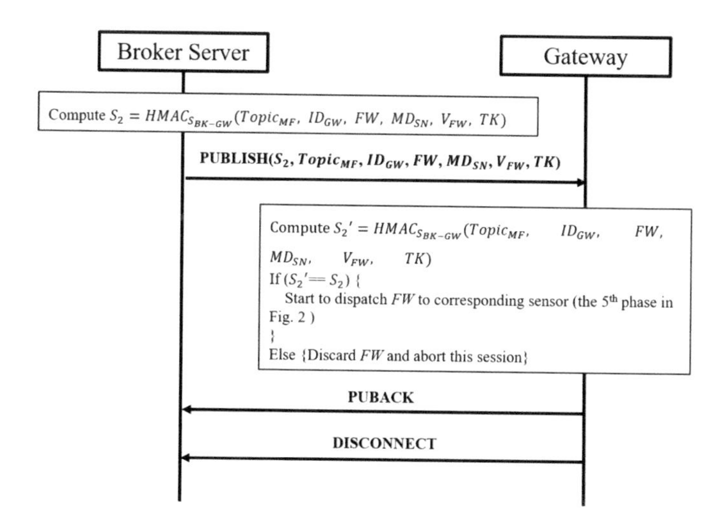

Fig. 7. The protocol to publish new released firmware *FW* from a firmware broker server to a gateway through MQTT scheme.

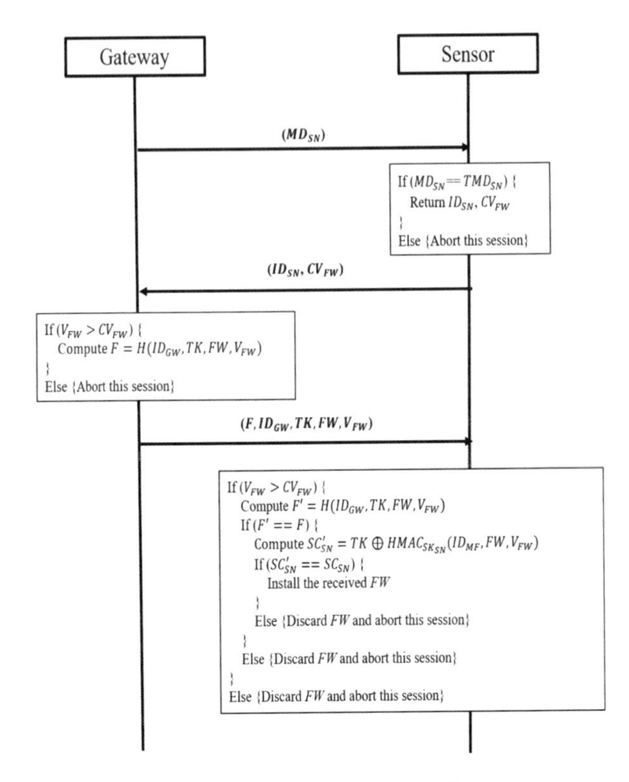

Fig. 8. The protocol to dispatch new released firmware *FW* from a gateway to a sensor device through wireless connection.

Figures 4 and 5 show the protocol design of the first and second phases of firmware update process (as shown in Fig. 2) in the proposed framework accordingly. Figures 6 and 7 show the protocol design of the third and fourth phases of firmware update process (as shown in Fig. 2) in the proposed framework accordingly. Notice that the first four phases of firmware update process utilize MQTT scheme to distribute new released firmware. Figure 8 shows the protocol design of the fifth phase of firmware update process (as shown in Fig. 2) in the proposed framework, in which a one-to-one wireless connection such as Wi-Fi connection or Bluetooth connection between a gateway and a sensor device is assumed.

4 Security Analysis

The proposed firmware update framework is strong enough to defend against general security threats such as eavesdropping, man-in-the-middle attack, replay attack and impersonation attack. If an adversary launches eavesdropping attack in order to get fresh session keys used in the first and third phases of firmware update process and insert malicious firmware later to brokers or gateways, the adversary cannot succeed since the session keys used in the first and third phases of firmware update process are constructed based on ECDH key exchange scheme. If the adversary can derive dynamically generated session keys through this ECDH scheme, then he/she can resolve ECDLP problem which in general takes exponential time complexity to get the correct answer. The proposed framework can defend against man-in-the-middle attack because our protocols mutually evaluate public certificate and digital signature for both communicating parties during session key agreement stages (the first and third phases of firmware update process). The attacker has to learn the long term private keys of gateways, brokers and patch servers, or the secret values pre-installed in IoT devices in order to impersonate as a legal gateway, broker, patch server or sensor to inject malicious firmware into the firmware update framework. With the same reason, the proposed framework can resist impersonation attack. Replay attack will be identified through executing digital signature verification operation $Verf_{key}()$ since all messages to be verified (M_1 to M_4) are associated with fresh random numbers (r_1 to r_4). At the fifth phase of firmware update process, replay attack still cannot succeed because each IoT device will first check if the received version number V_{FW} is larger than the current installed version number CV_{FW}.

5 Conclusion

In this study, a secure IoT firmware update framework using MQTT protocol is introduced. A general firmware update architecture is constructed with IoT devices, gateway devices, firmware distribution broker servers, and firmware deployment servers of IoT vendors. A secure firmware update mechanism using MQTT protocol is developed to help IoT devices authenticate the source of received firmware and verify the integrity of the received firmware. Security analysis shows that the proposed framework is secure against major security threats.

Acknowledgment. The authors gratefully acknowledge the support from the Taiwan Information Security Center (TWISC) and Ministry of Science and Technology, Taiwan, under the Grant Numbers MOST 105-2221-E-011-080-MY3, MOST 107-2218-E-011-012, MOST 107-2218-E-011-002, MOST 108-2221-E-011-063 and MOST 108-2221-E-011-065.

References

1. Choi, B.C., Lee, S.H., Na, J.C., Lee, J.H.: Secure firmware validation and update for consumer devices in home networking. IEEE Trans. Consum. Electron. **62**(1), 39–44 (2016)
2. Yadav, P., Vishwakarma, S.: Application of Internet of Things and big data towards a smart city. In: The 3rd International Conference On Internet of Things: Smart Innovation and Usages (2018)
3. Choi, H., Song, J., Yi, K.: Brightics-IoT: towards effective industrial IoT platforms for connected smart factories. In: IEEE International Conference on Industrial Internet (2018)
4. OASIS: MQTT Version 3.1.1 Plus Errata 01. http://docs.oasis-open.org/mqtt/mqtt/v3.1.1/mqtt-v3.1.1.html. Accessed 30 Apr 2019

5. Thota, P., Kim, Y.: Implementation and comparison of M2M protocols for Internet of Things. In: 2016 4th International Conference on Applied Computing and Information Technology/3rd International Conference on Computational Science/Intelligence and Applied Informatics/1st International Conference on Big Data, Cloud Computing, Data Science and Engineering (ACIT-CSII-BCD), pp. 43–48 (2016)
6. IETF: The Constrained Application Protocol. https://tools.ietf.org/html/rfc7252. Accessed 30 Apr 2019
7. Ranjan, A.K., Hussain, M.: Terminal authentication in M2M communications in the context of Internet of Things. Proc. Comput. Sci. **89**, 34–42 (2016)
8. Lavanya, Natarajan: Lightweight authentication for COAP based IOT. In: The 6th International Conference on the Internet of Things, pp. 167–168 (2016)
9. Bamasag, O.O., Youcef-Toumi, K.: Towards continuous authentication in Internet of Things based on secret sharing scheme. In: The 2015 Workshop on Embedded Systems Security (WESS 2015), Amsterdam, Netherlands (2015)
10. Butun, I., Erol-Kantarci, M., Kantarci, B., Song, H.: Cloud-centric multi-level authentication as a service for secure public safety device networks. IEEE Commun. Mag. **54**(4), 47–53 (2016)
11. Hernandez-Ramos, J.L., Pawlowski, M.P., Jara, A.J., Skarmeta, A.F., Ladid, L.: Toward a lightweight authentication and authorization framework for smart objects. IEEE J. Sel. Areas Commun. **33**(4), 690–702 (2015)
12. Kumar, P., Gurtov, A., Iinatti, J., Ylianttila, M., Sain, M.: Lightweight and secure session-key establishment scheme in smart home environments. IEEE Sens. J. **16**(1), 254–264 (2016)
13. Hassan, R., Markantonakis, K., Akram, R.N.: Can you call the software in your device be firmware? In: 2016 IEEE 13th International Conference on e-Business Engineering (ICEBE), pp. 188–195 (2016)
14. Chandra, H., Anggadjaja, E., Wijaya, P.S., Gunawan, E.: Internet of Things: Over-the-Air (OTA) firmware update in lightweight mesh network protocol for smart urban development. In: 2016 22nd Asia-Pacific Conference on Communications, pp. 115–118 (2016)
15. Nilsson, D.K., Sun, L., Nakajima, T.: A framework for self-verification of firmware updates over the air in vehicle ECUs. In: 2008 IEEE GLOBECOM Workshops, pp. 1–5 (2008)

Detection of Intrusions to Web System Using Computational Intelligence

Daniel Mišík(✉) and Ladislav Hudec

Faculty of Informatics and Information Technologies,
Slovak University of Technology in Bratislava,
Ilkovičova 2, 842 16 Bratislava, Slovakia
misik.fiit@gmail.com

Abstract. The paper focuses on usage of computational intelligence for detecting web system intrusions. We analyze existing researches in intrusion detection with computational intelligence and web attacks, primarily HTTP request and response-based web attacks. We propose and implement detection system with ensemble classification model that includes a set of classifiers. It's composed of LSTM autoencoder, text classifier, Linear SVM (Linear Support Vector Classification), Extreme Random Forest and Logistic regression, which have statistical and extracted text features of web communication as input, and Linear SVM, Extreme Random Forest, which have just extracted text features of web communication as input.

We designed flexible and extendable modular architecture and ensemble classification model where we can add another classification submodels in the future.

Keywords: Web IDS using computational intelligence · Text classifiers · Stacking ensemble

1 Introduction

Web systems are targets of many attacks. As a protection, we use intrusion detection systems that analyze the traffic and web system to capture attacks. Nowadays, detection systems are primarily based on signature detection, and therefore can't detect so-called zero-day attacks[1]. These detection systems need updates of attack signatures in the database. It is necessary to find an autonomous way of protecting data and web systems that could be adapted to new types of attacks without the need to update database. Solution can be the usage of computational intelligence for detecting web system intrusions [5].

Web attacks can be divided to HTTP request/response-base attacks or time series-based attacks. The HTTP request/response-base attacks are attacks like Cross Site Scripting, SQL injection, XML external entities injection and so on. On the another side, time series attacks are attacks like fingerprinting, brute force, web scraping and so on. In this paper we analyze and detect especially the first group of attacks, HTTP request/response attacks.

[1] New attacks that have not yet been discovered.

L. Borzemski et al. (Eds.): ISAT 2019, AISC 1050, pp. 199–208, 2020.
https://doi.org/10.1007/978-3-030-30440-9_19

In the second chapter we focus on the related researches. We write about state of art in the field of intrusion detection and about concrete papers, that affects our proposal. In the third chapter we describe in depth our proposed approach. In the end, we show experiments with CSIC 2010 dataset in the chapter four, test classification speed in the chapter five and make conclusion with future works in the chapter six.

2 Related Work

There are long-term attempts to use computational intelligence in IDS, this problem is still up to date, because computational intelligence solutions still don't achieve the results of signature-based IDS [14]. Nowadays is preferred in intrusion detection combination of computational intelligence algorithms, ensemble models, before classifying with individual classifiers [1, 9]. This is because this option allows using of different strengths of algorithms and combining them.

Kakavand et al. [12] focus on text mining for network intrusion detection. Firstly, they processed network HTTP requests and extract payload from requests. After that they perform text categorization to construct payload features, weighing to set impotence of features and after that perform text classification. For text classification they used SVM[2] and MDM[3] approach. For weighing of payload features they used TF-IDF[4]. Vector weight using TF-IDF scheme shows the importance of the specific document's feature in the collection of documents. Large weights are given to terms that are often used in a specific document but rarely in the entire collection of documents. Weight ifidf (d, t, D) for expression t in document d from document collection D is calculated by multiplying expression frequency tf (t, d) and the inverse document frequency idf (t, D) and describes the specificity of the expression inside the document collection. The expression frequency tf (t, d) describes how many times t is found in document d and idf (t, D) describes the inverse frequency of the expression t in the collection D documents.

$$tfidf(d,t,D) = tf(t,d) * idf(t,D) \quad (1)$$

Their proposal was confirmed with ISCX 2012 dataset. TP[5] was 97.45% a FP[6] was 0.4%.

Reutov et al. [4] experimented with text classification algorithms and detecting intrusions of web systems. They used specific recurrent neural network that has good results in sentiment classification. Sequence-to-sequence autoencoder was composed of multilayer LSTMs. As input was used HTTP request in text form that was encoded with few LSTM's layers and after that decoded with another LSTM layers. Input HTTP request was compared with decoded HTTP request. Result of comparison was used to

[2] Support Vector Machine.

[3] Mahalanobis Distances Map.

[4] Term frequency, inverse document frequency.

[5] True positive.

[6] False positive.

classify HTTP request as anomaly or normal communication. If the error was too high, it was anomaly, if not, it was normal communication.

Their solution was confirmed with their own dataset. They had precision and recall around 99%. You can see architecture of LSTM autoencoder in the figure (see Fig. 1).

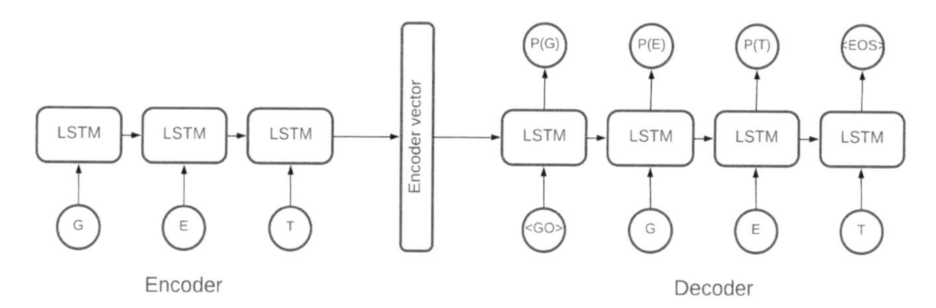

Fig. 1. Architecture of Reutov's LSTM autoencoder [4]

Min et al. [11] dealt with a deeper analysis of the body of network packets. Min's classifier, the Random Forest, was learning from aggregated statistical features and payload features. In order to extract the features from the payload, they transformed each byte from payload into a low-dimensional vector. These vectors were inserted into the slightly modified CNN[7] from Kim's work called Text-CNN [6]. This neural network has learned how to extract payload features. These extracted payload features were then aggregated with statistical features and give as input into a Random Forest classifier, which classified communication as anomaly or normal. Their proposal was confirmed with ISCX 2012 dataset. They had accuracy 99.13% and FAR[8] 1.18%.

3 Proposed Approach

Our main goal is to make classification model that is based on computational intelligence for detection intrusions to web systems and get better results than former works in this area. Because of popularity of this theme, we want also to create flexible and extendable modular architecture of IDS and ensemble classification model, where we can add another classification submodels in the future works.

Most of nowadays web systems have multilayer architecture. After extensive analysis we identify most important parts of communication flow between client and web server. We needs to monitor and analyze HTTP request, HTTP response and SQL query to the database, if it is part of communication flow. CPU and RAM monitoring and analyzing is also important. You can see important parts of communication flow in the figure (see Fig. 2).

---

[7] Convolution neural network.

[8] False acceptance rate.

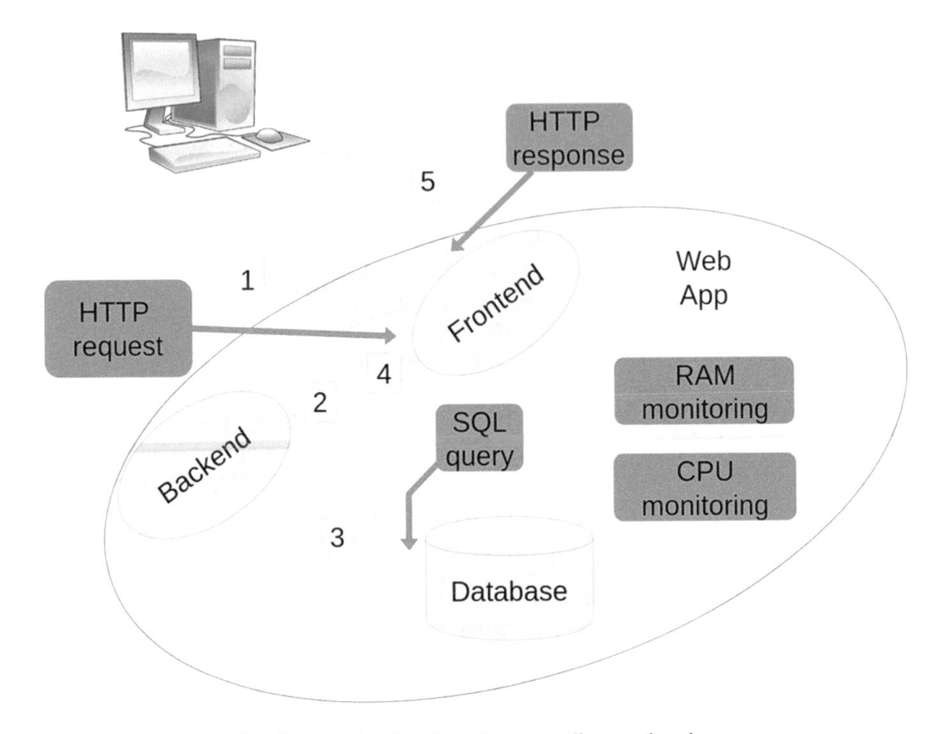

Fig. 2. Communication flow between client and web server

We created vulnerable web system which was monitored with our IDS. IDS is collecting HTTP request, response, SQL query and CPU, RAM data as in Fig. 2. Collected data are aggregated by clock stamp, because we need to join data which influence each other in certain communication flow between client and server. After aggregation we send data to the classification model which has two submodules: preprocessing module and classifier module. The output of classification model is binary classification of each aggregated block of data. You can see our classification model on the figure (see Fig. 3).

Our model can be divided into three types of classification that involved in ensemble model. Each type has specific data preprocessing and uses different classifiers at the end of the flow. These classifiers are named classifiers of the first layer.

The first type of classification, left flow in the picture, is classification with Reutov's LSTM autoencoder. The second type, flow in the middle of picture, is statistical and text features classification. The third, right flow in the picture, is text feature classification. The output of each classifier is an input for the second layer called Final classification decision. The Final classification decision processes outputs and decisions of individual classifiers and makes the final decision whether the data is classified as normal or anomalous.

Reutov et al. detected attacks from HTTP requests. The point is that the signs of HTTP request/response attacks can be also found in HTTP response and SQL query. For example XSS attacks signs can be found in HTTP response or SQL injection signs

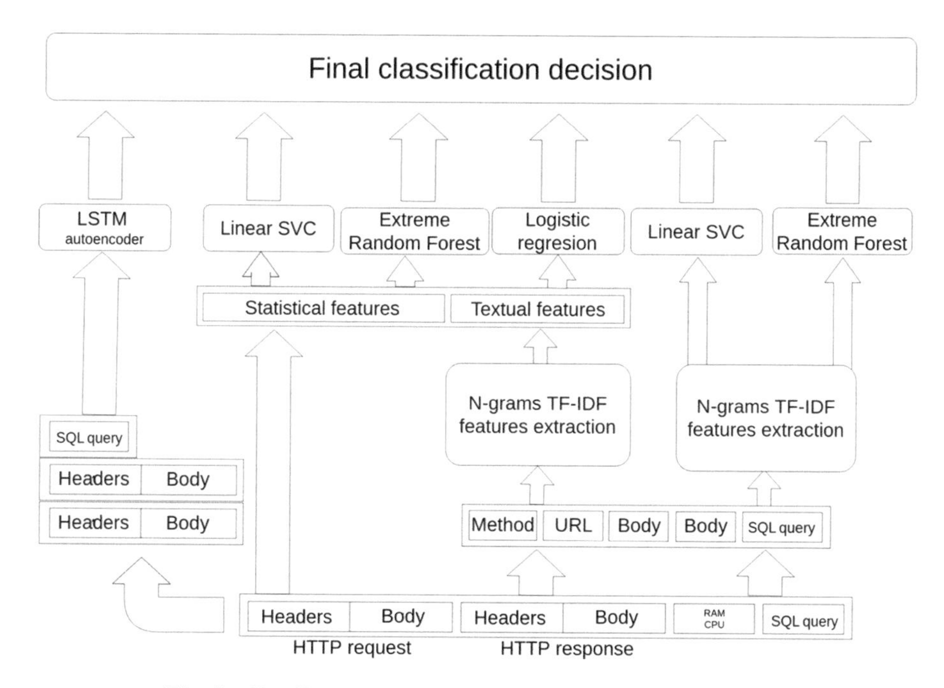

Fig. 3. Classification model for detecting web system intrusions.

can be found in SQL query. That's why we decided to detect from HTTP requests and also from HTTP responses and SQL queries. We just concatenate HTTP request, HTTP response and SQL query in textual form and sent it to the LSTM autoencoder. LSTM autoencoder is binary classifier and train just from normal data.

Signs of HTTP request/response attacks are multiple. These signs can be shown in statistical data, for example when length of HTTP request is too large, or in textual data, for example when HTTP payload contains text "UNION SELECT username, password, surname FROM users–" what can be sign of SQL injection. Based on this knowledge we create classifiers that are trained on statistical and text features. These statistical and text features classifiers have two subparts of preprocessing. In first subpart, we extract statistical features like number of alphanumeric signs in payload, length of HTTP message, length of HTTP payload, CPU and RAM measures. In second subpart, we extract textual features in two steps. Firstly, we process n-gram categorization from concatenated request method, request URL, payloads of request and response and SQL query. We used these textual data segments because in these segments is the most of textual signs of HTTP request/response attacks, for example XSS or SQL injection in payload or Path Traversal in URL, After categorization, we count weights to categorized text vectors with TF-IDF method. Then we aggregate these weighted vectored text features with statistical features and sent them to the Linear SVC, Extreme Random Forest and Logistic regression.

Text features classifiers are classifiers that are train just from text features. The extraction of text features is the same as we described above. Firstly, we extract textual features, we weight text vectors with TF-IDF and after that we sent these vectors to the Linear SVC and Extreme Random Forest.

Final classification decision is designed to have multiple different implementations. The output of this section is binary classification.

We implemented:

- Voting → Final classification decision is voting component that vote from outputs of subclassifiers for the final classification.
- Stacking[9] → Final classification decision is meta binary classifier that learns from outputs of subclassifiers trained with n-fold cross validation.

4 Evaluation

CSIC 2010 is one of the best know datasets for web system intrusions detection. We used this dataset for the evaluation of our approach. This dataset contains just HTTP requests and is delivered in 3 parts: normal train data, normal test data and anomaly test data. Peter Scully created well known preprocessed version of this dataset in csv format, which can be found on his website [2]. According to Peter Scully, entropy analysis shows very little difference between normal and normal test data and due to these similarities he does not recommend to use of normal test data in the evaluation [2].

In our published results we were working with original dataset, but we did not use normal test data. One of the reasons was also to use the same input data as former works so we have the comparable inputs. The few of former works [10, 15] were using Scully's csv format of dataset without normal test data.

We've validated our solution also on original dataset including normal test data and our results were even little bit better.

Due to limited computational power we have, LSTM autoencoder takes a lot of time for training and testing. We did stratified 5-fold cross validation[10] for models without LSTM autoencoder and stratified split[11] 80/20 of dataset for training/testing in models with LSTM autoencoder.

4.1 Evaluation Metrics

In evaluation we used metrics F1 score, precision, true positive rate[12], false positive rate. True positive, false positive, true negative, false negative are sign with TP, FP, TN and FN.

[9] Super Learner.

[10] Stratified cross validation is cross validation that use stratified folds. The folds are made by preserving the percentage of samples for each class.

[11] Stratified split of dataset works on same principle as stratified folds in stratified cross validation.

[12] Recall.

Precision is calculated:

$$\text{Precision} = \frac{\text{TP}}{\text{TP} + \text{FP}} \tag{2}$$

True positive rate, TPR, recall, is calculated:

$$\text{Recall} = \text{TPR} = \frac{\text{TP}}{\text{TP} + \text{FN}} \tag{3}$$

F1 score is calculated:

$$\text{F1} = \frac{2 \times \text{Precision} \times \text{Recall}}{\text{Precision} + \text{Recall}} \tag{4}$$

False positive rate, FPR, is calculated:

$$\text{FPR} = \frac{\text{FP}}{\text{FP} + \text{TN}} \tag{5}$$

4.2 Evaluation of Classifiers

Firstly, we will mention results of individual classifiers. Even if LSTM autoencoder gets great result on Reutov's dataset, on dataset CSIC 2010 got just precision 69.97% and recall 72.07%. We wasn't capable to get better results with Reutov's architecture of LSTM autoencoder, but we tried train LSTM autoencoder for certain URL path and we found out that for certain URL paths LSTM autoencoder has precision and recall around 99%. Each of individual statistical and text features classifiers had precision and recall above 91%. We have got the best results with 300 weighted text features and 50 statistical features. Each of text features classifiers had precision and recall above 98%. Among all individual classifiers has the best results text features Linear SVC with precision 99.92% and recall 99.74%.

Both ensemble models got better results as overall individual classifier. We experiments with Soft Voting and Hard Voting. Hard Voting was the better one with precision 99.94% and recall 99.56%. Neither of voting models beat the best individual classifier, text features Linear SVM. We experiments with multiple implementations of Stacking model. We implemented Soft and Hard Stacking ensemble models. Because of worse results of LSTM autoencoder, we tried to compare ensemble models with or without LSTM autoencoder. As meta classifier which learnt from output of individual classifiers, was chosen SVC algorithm. The best results has Soft Super Learner without LSTM autoencoder, precision 99.93% and recall 99.896%, and Hard Super Learner with LSTM autoencoder, precision 99.86% and recall 100%. These two models have also the best results among evaluated models. You can see comparison of our models in table (see Table 1).

Table 1. Comparison of our classification models

Detection method	F1	Precision	TPR	FPR
Voting + URL NN[a]	99.60%	99.92%	99.30%	0.06%
Hard weighted voting – NN[b]	99.749%	99.94%	99.56%	0.041%
Soft voting – NN	99.702%	99.848%	99.557%	0.105%
Hard Super Learner + NN	99.93%	99.86%	100%	0.10%
Text features linear SVM	99.82%	99.92%	99.74%	0.05%
Hard Super Learner – NN	99.828%	99.75%	99.908%	0.17%
Soft Super Learner – NN	99.91%	99.93%	99.896%	0.047%

[a]URL-based neural network, URL-based LSTM autoencoder
[b]Neural network, Reutov's LSTM autoencoder

We've also compared our results with the best results from other papers (see Table 2). We've chosen metrics TPR, recall and FPR for comparison.

Table 2. Comparison of our results with other researches on CSIC 2010

Detection method	TPR	FPR
Nguyen [13] – C4.5	94.49%	5.90%
Kozik – ELM[a] [8]	94.98%	0.79%
Kozik – REPTree[b] [7]	98%	1.50%
Löffler [10] – RF[c] + SVM	96.27%	14.38%
Šoltés [15] – dDCA[d]	87.88%	12.71%
Althubiti – RF [3]	99.90%	0.10%
Our results		
Text features Linear SVM	99.74%	0.05%
Soft Super Learner – NN	99.90%	0.05%
Hard Super Learner + NN	100%	0.10%

[a]Extreme Learning Machine Forest
[b]Reduced Error Pruning Tree
[c]Random Forest
[d]Deterministic dendritic cell algorithm

Our best classifiers have better results as any other detection method except Althubiti's Random Forest. Althubiti's Random Forest has TPR 99.90% and FPR 0.10%. Hard Super Learner with LSTM autoencoder has same FPR, but higher TPR value. Soft Super Learner without LSTM autoencoder has lower TPR and lower FPR.

For better comparison we can compare F1 score and precision values. Althubiti's Random Forest has F1 score 99.90% a precision 99.90%. Soft Super Learner has F1 score 99.91% and precision 99.93%. Hard Super Learner has F1 score 99.93% and precision 99.86%.

To summarize, our best detection methods are better than other detection methods except Althubiti's Random Forest. Depending on used metrics our methods are a little bit better or a little bit worse than Althubiti's solution.

5 Classification Speed of Our Classifiers

Nowadays, IPS is replacing IDS, so it is important to monitor time complexity of detection methods. Therefore, we had decided to test the classification speed of our classifiers in relation to the amount of data which they had to classify. We compared Stacking ensemble model with and without LSTM autoencoder, text features Linear SVM and Voting ensemble model with URL-based LSTM autoencoder. Results of test show that classification of classifiers without LSTM autoencoder is much faster than classification of classifiers with LSTM autoencoder. Text features SVM needs less than 3 seconds to classify 8500 samples. Stacking ensemble model without LSTM autoencoder needs less than 31 seconds to classify 8500 samples. Voting ensemble model with URL-based LSTM autoencoder needs up to 301 seconds to classify 8500 samples. Slowest one was Stacking ensemble model with LSTM autoencoder that needs up to 500 seconds to classify just 4500 samples.

6 Conclusion and Future Work

In this work we were trying to create ensemble classification model that combine strengths of individual models for getting better overall results. We've implemented prototype that was trained on CSIC 2010. Our results were comparable with state of art results. We've also implemented extensible generic IDS where we plug in our detection methods. This IDS can be extended via plug in new type of data monitors or new type of classifiers. We can add new evaluation methods in the future works. We also test this detection system by monitoring of vulnerable web system that we implemented for this purpose.

Our detection system detects currently HTTP request/response attacks. It's considered to be used with other detection systems that detect time series attacks.

Unfortunately there are still some issues. Dataset CSIC 2010 contains just HTTP request, but our proposed approach also consider HTTP response, SQL query, RAM and CPU monitoring as valuable data for analysis. Because of there is no comparable dataset with all of these data types, we were not able to evaluate effect of another data types except HTTP requests on final results. Our only option in the future is to create our own dataset and test effect of additional data types on the final results.

Even if there is no proper dataset for our full proposal, in the future we should test our methods on other valuable datasets which contains just HTTP requests because we need to know how good methods on another training data are.

We also think that there is space to improve our methods. We can improve Stacking ensemble model with changing parameters, changing meta classifier and so on. We can also try to change Reutov's LSTM autoencoder architecture to get better individual results. The speed of classification is also valuable feature of detection methods. We would like to focus also on improving performance of our methods.

Acknowledgement. This work was partially supported by Eset Research Centre.

References

1. Aburomman, A.A., Reaz, M.B.I.: A survey of intrusion detection systems based on ensemble and hybrid classifiers. Comput. Secur. **65**, 135–152 (2017)
2. Scully, P.: Csic 2010 http dataset in csv format (for weka analysis). https://petescully.co.uk/research/csic-2010-http-dataset-in-csv-format-for-weka-analysis/. Accessed 16 Apr 2019
3. Althubiti, S., Yuan, X., Esterline, A.: Analyzing HTTP requests for web intrusion detection (2017)
4. Reutov, A., Stepanyuk, I., Sakharov, F., Murzina, A.: Seq2Seq for Web Attack Detection. https://github.com/PositiveTechnologies/seq2seq-web-attack-detection. Accessed 8 Dec 2018
5. Bhuyan, M.H., Bhattacharyya, D.K., Kalita, J.K.: Network anomaly detection: methods, systems and tools. IEEE Commun. Surv. Tutor. **16**(1), 303–336 (2014)
6. Kim, Y.: Convolutional neural networks for sentence classification. In: Proceedings of the 2014 Conference on Empirical Methods in Natural Language Processing (EMNLP), Association for Computational Linguistics, pp. 1746–1751 (2014)
7. Kozik, R., Choras, M.: Adapting an ensemble of one-class classifiers for a web-layer anomaly detection system. In: 2015 10th International Conference on P2P, Parallel, Grid, Cloud and Internet Computing (3PGCIC), pp. 724–729 (2015)
8. Kozik, R., Choraś, M., Holubowicz, W., Renk, R.: Extreme learning machines for web layer anomaly detection. In Choraś, R.S. (ed.) Image Processing and Communications Challenges 8, Cham, Springer International Publishing, pp. 226–233 (2017)
9. Langin, C., Rahimi, S.: Soft computing in intrusion detection: the state of the art. J. Ambient Intell. Humaniz. Comput. **1**(2), 133–145 (2010)
10. Löffler, M.: Improvement of intrusion detection using multiple classifier model. Diploma thesis. FIIT STU (2017)
11. Min, E., Long, J., Liu, Q., Cui, J., Chen, W.: TR-IDS: anomaly-based intrusion detection through text-convolutional neural network and random forest. Secur. Commun. Netw. **2018**, 1–9 (2018)
12. Kakavand, M., Mustapha, N., Mustapha, A., Abdullah, M.T.: A text mining-based anomaly detection model in network security. Glob. J. Comput. Sci. Technol. **14**(1), 22–31 (2015)
13. Nguyen, H.T., Torrano-Gimenez, C., Álvarez, G., Petrovic, S., Franke, K.: Application of the generic feature selection measure in detection of web attacks. In: CISIS (2011)
14. Wu, S.X., Banzhaf, W.: The use of computational intelligence in intrusion detection systems: a review. Appl. Soft Comput. **10**(1), 1–35 (2010)
15. Šoltés, F.: Improving security of a web system using biology inspired methods. Diploma thesis. FIIT STU (2016)

New Encryption Method with Adaptable Computational and Memory Complexity Using Selected Hash Function

Grzegorz Górski[1,2](✉) and Mateusz Wojsa[1,2]

[1] Koszalin University of Technology, Śniadeckich 2, 75-453 Koszalin, Poland
{grzegorz.gorski,mateusz.wojsa}@tu.koszalin.pl
[2] Transition Technologies S.A., Pawia 55, 01-030 Warsaw, Poland

Abstract. In this paper, we describe the new method of data encryption/decryption with selectable block and key length and hash function. The block size directly improves the efficiency but also impacts on memory complexity of the presented solution. The choice of hash function length results in computational complexity and complicates attacks using rainbow tables. The algorithm is a modification of SP (Substitution-Permutation) concept with the use of static S-blocks and dynamically indexed block permutation. In further step, usage of chosen operational modes of the method (ECB, CBC, CTR) is presented. In the article there is the example with all the algorithm parameters i.e. input and output data, key value, hash function type and set of all method internal states. The efficiency of the solution was experimentally examined and compared with the most popular bock cipher algorithms e.g. AES, Serpent, BlowFish, TwoFish. The obtained initial results indicate that the new method can be dedicated especially for systems with high security requirements. The paper conclusions contain propositions of algorithm implementation optimizations and further research.

Keywords: Data encryption · Block cipher · Hash function · Computational and memory complexity · Cipher operational modes

1 Introduction

To start with, it should be emphasized that data confidentiality and security are regarded as crucial requirements for contemporary distributed information systems. Therefore, there are several scientific articles presenting different cipher algorithms [1–4]. Large collections of data are usually encrypted with symmetric methods using permanently specified block size. These iteration solutions allow to achieve required security level by increasing or decreasing cipher key length what results in different number of algorithm internal rounds. In spite of being flexible, multi-round algorithm is also believed to be complicated and is still examined in hope some back-door capabilities are discovered. Despite the fact that the ciphers are believed to be worldwide standard they are selected by means of competition. However, national secret services agencies constantly prefer exclusively designed and approved encryption methods. The same agencies require exclusive traffic and content control with the use of

L. Borzemski et al. (Eds.): ISAT 2019, AISC 1050, pp. 209–218, 2020.
https://doi.org/10.1007/978-3-030-30440-9_20

contemporary 4G and future 5G telecom network as far as the investigation concerns a given county or its citizens. Encryption/decryption solutions are usually based on SP network i.e. contain static substation blocks (S-blocks) and static permutation blocks for data processing which are usually before bite XOR operation with key subset [5]. The other researchers focus on optimization of cipher algorithms using internal cache of processor, possibilities of parallel computing with graphic multicore processor or ASIC integrated circuits e.g. FPGA [6, 7]. It results in improvement in coding efficiency or integration scale. However, there is everlasting necessity of new cipher algorithms especially in point-to-point communication i.e. solutions designed for presentation or application layers of OSI reference model. The purpose was to define a new encryption/decryption method with adaptable computational and memory complexity using selected hash function. The algorithm should have a single processing round and offer higher security level than referential encryption block algorithms. The achieved level of security could be evaluated with computational complexity of brute-force attack.

Presented paper structure begins with the new GW method description. It contains initial requirements and assumptions, step by step explanation of encryption and decryption procedure for the algorithm. Next, the complexity analysis and operational modes are presented. The final part of the article is about experiment assumptions and discussion on obtained results.

2 The Method Description

Contrary to other symmetric block ciphering algorithms the authors assumed that the new method must have adaptable block size and key length [8]. It is suggested that the amount of data that is processed in the time of one cycle and is multiplicity of a single byte. However, it is not required. The new method referred as GW (Gorski-Wojsa) algorithm is not a multi-round solution i.e. the whole coding and decoding process is contained in a single step. It makes a whole algorithm much more simple and allows a user to point out the operations which directly impacts achieved security level. The software implementation of the algorithm should guarantee a user the possibility to choose the crucial security operation from predefined set of functions or create their own one according to general specification. The whole algorithm based on crucial security operation should guarantee that there is no possibility to hide inside the back-door functionality. The key length should be flexible what means that it does not need to belong to the typical collection (128b, 192b, 256b etc.). Constant memory complexity of the algorithm is not required. Necessary amount of memory can depend on assumed block size. Finally, the method should offer significantly higher computational complexity than the other symmetric block solutions.

Following the requirements mentioned above, hash function was chosen as a crucial security operation of the GW algorithm. This type of operation referred to one directional compress function irrevocably converts plain data into unique data. As a result, there is no correlation with the corresponding plain data [9]. It can be proved that the fact of hash function backdoor existence is equivalent to retrieve input data form hash value without searching the whole collection of hash function arguments. In GW algorithm case, achieved security level stands to computational complexity of the best attack against hash function. The user of GW algorithm can choose any hash function, which he finds trustworthy. However, dimension of hash value collection directly impacts the computational complexity achieved by GW algorithm. Summarizing, GW algorithm hash function irrevocably converts encryption key into a pseudorandom number used to dynamically indexed block permutation [10, 11].

2.1 The GW Algorithm

The presented encryption method runs two separate procedures i.e. one for encryption and the second one for data decryption. The first one is given in Fig. 1. The encryption procedure of GW algorithm starts with some initial assumptions. First of them is to determine cipher key k_S. The key length can vary from the single bit to maximum amount accepted by the chosen hash function what fulfills the adjustable key length requirement. If one assumes that the k_s length is contained within <1, n> the potential multitude of cipher key collection K_x is given bellow:

$$Card\ (Ks) = 2^n! \tag{1}$$

This value does not influence computational complexity of the method because of constraint derived from maximum number of hash function values. If the length of chosen hash function H() is H_l than the multitude of hash value collection declares the formula:

$$Card\ (H(k_s)) = 2^{Hl} \tag{2}$$

The second assumption of GW algorithm is the block size m given in bits. It is mentioned above that preferred but not required is multiplicity of the single byte. Block length m directly impacts on memory complexity of the presented solution. Formula (2) also defines maximum number of possible block values if one exchanges H_l with m. The following initial parameter is encryption table $T_e[]$ that is used to convert each data block value B_i into a random substation value S_i. The index of $T_e[]$ is contained within <0, 2^m-1> and the substation values stored in $T_e[]$ are unique and must overlay whole collection of data block values. $T_e[]$ table has random order. There is an additionally defined table defined as decryption table $T_d[]$ used during reverse conversion i.e. from substitution value S_i to data block B_i during decoding procedure. Both tables $T_e[]$ and $T_d[]$ are obviously correlated and some programming languages accept theirs definition as one inverted table. Memory complexity given in bites of GW algorithm depends on the block size m which is presented below:

$$O(m) = 2^{m+1}m \tag{3}$$

Block size multiplication k is the last initial parameter that should be determined. It is calculated as division hash function length H_l and block size m rounded down to the whole i.e. without fractional part. This value is used in XOR operation during calculation of subblock indexes of hash value H_i. Next couple of steps are devoted to auxiliary indexes calculation. The calculation schema with all details is presented in Fig. 2.

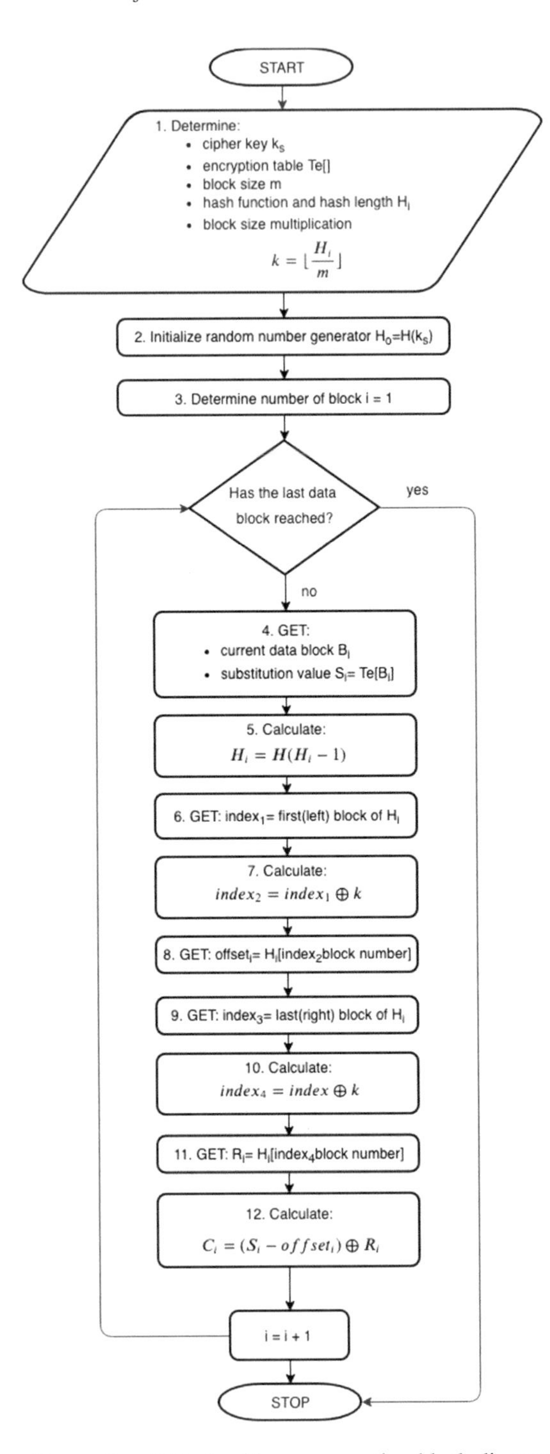

Fig. 1. The GW algorithm – encryption block diagram

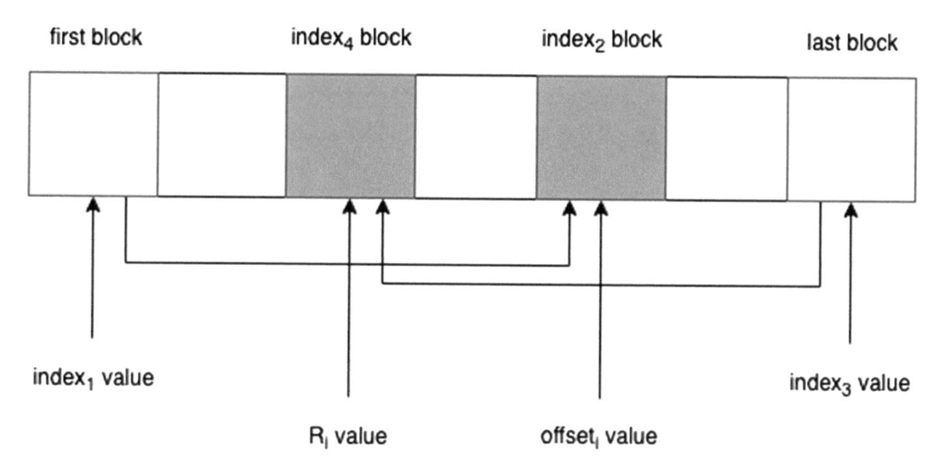

Fig. 2. The GW algorithm – Offset_i and R_i calculation scheme based on random nuber H_i

The index_1 is equal to the highest (on the left) subblock of hash value H_i. Following, the index_2 is calculated by basing on index_1, XOR operation and multiplication parameter k. Index_2 represents number of subblock where the offset_i value is collected from. Thanks to modulo operation regardless index_1 value index_2 never reaches out the H_i vector. The similar mechanism is used for index_3 and index_4 calculation. However, index_3 is received from the opposite side of hash value H_i i.e. from the lowest subblock (on the right).

Index_4 represents subblock number of hash value H_i where the random number R_i is stored. The last step of the sequence repeated for each block is designed to calculate cryptogram block based on substitution value S_i, offset_i and random number R_i. The auxiliary indexes mechanism applied in the GW algorithm could be referred as dynamically indexed block permutation. It causes random parameters offset_i and R_i are dynamically spread out inside the whole hash vector H_i what significantly complicates attacks against hash function using collected rainbow tables.

Decryption procedure is similar to encryption and it is presented in Fig. 3. It also starts with some initial assumption. However, on the contrary to encryption process decoding phase uses decryption table $T_d[]$. Initialization step (point no 2 and 3) is identical in both procedures. Following sequence of operations is repeated for each cryptogram block instead of data block in first procedure. Following, calculation of current hash value H_i and auxiliary indexes from index_1 to index_4 is proceeded. Two additional parameters i.e. offset_1 and R_i are established in the same way. In contrast to encryption process decoding process cryptogram block C_i is obtained first and then random number R_i and offset_i values are calculated and at the end of repeated sequence substitution value S_i is received. The one before last operation uses decryption table $T_d[]$ and converts substitution value S_i into data block B_i. Finally, the internal block counter i is incremented.

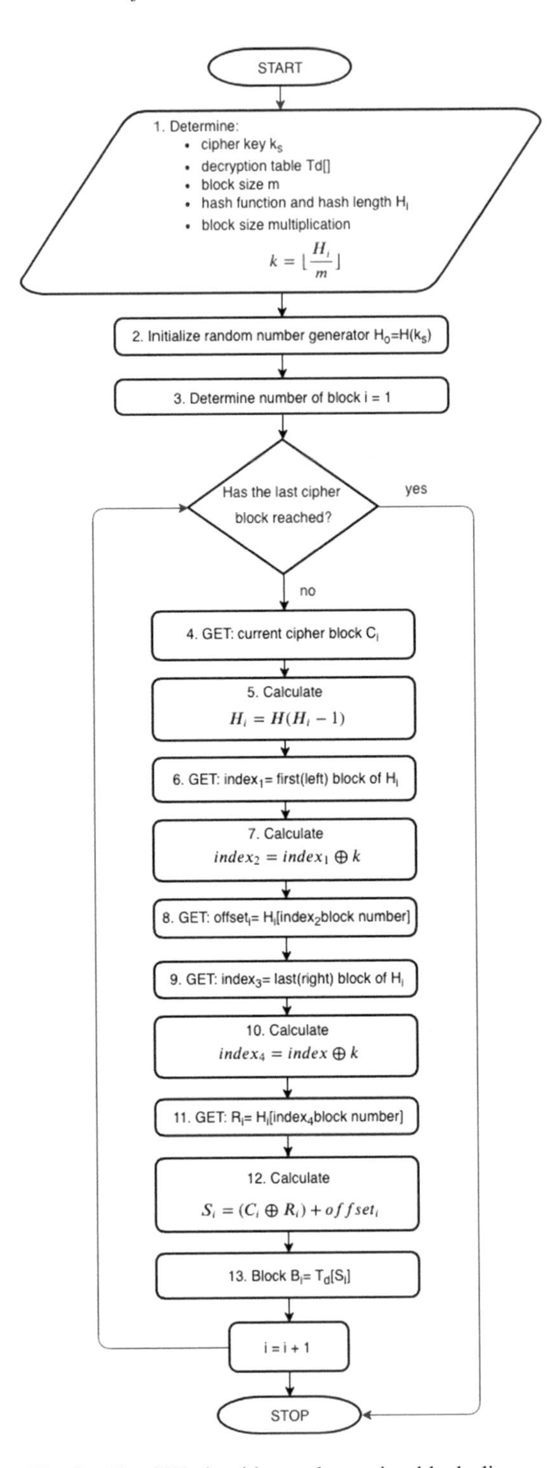

Fig. 3. The GW algorithm – decryption block diagram

Computational complexity of GW algorithm strongly depends on which method parameters can be regarded as secret i.e. exclusively shared by users sending and receiving encrypted messages. If one assumes that only the cipher key k_s is a confidential than algorithm complexity O_1 is given below with limitation described below formula (1).

$$O_1(n) = 2^{Hl} \tag{4}$$

If cipher key k_s and both tables $T_e[]$ and $T_d[]$ are secret than the GM algorithm gains significantly higher computational complexity O_2 what presents the following formula

$$O_2(n) = (2^m)!\, 2^{Hl} \tag{5}$$

where m refers to data block size given in bits. Considering Stirling's approximation the algorithm bit complexity O_2 is given as follows:

$$O_2(n) = \frac{\sqrt{2\pi}}{e^{2^m}} * 2^{m(2^m + 0,5) + H_l} \tag{6}$$

where H_l refers to hash value length given in bits. Formula (6) considering the fact that decryption table $T_d[]$ deterministically depends on encryption table $T_e[]$.

Example 1. Let us assume that the block size m = 8b, SHA-512 is used hash function than memory complexity of GW algorithm is equal to 512 bytes. If only cipher key k_s is confidential computational complexity O_1 is equal to 2^{512}. If both tables i.e. encryption and decryption are additionally secret computational complexity O_2 can be roughly approximated:

$$O_2(n) > 2^{2053} \tag{7}$$

Application of larger block sizes for instance 16 or 32 bits long is still possible. It requires respectively 256 kB or 16 GB amount of memory. However obtained bit complexity O_2 is hard to image and due to efficiency reasons out of reach for typical symmetric block cipher algorithms.

3 GW Operational Modes

Block cipher algorithms can process data in different operational modes that are to some extent algorithm application schemas. The modes define how the input data is divided into single blocks and if there are any dependencies between successive cryptogram block and other algorithm parameters. Electronic Code Book (ECB) is the most popular mode that assumes fully independent block coding and decoding. In fact, both procedures of GW algorithm are described above and operated in ECB mode. If any relationships between successive cryptogram block are involved then Cipher Block

Chaining mode (CBC) is the most probable. In GW algorithm CBC can be easily implemented assuming new formulas for hash value calculation presented below:

$$H_0 = H(k_s \,||\, IV),$$
$$H_i = H(H_{i-1} \,||\, S_{i-1}) \tag{8}$$

where IV refers to Initialization Vector i.e. constant algorithm parameter, cipher key k_s and cryptogram block calculated in previous iteration. The set of formulas (8) is the same for coding and decoding procedure. Counter Mode (CTR) is the last operational mode presented in the paper [12]. Similarly to ECB CTR schema supports independent block processing. However, it allows for achieving in different encryption procedures unique cryptogram block form the same algorithm, cipher key and input data block. In that case CTR mode for GW algorithm can be obtained with modified set of arguments of hash function formula:

$$H_0 = H(k_s \,||\, nonce) \tag{9}$$

$$H_i = H(H_{i-1} \,||nonce||\, i)$$

where nonce is random number usually determined for whole communication or encryption/decryption session.

4 Initial Results

4.1 Experiment Assumptions

Experiments were performed using laptop (hard drive: SSD type, RAM 16 GB, processor: Intel Core i7-8650U 7th generation 1.9-4.2 GHz, Windows 8). The implementation of GW algorithm was made in Java with two parameter sets. The first version (GW-1B) had the block size equal to 1 byte, 128 bit key length and SHA-512 hash function. The other one (GW-2B) utilized 2 byte block size, 256 bit key length and SHA-1024 hash function. For both mechanism of object collection was involved. There was no code optimization considering hardware architecture [13]. AES, Serpent, BlowFish, TwoFish are chosen as referential symmetric block cipher algorithms. Bouncy Castle version 1..59 was used as crypto library containing Java implementation of cryptographic algorithms. All encryption and decryption procedures were repeated one hundred times and the results were averaged to reduce impact of singular errors.

4.2 Efficiency Benchmark

For all examined algorithms both encryption and decryption procedures were involved. Results of experiments are given in Table 1. For both operations GW-1B algorithm processing time is approximately from 6 to 22 times longer than other referential algorithms. GW-2B version requires approximately from 11 to 42 times longer to encrypt or decrypt the same amount of data. However, for all referential algorithms the same computational complexity O_1 can be obtained with cascade system by connecting several blocks with shorter key length. The overall processing time of cascade systems is a single block processing time multiplied by the number of block (4 or 8). Initial results seem to be promising if one considers much higher computational complexity of GW methods in O_2 mode (with secret encrypting and decryption tables). Available programming APIs for referential algorithms do not support such long encryption keys to achieve complexity compared with O_2. In that case, multistage processing with shorter keys could not be a solution.

Table 1. Algorithm benchmark

Algorithm	Procedure	Key length	Time [msec]	Complexity O_1	Complexity O_2
GW-1B	Encryption	128b	2.891463	2^{512}	$>2^{2053}$
GW-1B	Decryption	128b	3.984224	2^{512}	$>2^{2053}$
GW-2B	Encryption	256b	4.171052	2^{1024}	$>2^{984073}$
GW-2B	Decryption	256b	5.215927	2^{1024}	$>2^{984073}$
AES	Encryption	128b	0.123628	2^{128}	–
AES	Decryption	128b	0.126837	2^{128}	–
Serpent	Encryption	128b	0.240379	2^{128}	–
Serpent	Decryption	128b	0.261113	2^{128}	–
BlowFish	Encryption	128b	0.479701	2^{128}	–
BlowFish	Decryption	128b	0.466195	2^{128}	–
TwoFish	Encryption	128b	0.394649	2^{128}	–
TwoFish	Decryption	128b	0.394579	2^{128}	–

5 Summary

In this paper, the new concept of an encryption method using selected hash function was presented. The GW algorithm allows to achieve assumed level of computational complexity with selecting appropriate block size and length of hash value. The block size has significant impact on the amount of memory required by the solution. Initial efficiency tests confirm that new solution is several times slower than referential symmetric block algorithm but its computational complexity is corresponding several times higher. Further research will focus on optimizing the code implementation. Insertion of encryption and decryption tables into processor cache memory will be put into consideration. Hardware implementation of GW method using FPGA integrated circuit will be also put to the test.

References

1. Noura, H., Sleem, L., Noura, M., Mansour, M., Chehab, A., Couturier, A.: A new efficient lightweight and secure image cipher scheme. Multimed. Tools Appl. **77**(12), 15457–15484 (2018)
2. Zhang, X., Heys, H., Li, C.: Energy efficiency of encryption schemes applied to wireless sensor networks. Secur. Commun. Netw. **5**(7), 789–808 (2012)
3. Sallam, S., Beheshti, B.D.: A survey on lightweight cryptographic algorithms. In: Proceedings of IEEE Region 10 Annual International Conference on TENCON 2018, pp. 1784–1789. TENCON, October 2018
4. Bogdanov, A., Isobe, T., Tischhauser, E.: Towards practical whitebox cryptography: optimizing efficiency and space hardness. In: Cheon, J., Takagi, T. (eds.) Advances in Cryptology - ASIACRYPT 2016. LNCS, vol. 10031, pp. 126–158. Springer, Heidelberg (2016)
5. Shibutani, K., Bogdanov, A.: Towards the optimality of Feistel ciphers with substitution-permutation functions. Des. Codes Cryptogr. **73**(2), 667–682 (2014)
6. Sugier, J.: Implementation efficiency of BLAKE2 cryptographic algorithm in contemporary popular-grade FPGA devices. In: Kabashkin, I., Yatskiv, I., Prentkovskis, O (eds.) Lecture Notes in Networks and Systems, vol. 36 pp. 456–465. Springer, Heidelberg (2018)
7. Szefer, J., Chen, Y.Y., Lee, R.B.: General-purpose FPGA platform for efficient encryption and hashing. In: Charot, F., Hannig, F., Teich, J., Wolinski, C. (eds.) IEEE International Conference on Application-Specific Systems Architectures and Processors. IEEE (2010)
8. Hei, X.L., Song, B.H., Ling, C.J.: SHIPHER: a new family of light-weight block ciphers based on dynamic operators. In: IEEE International Conference on Communications. IEEE (2017)
9. Jithendra, K.B., Shahana, T.K.: Elastic serial substitution box for block ciphers with integrated hash function generation. In: Bi, Y., Kapoor, S., Bhatia, R. (eds.) Proceedings of Sai Intelligent Systems Conference (INTELLISYS) 2016, vol 2. LNNS, vol. 16, pp. 658–671. Springer, Heidelberg (2018)
10. Noura, H., Courousse, D.: Lightweight, dynamic, and flexible cipher scheme for wireless and mobile networks. In: Mitton, N., Kantarci, ME., Gallais, A., Papavassiliou, S. (eds.) Lecture Notes of the Institute for Computer Sciences Social Informatics and Telecommunications Engineering, vol. 155, pp. 225–236. Springer, Heidelberg (2015)
11. Kumar, M., Dey, D., Pal, S.K., Panigrahi, A.: HeW: a hash function based on lightweight block cipher FeW. Defence Sci. J. **67**(6), 636–644 (2017)
12. Alvarez-Sanchez, R., Andrade-Bazurto, A., Santos-Gonzalez, I., Zamora-Gomez, A.: AES-CTR as a password-hashing function. In: Garcia, H.P., Alfonso-Cendon, J., Gonzale, L.S., Quintian, H., Corchado, E. (eds.) Advances in Intelligent Systems and Computing, vol. 649, pp. 610–617. Springer, Heidelberg (2018)
13. Takahashi, H., Nakano, S., Lakhani, U.: SHA256d hash rate enhancement by L3 cache. In: IEEE Global Conference on Consumer Electronics. IEEE (2018)

OSAA: On-Demand Source Authentication and Authorization in the Internet

Bartłomiej Dabiński(✉)

Wroclaw University of Science and Technology, Wroclaw, Poland
bartlomiej.dabinski@pwr.edu.pl

Abstract. Lack of ability to control inbound traffic is one of the essential security vulnerabilities of the present Internet. It is the consequence of the fundamental fact that the Internet was built as a highly distributed public network, in which every node may freely send arbitrary traffic to any other node. This vulnerability can be exploited by a variety of DoS attacks (with a volumetric DDoS attack being the most prominent example) or non-malicious phenomena like flash crowds. In this paper, state-of-the-art solutions aiming to mitigate these risks have been discussed and a novel proposal, On-demand Source Authentication and Authorization (OSAA), has been presented. OSAA does not target a particular threat but addresses the root cause of the vulnerability. The proposed architecture enables Internet end nodes to authenticate traffic sources and facilitates cost-effective filtering of unauthorized traffic. The solution is based on a capability-based security model and public key infrastructure. Key characteristics of OSAA are strong security of provided services and a viable business case with clear economic incentives for parties bearing the workload.

Keywords: Capability-based security · Denial-of-service attacks · Internet · Public key infrastructure

1 Introduction

Three elementary attributes defining security are confidentiality, integrity, and availability [1]. While the Internet does not assure any of them, there have been a number of techniques developed that provide confidentiality and integrity protection on top of the Internet Protocol (IP) layer, like Transport Layer Security (TLS) or Internet Protocol Security (IPsec). On the other hand, there are no universal measures protecting from attacks against availability. Currently existing solutions suffer from shortcomings with respect to modern, sophisticated denial-of-service (DoS) and distributed DoS (DDoS) attacks. Moreover, their cost is in many cases prohibitive.

The reason for this status is that the Internet relies on a destination-oriented routing and connectionless transmission, not providing any means to authenticate or authorize traffic sources. The black-hat community has a head start because it is moderately easy to build a *botnet* (a network of exploited hosts used for an attack) and DDoS attacks scale very well in opposite to defense systems. To make matters worse, finding the culprits of well-prepared, multi-tiered attacks exploiting *IP spoofing* (as described in [2]) is practically infeasible.

L. Borzemski et al. (Eds.): ISAT 2019, AISC 1050, pp. 219–230, 2020.
https://doi.org/10.1007/978-3-030-30440-9_21

Although researches already pointed in the late '90s that these intrinsic weaknesses of the Internet would pose a serious issue (and research concerning a related topic of QoS started even a decade earlier), the architecture of the Internet has not changed in that regard and the problem remains open. Neither DDoS attacks have faded. On the contrary: they have become more intensive in all dimensions: frequency, capacity, and complexity [3, 4]. A prominent example of this trend is the rise of the largest peak attack size from 24 Gb/s in 2007 to 309 Gb/s in 2013 to 1.7 Tb/s in 2018 [4].

In this paper, a novel solution, On-demand Source Authentication and Authorization (OSAA) has been proposed to address the root cause of the problem. OSAA explores new areas of the solution space and provides the unique set of properties:

- on-demand activation of the control mechanisms in a timely manner,
- preventative capability setup (proactive requests, pushing capabilities),
- insensibility to routing path changes,
- fast response to an enforcement decision (addressing *hit-and-run* [5] attacks),
- resistance to *denial-of-capability* [6] attacks.

Moreover, OSAA features implementation feasibility. The concept has a viable business case, allows gradual implementation and cooperation with legacy systems, and enables stateless, cost-effective traffic filtering.

2 Related Work

The related work may be divided into three areas:

- Ensuring accountability by identification of traffic sources. The problem is known as *IP traceback*. Identification as such does not build any defense but facilitates other techniques and is a prerequisite for legal prosecution, then acts as a deterrent.
- Attack detection, traffic classification (wanted/unwanted) and constructing signatures. This area is beyond the scope of the paper.
- Providing authorization for traffic sources (limiting unwanted incoming traffic).

The scope of this chapter is limited to the solutions that are currently in use and the groundbreaking ideas that brought the most attention of the research community.

The simple solution to solve the IP traceback problem is *ingress/egress filtering*, recommended in RFC 2827 [7]. The filtering should be performed by Internet service providers (ISPs) to ensure that traffic entering the public Internet uses only valid source IP addresses. Alas, many ISPs do not comply.

Currently employed techniques for IP traceback includes traffic tracking on subsequent nodes [8] or log analysis. These operations require manual intervention and cooperating with other ISPs (who in general case are not obliged to offer any support). In the case of distributed attacks, this approach is mostly infeasible.

Among novel techniques related to IP traceback, *packet marking* is a leading concept. It consists in adding to a packet header a mark identifying the path the packet traverses. In *deterministic packet marking* (DPM), each node appends its identifier to a processed packet. To solve the issues caused by a variable length of the mark, Shokri et al. proposed in [9] to limit the number of marking devices to edge nodes.

Probabilistic packet marking (PPM) presented first by Savage et al. in [10], uses a fixed-length mark that is overwritten by subsequent nodes with probability $p > 0.5$. After collecting enough packets, an end host can calculate the path from the distribution of the marks. Song and Perrig presented in [11] significant enhancements to the PPM scheme, but their solution still does not cope with highly distributed attacks involving thousands of sources. This pitfall is avoided by Pi (path identification mechanism) proposed in [12] by Yaar et al. However, the advantage of Pi was achieved by sacrificing accountability – the aim of Pi is only to distinguish flows on per path basis for filtering purposes.

IP blackholing (RFC 3882 [13]) is a simple technique widely used by ISPs to mitigate floods of unwanted traffic. In case of an attack, the victim requests (usually automatically) their upstream provider to inject a black hole route for an attacked IP address. Legitimate traffic is dropped as well, so the affected host is sacrificed to save the rest of the network. The method is not effective if the whole network has been attacked.

A more sophisticated approach is to use *sink-hole routing* to redirect all suspicious traffic to a *scrubbing center*, which performs filtering. Distributing scrubbing centers over the Internet and utilizing anycast addressing improves scalability but, on the other hand, the reaction time becomes an issue (caused by BGP convergence time), which is exploited by hit-and-run attacks. Lastly, this kind of protection is relatively expensive because specialized equipment and large bandwidth are required for scrubbing centers. This is not uncommon that a DDoS protection service for an Internet connection costs even more than the base connectivity service itself.

A problem akin to source authorization is *quality of service* (QoS) assurance. Although perspectives are different, the common goal is to ensure that the right traffic gets priority in an overload condition. Integrated Services (IntServ), described in RFC 1633 [14], guarantees QoS on a per flow basis. Because of many implementational concerns of IntServ (scalability, charging model, etc.), the focus has shifted towards Differentiated Services (DiffServ), described in RFC 2474 [15]. DiffServ operates on per traffic class basis, hence offers decent scalability. In respect to volumetric DoS attacks, DiffServ is only limiting the damage to an attacked class.

Among novel proposals, *pushback*, presented by Ioannidis and Bellovin in [16], is a promising concept. It consists in forwarding the information about an incident to upstream nodes, which are supposed to take action (limit or drop specified traffic) or further delegate the action towards the source. A distributed response of pushback provides scalability necessary to handle massive DDoS attacks. The drawbacks lay in (1) difficulties with describing sophisticated attacks in a standardized form, (2) high cost of traffic filtering or limiting in transit networks, in particular for complex signatures.

Another concept that attracted much attention of the research community employs a capability-based security model and was proposed first by Anderson et al. in [17]. Yaar et al. proposed SIFF [18], a stateless Internet flow filter, which features low computational and memory overhead. In this approach, traffic is classified by a destination node in two groups: privileged (packets with obtained capabilities) and unprivileged. FGC (fine-grained capabilities) [19] extend the number of classes allowing the prioritization of sources with different trust levels.

Argyraki and Cheriton coined the term *denial-of-capability* (DoC) pointing in [6] that network capabilities are susceptible to DoS attacks targeting the capability service itself. Portcullis [20] attempts to fix the vulnerability using a *proof of work* concept. Wang et al. argue [21] that support from source networks is necessary to eliminate DoC.

The taxonomy presented in [22] studies the solution space of capability-based DDoS defense architectures and identifies unexplored areas. Some of these areas, like on-demand packet marking and capability initialization, pushing a capability to a source, and susceptibility to routing path changes, are addressed by OSAA.

3 Problem Statement and Assumptions

The goal of this work is to propose a solution that enables recipients to authenticate senders and authorize the usage of their network bandwidth. These security features – missing in the current architecture of the Internet – allow easily mitigating detrimental effects of malicious (like DDoS) and non-malicious (like *flash crowds*) traffic.

The proposed solution should meet the following requirements:

1. Scalability to the size of the Internet.
2. Interoperability with legacy systems.
3. Significant effectiveness without full implementation in the Internet.
4. Support for IPv4 and IPv6.
5. A viable business case for all parties expected to implement the solution.
6. Strong security of the authentication and authorization services.
7. Enforcing authorization decisions in a timely manner.
8. Negligible impact on legitimate user traffic (throughput, packet loss, latency).

While most research proposals comply with requirements 1–4, the other four are often neglected and thus deserve elaboration. The importance of requirement 5 arises from the fact, that the Internet is composed of networks managed by independent companies, often competitors. Therefore, the decisions are not made on technical merits only, but a strong business case is needed in the first place. Ingress/egress filtering is a solid example underlining the validity of this point. Although complete implementation in the Internet could easily eliminate IP spoofing, many ISPs do not comply just because there is no value from their particular perspective (neither a penalty nor an incentive). In this context, the assumption (implicitly present in many proposals) that ISPs will implement much more complex techniques (marking, filtering, etc.) without any economic justification, is clearly unfounded. The lack of a valid business model makes a significant portion of proposed solutions infeasible in the real world.

Requirement 6 is crucial as well. Each novel system that requires collaboration introduces new interfaces and thereby new attack vectors. If the interfaces are not secure, the system may be not effective or even may facilitate attacks.

Requirements 7–8 are particularly important for techniques that require establishing some kind of a protected channel (capability-based systems fall into this category). These techniques tend to either impose an initial delay for each new flow [18, 19] (which is detrimental for interactive, transaction-oriented services) or face issues with activating the protection on-demand in a timely manner [20, 21] (which makes them vulnerable to hit-and-run attacks).

The solution proposed in this paper is based on the following assumptions:

1. Attackers may generate arbitrary packets.
2. Many attackers may cooperate.
3. Attackers know in-detail the principles of operation of the defense mechanisms.
4. A recipient (and only a recipient) is able to distinguish between sources of wanted and unwanted traffic, with the following reservations:
   a. Analysis of multiple packets may be required to classify a traffic source.
   b. A recipient has means to determine if a given source address is authentic.

4 The Design of OSAA

4.1 Network Model

In order to describe and further explore the proposed solution, the following network model is assumed: a tree rooted in *D* (destination), with nodes T_i (trusted) and U_i (untrusted), and leaves S_i (source), as depicted in an example in Fig. 1. Each vertex is identified by a 32-bit address.

A message (also referred to as a packet) transmitted in the network consists of a variable-length payload and a fixed-length header. The header contains source and destination nodes' addresses and a 16-bit field reserved for a *token* representing the capability. There are two types of packets: *user plane* (u-plane) and *control plane* (c-plane). U-plane packets are transmitted only from S_i towards D. A leaf S_i may transmit either wanted or unwanted u-plane traffic. C-plane packets carry OSAA control data. Packets may contain either a genuine or spoofed source address. An intermediate node (T_i or U_i) is responsible for traffic routing, i.e. forwarding packets to a destination indicated in a packet header or to another T_i/U_i node that is closer to the destination. T_i nodes perform additionally functions related to OSAA (e.g. token issuing/verification).

It is assumed that a public key infrastructure (PKI) is in place. D and S_i have public certificates (issued for their network addresses) and private keys capable to generate digital signatures. All certificates are issued and signed by globally trusted entities.

4.2 OSAA Description

OSAA specifies the following roles:

- Destination Node (D) – the protected node; D must implement OSAA,
- Trusted Node (T) – an intermediate node implementing OSAA and either (1) neighboring D and trusted by D or (2) neighboring and trusted by another T,
- Untrusted Node (U) – an intermediate node without an established trust relationship neither with D nor with any of T nodes,
- Source Node (S) – a node sending traffic to D; S may or may not implement OSAA.

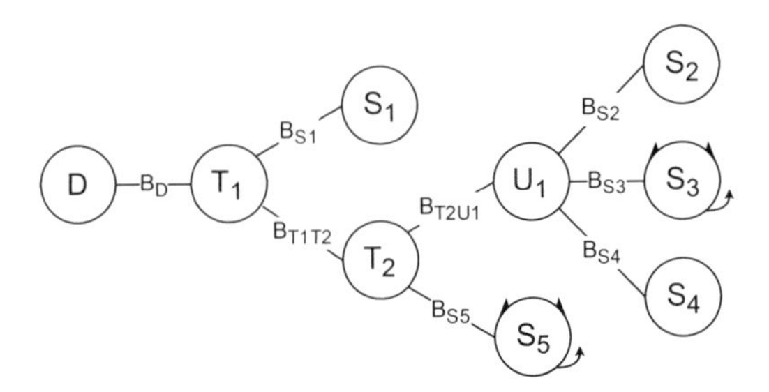

Fig. 1. The network is represented as tree $G = (V, E)$, where V is a set of vertices (representing network nodes) and E is a set of edges with assigned weights. Edges between nodes represent connections with fixed bandwidth (expressed in bits per second). It is assumed the bandwidth is equal and separate for both directions (it reflects symmetric full-duplex transmission). Source nodes S_3 and S_5 are malicious and send unwanted traffic.

OSAA defines the following message types:

- Authentication Request (AR) – a message sent by D or T (on behalf of D) to S informing that S is required to prove its identity and obtain a token,
- Authenticity Proof (AP) – a message from S to D containing a public certificate of S, timestamp (for replay-attack protection) and a digital signature of the message,
- Token Offer (TO) – a message from D to S indicating short-term authorization, TO contains a token (in the header), the token validity time, a public certificate of D, and the digital signature of the message,
- Key Offer (KO) – a message from D to S indicating long-term authorization, KO contains a *KS* key (that allows generating tokens), setup parameters, and the digital signature of the message,
- IR (Incident Report) – a message from D to S informing about malicious traffic received from S,
- Token Refresh (TR) – a message from S to D containing a valid token and informing that S needs a new token,
- Protection Control (PC) – a message from D or T to T controlling protection mechanisms, i.e. enabling/disabling filtering or enabling/disabling AR messages; PC contains an affected address or a range of addresses,
- PU (Parameters Update) – a message from D or T to T containing a *KM* key or parameters update.

Token validity is time-limited. The generation of a chain of tokens requires two keys (KM and KS). The master key (KM) is generated by D and is shared with T. Key KM is persistent. The secondary key (KS) is a cryptographic hash generated from key KM, S and D addresses, and counter *C1* based on the current date and time. Counter C1 is incremented with a fixed period equal to the validity time of KS. Key KS is directly used to generating tokens and may be shared by D with S.

The token is a 16-bit value consisting of a tag (1 bit) and a cryptographic hash (15 bits). The cryptographic hash is calculated from 4 inputs: source address, destination address, key KS, and counter *C2* based on the current date and time. The tag, denoted as *C2t*, allows changing the key seamlessly, as shown in Table 1. The token validity time equals the double value of the counter C2 period. Thus, the C2 period should be set to a value small enough to make attacks against a 15-bit token hash infeasible.

Table 1. Chain of tokens generated from master key *KM* in time range 0–15 [time unit]. *Src* and *Dst* are S and D network addresses, respectively. *H* is a hash function. ^ is a bitwise concatenation operator. Column "Valid tokens" indicates the time range when a token is valid.

Counter C1	Key KS	Counter C2	Tag C2t	Token	Valid tokens		Timer
$C1_0$	$KS_1 = H(C1_0, Src, Dst, KM)$	$C2_0$	0	T_0 = '0'^$H(C2_0, Src, Dst, KS_1)$	T_0		0
						T_1	1
		$C2_0+1$	1	T_1 = '1'^$H(C2_0+1, Src, Dst, KS_1)$			2
					T_2		3
		$C2_0+2$	0	T_2 = '0'^$H(C2_0+2, Src, Dst, KS_1)$			4
						T_3	5
		$C2_0+3$	1	T_3 = '1'^$H(C2_0+3, Src, Dst, KS_1)$			6
					T_4		7
$C1_0+1$	$KS_2 = H(C1_0+1, Src, Dst, KM)$	$C2_0 + 4$	0	T_4 = '0'^$H(C2_0+4, Src, Dst, KS_2)$			8
						T_5	9
		$C2_0+5$	1	T_5 = '1'^$H(C2_0+5, Src, Dst, KS_2)$			10
					T_6		11
		$C2_0+6$	0	T_6 = '0'^$H(C2_0+6, Src, Dst, KS_2)$			12
						T_7	13
		$C2_0+7$	1	T_7 = '1'^$H(C2_0+7, Src, Dst, KS_2)$			14
							15

D generates a token and shares it with S in a TO message. In such case, the remaining validity time is attached. This way it is ensured that S is equipped with capability valid at least for a token validity period and at maximum for a double token validity period. A token may be also generated by S. In this case D has to send KO including key KS, current tag C2t, C1 and C2 periods, and C1 and C2 remaining counter times.

Owing to overlapping validity times, the system tolerates a clock drift and timing skew not exceeding a half of C2 period. Because in each point of time two valid tokens exist, tag C2t is used to distinguish which token was used by S.

4.3 Typical Defense Scenario

A typical scenario of an incident and the OSAA response comprises the below-mentioned steps. The incident used in the example consists in flooding node D with unwanted traffic from multiple rogue S nodes, while other, legitimate S nodes are trying to communicate with D at the same time. The flooding, in the initial phase, progressively gains momentum. Figure 2 shows the c-plane communication sequence of the scenario using the network described in Sect. 4.1 above.

1. (Optional) S nodes may proactively send AP along with their first message in a flow[1]. In such cases, D responds with TO or KO to trusted (legitimate) S nodes.
2. D detects the incident in its initial phase (unwanted traffic hits a warning threshold).
3. D transmits KO messages to trusted S nodes (pushes capabilities).
4. Trusted S nodes that support OSAA start marking their messages with valid tokens.
5. D decides that filtering of unwanted traffic is required (critical threshold reached).
6. D issues PC to neighboring T nodes to enable traffic filtering (discard messages without a valid token). Discarded messages are responded with AR.
7. T may further delegate its tasks by sending PC to its trusted T nodes (towards S).
8. S that supports OSAA and does not have a capability responds to AR with AP.
9. T nodes verify AP messages and forward the valid ones to D. T limits the rate of forwarded AP messages per an S node, which mitigates potential DoC attacks.
10. D decides whether to respond or not to AP. The response may be TO (short-term authorization), KO (long-term authorization) or IR (reject).
11. S sends a TR message if the validity time of its token is about to expire and a new token is needed.
12. D decides whether to respond or not to TR. The response may be a TO (short-term authorization), KO (long-term authorization) or IR (reject).

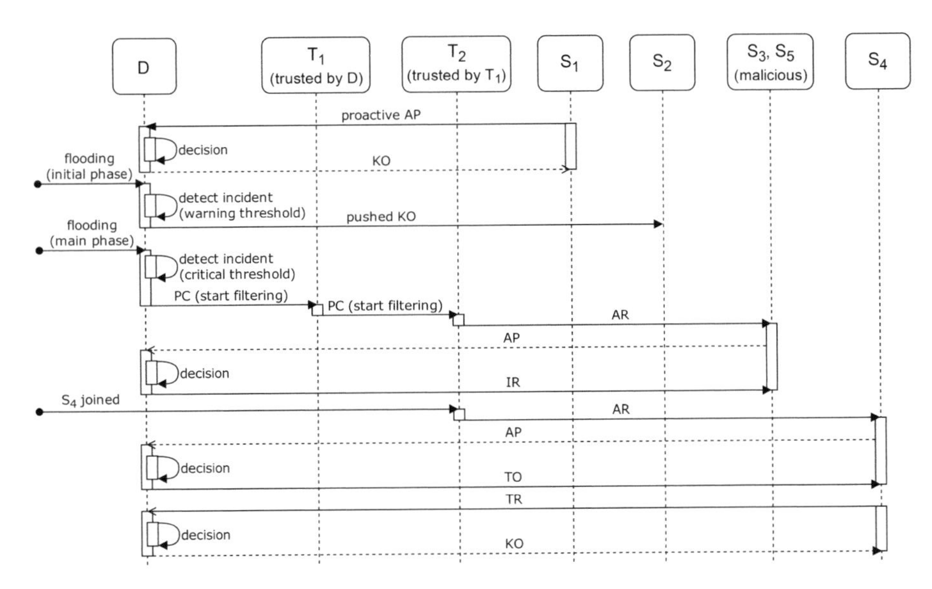

Fig. 2. Example of a defense scenario. Trusted nodes S_1 and S_2 communicate with D before the attack starts. S_1 gets a token using proactive AP, S_2 benefits from the pushing technique. In a result, both S_1 and S_2 gain capabilities before the flooding momentum reaches a critical threshold. Trusted node S_4 starts the transmission while D is under attack, then it requires authorization before it can reach D with u-plane traffic.

[1] A *flow* is defined by a pair of network addresses (source and destination).

4.4 OSAA Security Analysis

Confidentiality. OSAA does not provide confidentiality for c-plane, which potentially allows an eavesdropper to obtain valid tokens with no authorization. Compared to a no-defense scenario, an adversary intending to cause DoS is required to adopt additional measures to perform eavesdropping of victim's traffic, which is considered much harder than flooding itself. To some extent, the issue is mitigated by limiting the lifetime of tokens.
Confidentiality for OSAA c-plane can be ensured utilizing the public-key cryptography. A promising approach is to replace explicit tokens with hash message authentication codes (HMACs) generated upon secret tokens. Further research is needed to verify the performance impact of this scheme.

Integrity. Integrity of sensitive c-plane messages is protected by digital signatures. Integrity is not assured, for (1) communication between trusted neighboring nodes, (2) AR, IR, TR messages. The risk attached to this vulnerability is considered low.

Digital signatures can be employed for all c-plane message types, but pros and cons need to be weighed carefully to make an adequate trade-off between performance and security.

Availability. Availability is a critical attribute for a DoS-mitigating system. Like any capability-based solution, OSAA is subjected to DoC. As noticed in [20], best the system can do under flooding DoC attack (trying to saturate the network bandwidth) is to share the bandwidth equally between all requesters. Since OSAA ensure source authentication, only nodes with non-spoofed addresses may attempt to flood D with AP messages, which makes solving the problem trivial, yet performance impact of the signatures' verification of attack's packets is not negligible and is an open field for further research. OSAA proposes rate filtering on per source address basis, but round-robin scheduling would be a possible approach as well.

The system is protected against replay attacks by including a time-related parameter (counter C2) in the token generation process.

The resistance to brute-force attacks against a token is based on a 15-bit hash value. A token is valid for double C2 period time. The C2 period is intended to be a configurable parameter, so it needs to be specified during the deployment phase considering the victim bandwidth and assumed adversary bandwidth for a single malicious source.

The system does not offer protection against an authorized S node that converted to a malicious one (or was incorrectly classified). In such a situation, the node can flood D until the token expires. Extending OSAA with revocation features is currently under the author's investigation.

Other DoS attack types involve overloading CPU or memory resources of D or T nodes. Performance characteristics of OSAA, which are currently being investigated, are essential to assess the risk of such attacks.

4.5 Time Characteristics Analysis

In terms of u-plain delays, OSAA has no impact unless the protection has been activated. This is an important property because in the real world the Internet end nodes for most time do not need the protection.

If the protection is active, the delay of capability establishment will be imposed (refer to the communication of S_4 in Fig. 2). To reduce the delay, the capability setup may be piggybacked e.g. in TCP three-way handshake, as other researchers suggest [6, 20].

OSAA is considered to provide activation of the control mechanisms in a timely manner, because of the following features:

- Proactive capability requests and pushed capabilities allow obtaining a capability in advance. Enabling protection does not affect the traffic with granted capabilities.
- Lightweight process of granting capabilities. No communication with external systems is needed as opposed to other capability-based proposals that require e.g. DNS queries [20]. Standard hashing/signing operations are sufficient for OSAA, which has lower computation costs than proof of work schemes [20].
- The victim does not need to change its network address when ten protection is activated (as opposed to PATRICIA [21] and related proposals).

5 Implementation Considerations

OSAA assumes that a trust relationship is established between neighboring nodes. This may be ensured fairly easily because neighboring ISPs either have a contractual relationship or communicate via an IX (Internet exchange point) whose operator may act as a mediator (similarly to the way BGP peering is facilitated).

OSAA assumes also that a PKI is already in place. Regional Internet Registries (RIRs) are expected to act as Certification Authorities (CAs). The changes towards this idea have been already initiated in recent years. For example, as of today RIPE NCC (one of five RIRs) issues standard X.509v3 certificates for certain types of address blocks and extending of this service is planned [23].

Regarding the token placement in IPv4 or IPv6 packet headers, Identification Field (16 bits) or Flow Label (20 bits) are to be used, respectively.

OSAA provides a valid business case for all involved parties:

- For destination end hosts, OSAA offers anti-DoS features.
- Regarding transient nodes, it is assumed the cooperation will base on business contracts between neighboring operators (OSAA as an add-on service).
- For source hosts, implementing OSAA provides privileged access to resources in case the destination end suffers from DoS or a flash crowd. Additionally, reports about attacks coming from own network (IR messages) support identification of infected machines, thus provide an opportunity to fix security vulnerabilities.
- Critical systems may mandate OSAA for all connecting hosts, thus enhancing the service availability. This is valid in particular for closed systems that use the Internet as a communication medium.

OSAA design has the following limitations:

- Coarse level of granularity (capabilities assigned on per host basis). It is not possible to drop only certain traffic of a given host. It is believed that this limitation is not practically significant because groups of legitimate users and attackers are mostly disjoint. Further research is needed to evaluate this assumption.
- OSAA is not compatible with network address translation (NAT).

6 Future Work

The following challenges and issues are to be studied in the future:

- Ensuring confidentiality for OSAA messages.
- Capability revocation feature.
- Packet-limited (instead of time-limited) tokens for short-term authorization.
- Whitelisting of trusted legacy sources.

There is also a need to verify the end-to-end performance characteristics and compare OSAA with other solutions. Because of the architectural diversity of systems proposed by the research community, a common platform is required to perform a credible comparative analysis. The benchmarking present in many papers (like [20]) does not consider computational and memory costs for different components of the system, which may be significantly different for different architectures.

7 Summary

In this work, security vulnerabilities of the Internet in terms of availability have been elaborated. Contemporary solutions for this problem as well as state-of-the-art research proposals have been analyzed. The paper has presented OSAA, On-demand Source Authentication and Authorization. The design is based on a capability-based security model.

The main contributions of the work are (1) exploration of pushed capabilities, (2) a practical application of public key infrastructure to solve DoC problem, (3) presenting the holistic solution with a unique set of features regarding performance, security, and implementation feasibility.

Acknowledgments. The author would like to thank Professor Grzegorz Kolaczek for his insightful comments and technical assistance. Words of gratitude are owed also to Nokia Networks for financial support of the research.

References

1. ISO/IEC: International Standard 27000. Information technology – Security techniques – Information security management systems – Overview and vocabulary (2016)

2. Hoque, N., Bhattacharyya, D.K., Kalita, J.K.: Botnet in DDoS attacks: trends and challenges. IEEE Commun. Surv. Tutorials **17**(4), 2242–2270 (2015)
3. Akamai Technologies: State of the Internet/Security Report. Q1 2017 (2017)
4. NETSCOUT Systems: NETSCOUT's 14th Annual Worldwide Infrastructure Security Report (2018)
5. Imperva: The Top 10 DDoS Attack Trend (2015)
6. Argyraki, K., Cheriton, D.: Network capabilities: the good, the bad and the ugly. In: Proceedings of the ACM Workshop on Hot Topics in Networks, College Park, ACM (2005)
7. IETF: RFC 2827 - Network Ingress Filtering: Defeating Denial of Service Attacks which employ IP Source Address Spoofing (2000)
8. Cisco Systems: Characterizing and Tracing Packet Floods Using Cisco Routers. http://www.cisco.com/c/en/us/support/docs/security-vpn/kerberos/13609-22.html. Accessed 02 May 2019
9. Shokri, R., Varshovi, A., Mohammadi, H.: DDPM: Dynamic Deterministic Packet Marking for IP traceback. In: 14th IEEE International Conference on Networks. IEEE, Singapore (2006)
10. Savage, S., Wetherall, D., Karlin, A., Anderson, T.: Practical network support for IP traceback. In: Proceedings of the Conference on Applications, Technologies, Architectures, and Protocols for Computer Communication, ACM, Stockholm (2000)
11. Song, D.X., Perrig, A.: Advanced and authenticated marking schemes for IP traceback. In: Proceedings on Twentieth Annual Joint Conference of the IEEE Computer and Communications Societies, INFOCOM 2001. IEEE, Anchorage (2002)
12. Yaar, A., Perrig, A., Song, D.: Pi: a path identification mechanism to defend against DDoS attacks. In: Proceedings of the Symposium on Security and Privacy. IEEE, Berkeley (2003)
13. IETF: RFC 3882 - Configuring BGP to Block Denial-of-Service Attacks (2004)
14. IETF: RFC 1633 - Integrated Services in the Internet Architecture: an Overview (1994)
15. IETF: RFC 2474 - Definition of the Differentiated Services Field (DS Field) in the IPv4 and IPv6 Headers (1998)
16. Ioannidis, J., Bellovin, S.: Implementing pushback: router-based defense against DDoS attacks. In: Network and Distributed System Security Symposium. Proceedings, Internet Society, San Diego (2002)
17. Anderson, T., Roscoe, T., Wetherall, D.: Preventing Internet denial-of-service with capabilities. ACM SIGCOMM Comput. Commun. Rev. **34**(1), 39–44 (2004)
18. Yaar, A., Perrig, A., Song, D.: SIFF: a stateless Internet flow filter to mitigate DDoS flooding attacks. In: Proceedings of the IEEE Symposium on Security and Privacy. IEEE, Berkeley (2004)
19. Natu, M., Mirkovic, J.: Fine-grained capabilities for flooding DDoS defense using client reputations. In: Proceedings of the 2007 Workshop on Large Scale Attack Defense. ACM, Kyoto (2007)
20. Parno, B., Wendlandt, D., Shi, E., Perrig, A., Maggs, B., Hu, Y.C.: Portcullis: protecting connection setup from denial-of-capability attacks. ACM SIGCOMM Comput. Commun. Rev. **37**(4), 289–300 (2007)
21. Wang, L., Wu, Q., Dung, L.D.: Engaging edge networks in preventing and mitigating undesirable network traffic. In: 3rd IEEE Workshop on Secure Network Protocols. IEEE, Beijing (2007)
22. Kambhampati, V., Papadopoulos, C., Massey, D.: A taxonomy of capabilities based DDoS defense architectures. In: 9th IEEE/ACS International Conference on Computer Systems and Applications (AICCSA). IEEE, Sharm El-Sheikh (2011)
23. FAQ: Certification – RIPE Network Coordination Centre. https://www.ripe.net/manage-ips-and-asns/resource-management/faq/certification. Accessed 05 Oct 2019

Analysis of Blockchain Selfish Mining Attacks

Michal Kędziora(✉), Patryk Kozłowski, Michał Szczepanik, and Piotr Jóźwiak

Faculty of Computer Science and Management, Wroclaw University of Science and Technology, Wroclaw, Poland
michal.kedziora@pwr.edu.pl

Abstract. The paper presents vector of attack on the mechanism of achieving a proof of work consensus, such as the application of selfish mining strategy. The aim of the work was to analyze to what extent these risks may affect the fairness of the cryptocurrency extraction process. This goal was achieved by performing simulations using appropriate models. The aim of the experiment was to check how the choice of parameters affects the distribution of block extraction between network users. We use an algorithm of persistent mining in the pursuit of a longer chain with a modified selfish extraction algorithm.

Keywords: Blockchain · Selfish mining · Cryptocurrency · Proof of work

1 Introduction

The Bitcoin protocol requires that the majority of networks be fair, i.e. it adheres to the cryptocurrency protocol in a way that does not diverge from the implemented one. In a situation when most nodes join the group and start cooperating with each other, the cryptocurrency ceases to be decentralized and begins to be controlled by miners who are in collusion with each other [1]. It is important, therefore, that the consensus algorithm be designed in a way that discourages nodes from diminishing in order to gain benefits [8, 9]. Unfortunately, experience shows that in the case of the proof of work algorithm, there is a threat of miners collusion in order to achieve the greatest benefits. In order to reduce the variance of their income, as well as due to randomness and uncertainty of making a profit, individual nodes extracting blocks merge into larger groups called mines [7]. As part of the mine, all miners have their contribution to solving the cryptographic problem, which is finding the right value of nonce, and the reward for their effort is to receive a virtual currency, in proportion to the work of the entire mine. The block chain is vulnerable to attacks by miners who are in collusion. It is widely believed that a successful attack on a network of cryptocurrencies should have 51% of the computing power, however, even nodes controlling a smaller part of the resources can be dangerous and pose a threat to the safety of the blockchain [2].

Selfish mining is a strategy of generating income from block mining in a way that deviates from the standard protocol, contributing to an increase in mine income, rather than the amount of work done for the network [11]. As a result of applying this strategy, fair nodes that adhere to the cryptocurrency protocol use their resources

L. Borzemski et al. (Eds.): ISAT 2019, AISC 1050, pp. 231–240, 2020.
https://doi.org/10.1007/978-3-030-30440-9_22

aimlessly. The set of strategies that allow such action can be called malicious extraction. In this work we present simulation and analysis of selfish mining attacks.

2 Blockchain Selfish Mining

The cryptocurrency system consists of a set of miners 1, 2, 3..., n. Each of the miners has computing power m_I equal to that of $\sum_{1}^{n} m_i = 1$ and each of them chooses a block for which he will perform a specific job. In addition, each of the miners operates in a rational way, i.e. trying to maximize their income. The strategy of selfish mining allows to extract more blocks than it would result from the proportional power of the mine. The main aspect of the strategy is forcing honest nodes to do purposeless calculations on an obsolete, public block chain for blocks that will not be eventually included. Achieving this goal is based on the selective publication of individual blocks by an unfair mine, in order to discredit the work done by the rest of the network. The mine keeps the extracted block secret from the entire network, simultaneously creating branching chains and creating a private copy while the rest of the network is working on a public, shorter version. The consequence of such action is the futility of calculations made by the honest part of the network at the moment when the dishonest person decides to make his block public. However, such a state of affairs cannot persist indefinitely, because dishonest miners have only a part of the network's mining power, which prevents them from arbitrarily managing the chain, so there is also a need to abandon work on the private chain and start working on the public version when only it will last longer [13].

On the basis of the selfish extraction algorithm, it can be noted that decisions are made on the basis of events related to the formation of a block, both extracted by the dishonest mine and the rest of the network. In a situation where the public chain of blocks is longer than the private one over which the mine works, it becomes useless due to the power difference between a fair and dishonest network. This results in a poor chance of chasing off the public chain, therefore the public version is adopted and working on it in hiding in the hope of obtaining disproportionately higher than normal incomes. However, in a situation where a dishonest mine finds a block, it is in a favorable position in relation to the remaining, fair part of the network. Instead of spreading the newly mined block immediately, the selfish extraction algorithm holds it only on the private version of the chain. At this moment, two scenarios are possible:

In the first case, when a dishonest mine increases its advantage, finding another block (two blocks or more advantage over a public blockchain), it has a comfortable position, allowing free broadcasting of the first private block of private chains to anyone found by the honest part of the network. The advantage at some point will melt to one block due to the power disparity between the dishonest mine and a fair part of the network, making the length of the public chain dangerously approaching the private version. In this situation, the mine decides to make its entire chain public and the state of the cryptocurrency system returns to the initial one when there is only one version of the chain.

In the second case, an honest part of the network extracts the block and eliminates the advantage of an inaccurate mine, making it public its private chain. This can lead to a situation where there is no certainty as to which chain will become the main and Work will be undertaken by honest miners, as the dishonest part of the network will continue to work on its private version in the hope of creating a block and gaining an advantage. If you gain an advantage over the public chain, the mine will receive income from the creation of two blocks. However, if it is the first block that is found in favor of a public blockchain, then it will receive income from two blocks, while the mine will not receive any income [1].

2.1 A Selfish Blockchain Mining Model

The theoretical analysis of the selfish extraction strategy is based on breaking it into several cases, which after the merger show the operation in all possible scenarios. The result is an equation describing the expected income of an unfair mine depending on the percentage share in the power of the network, as well as on how much of the honest community will work on the chain received from the selfish miners.

$$R_{pool} = \frac{r_{pool}}{r_{pool} + r_{others}} = \cdots = \frac{\alpha(1-\alpha)^2(4\alpha + \gamma(1-2\alpha)) - \alpha^3}{1 - \alpha(1 + (2-\alpha)\alpha)} \tag{1}$$

R_{pool} is the ratio of blocks extracted by the dishonest mine to all extracted blocks,
r_{pool} is the income of selfish miners,
r_{others} is the income of the rest of the network,
α is the share of mine power in the total power of the network,
$(1 - \alpha)$ is the remaining power of the network,
γ is part of honest nodes that decide to work on a chain originating from an unfair mine, $(1 - \gamma)$ is the remaining part of honest miners, performing work on a chain version originating from honest nodes.

A simulation can be created that allowed for experimental verification of the strategy. In a paper of Eyal and Siter [1], it can be seen that the simulation results coincide with the theoretical predictions, as well as the fact that it is justified to use the selfish strategy of mining in place of the standard protocol above certain threshold values γ.

3 Experiments

The research used a previously used environment in our paper [3] to simulate the cryptocurrency extraction process based on the proof of fork consensus mechanism. Appropriate changes were introduced regarding the method of unfair mining by a selfish miner who is represented as one user of the cryptocurrency network, which corresponds to the situation when the mine has a central coordinator, while the rest of the network is a cooperating, honest coal mine. A simulation environment was written in C# 6.0 on Microsoft .NET version 4.5 platform, based on the technical specification the implementation of the Bitcoin client [4]. Each of the network nodes performed work in the search for the value of the hash function corresponding to the corresponding one assumption about the difficulty of finding a block, and its quantity could be easily chosen because of the multi-threaded architecture. All statistical tests and charts are generated in the R-Studio version 1.0.136 using the language R.

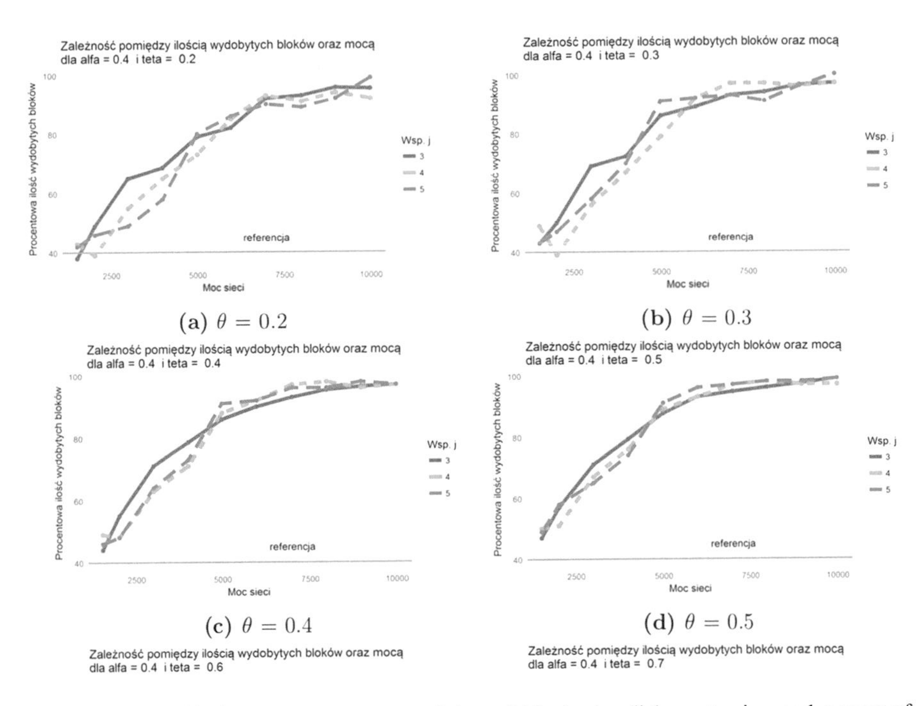

Fig. 1. Relationship between percentage mining of blocks by dishonest mine and power of network extraction for $\alpha = 0.45$

The aim of the experiment is to check how the choice of parameterd affects the distribution of block extraction between network users. We use an algorithm of persistent mining in the pursuit of a longer chain [5] with a modified selfish extraction algorithm [1], which fits into the family of hybrid strategies, combining several approaches, depending on the situation in which a dishonest mine is located in relation to the rest of the network. Parameters that changed during the simulation are: α - percentage share of dishonest mine power in the entire network capacity, j - difference in the length of the public and private chain to which it is unfair the mine will try to catch up with the public chain, θ-indicator of connection with the fair part of the network, expressed as a percentage of connected nodes, P - network power expressed in hash per s.

The simulation was carried out for one dishonest mine that was represented by the central coordinator as one user in the cryptocurrency network. For all the tests carried out, the number of nodes in the network n = 50 and the difficulty of extraction D = 3.

In addition, a change was introduced to the original Bitcoin protocol, whereby the dishonest mine did not transmit messages from honest nodes. The tests were carried out for the following variable values:

$P = \{1500, 2000, 3000, 4000, 5000, 6000, 7000, 8000, 9000, 1000\}$
$\theta = \{0.2, 0.3, 0.4, 0.5, 0.6, 9, 0.7, 0.8, 0.9\}$
$\alpha = \{0.25, 0.3, 0.35, 0.4, 0.45\}$

For each of the graphs, a line with a reference value for the percentage value of extracted blocks was plotted, which should result from the proportional share of dishonest mine in the power of the network.

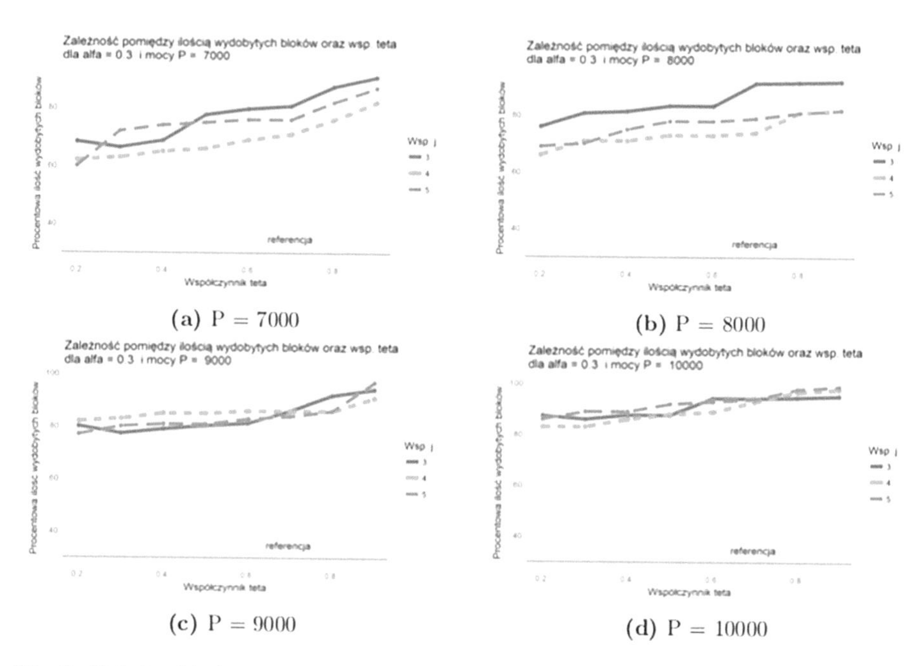

(a) P = 7000 (b) P = 8000

(c) P = 9000 (d) P = 10000

Fig. 2. Relationship between the percentage mining of blocks by the dishonest mine and the coefficient θ for $\alpha = 0.3$

The First simulated case was to analyze the relationship between the power of the network and the number of blocks extracted. Simulation results were collected for various parameters α, θ, P in the form of graphs of dependencies between the power of the network and the number of blocks extracted by the dishonest mine depending on the percentage share in the power of the network (Fig. 1). An orange reference line was added to indicate the number of blocks extracted, which results from the honest observance of the protocol and is directly proportional to the share of dishonest mine power in the total power of the network.

Second part of simulations where performed to research the relationship between the coefficient θ being excavated and the percentage of blocks. Simulation results were collected for various parameters α, θ, P in the form of diagrams of relationship between the connection factor with a fair share of the network and the number of extracted blocks by an unfair mine depending on the percentage share in the network power (Fig. 2). The reference line indicating the number of extracted blocks, which results from honest protocol compliance, has been added in orange, and is directly proportional to the share of dishonest mine power in the total power of the network.

3.1 Analysis of the Results

Analyzing the malicious extraction algorithm used in this work, one can notice the impact of the j parameter selection on the results obtained by the dishonest mine, however, it is impossible to clearly determine which one would best be executed regardless of the situation. It can only be determined which parameter j will be the best for the given parameters α, θ and the computing power of the P network.

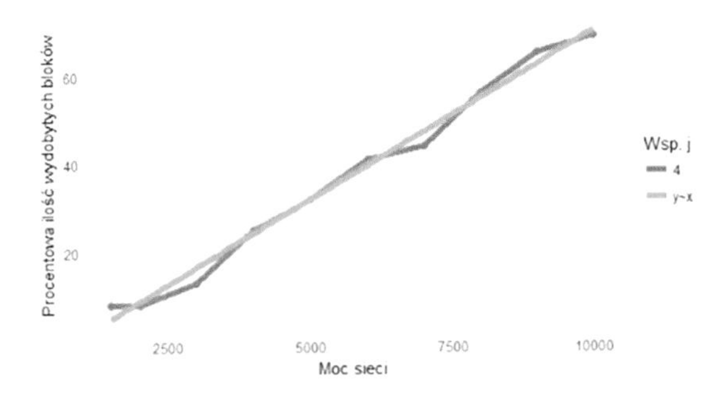

Fig. 3. Adjustment of the line to the relationship between the number of extracted blocks and the power of the network for $\alpha = 0.25$ and $\theta = 0.3$.

In the part concerning the relationship between the power of the network and the number of extracted blocks, one can notice two tendencies for the course of the curves depending on the parameter α. For the value $\alpha = \{0.25, 0.3\}$ there is a direct relationship between the power output and the number of extracted blocks (Fig. 3), which was noted in the case of the analysis of the possibility of a Sybil attack [3]. However, with the increase of the dishonest share of the mine in the total power of the network, this trend has changed into a logarithmic one (Fig. 4). The reasons for this state of affairs are to be found in the increasing number of branches of the block's chain, resulting from the low difficulty in extracting the block and the increasing power of the network.

The implementation of the dishonest mine operation assumed that it would not participate in solving emerging branches, all the while working on a private copy, however, the public part of the network had entered the protocol in accordance with the protocol, so its power was divided between branchings from a fair mine also produced accidentally by honest miners.

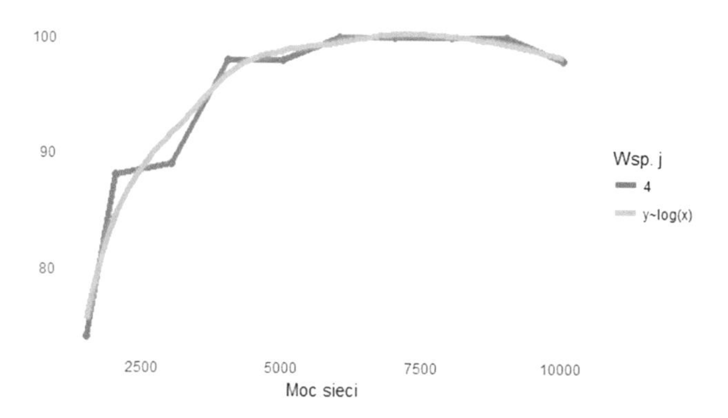

Fig. 4. Adjustment of the logarithmic curve to the relationship between the number of extracted blocks and network power for $\alpha = 0.45$ and $\theta = 0.9$

In order to verify the hypothesis concerning the distribution of fair grid power between existing branches, the average time to find a solution for the block was checked depending on the power of the network (Fig. 5).

The correlation value for the average solution finding time for the block and the number of resulting branches is r = −0.81, which indicates a strong inverse correlation between them. This shows that as the power of the grid increases, the emergence of branching is an increasingly frequent occurrence. For network power above 5000 hash average the time needed to find a block solution is less than 1 s, which means that at least 2 blocks should be found on each block chain length, but for data from simulation it is an average of 1.5 blocks. However, it should be taken into account that when a block is found by one of the users, the remaining part of the network stops working on its current block and starts working on the new one. This explains the existing state of affairs of the formation of a smaller number of branches, than it would result from the average time needed to find the appropriate value of the shortcut function depending on the power of the network.

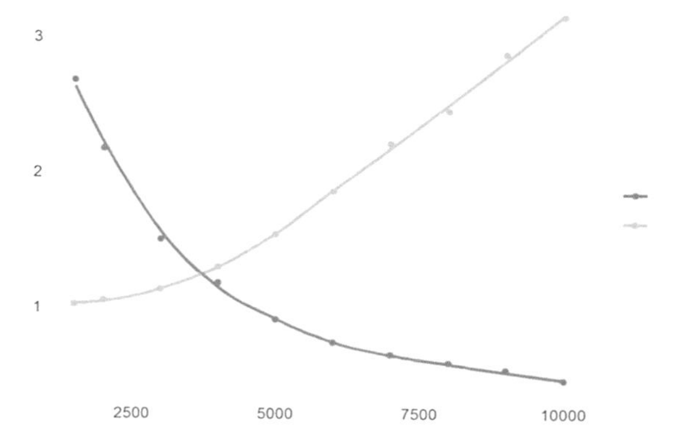

Fig. 5. Chart showing the average number of chain branches and the average time needed to solve the block for difficulties D = 3 depending on the power of the network. The mean amount should be understood as the number of concurrent blocks at one chain height during simulation.

With the increase of θ for the constant α parameter, the power of the network decreases, for which it exceeds the reference line resulting from the proportional share of dishonest mine power in the total network capacity, however, for the first time independently of P and θ power, used in this paper algorithm, scores better than the reference values for $\alpha = 0.45$, while for $\alpha = 0.4$ for $\theta > 0.3$. In addition, for the parameter $\alpha = \{0.25, 0.3.0.35\}$, the best algorithm was to use j = 3, while the increase in the share of dishonest mine in the total power of the network can not be distinguished by a specific value. However, it can be noticed that for $\alpha = \{0.4, 0.45\}$ a higher slope to the horizontal axis for j = {4, 5} compared to j = 3, which means that the change in the value of θ has a greater impact on them, which is important information when it is not possible to model the parameter θ and it is at an unknown level. In this case, the adversary may be tempted to use the j = 3 parameter as the safest and most effective in an unknown network environment (Fig. 6).

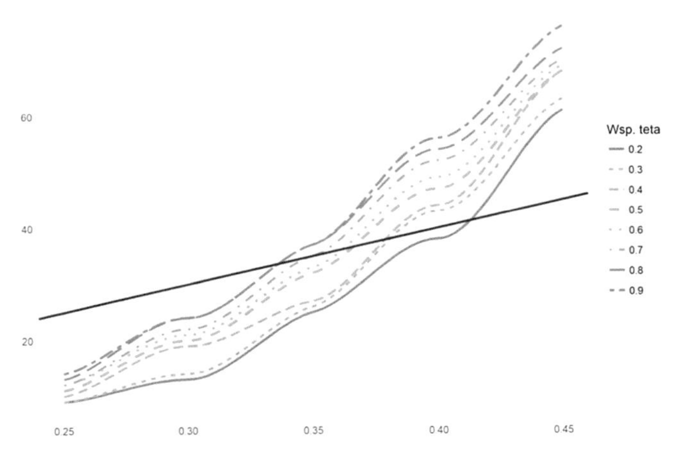

Fig. 6. Comparison of the achievements of the algorithm used in the work for different values θ for power 1500 hash/s and j = 3

In the part, where the dependence of parameter θ on the number of extracted blocks was examined, it was shown that it is directly proportional, regardless of the power of the network and the share of unfair mine power in the power of the entire α network. For the smaller network powers P = {1500, 2000, 3000, 4000}, the best strategy is to use the j = 3 parameter. As the total power increases, it is not possible to clearly identify the best-performing strategy depending on the decision a concrete application should be taken, knowing the other parameters α, P and θ.

Similarly to the previous situation, when the dependence of network power on the number of extracted blocks was examined, as now, the apparent dependence can be explained by the number of branches created depending on the power of the network - at the moment of obtaining an advantage by an unfair mine, the number of blocks that must be abandoned through the fair part of the network is j, which results in a situation in which a considerable amount of computational resources used to find rejected blocks is wasted.

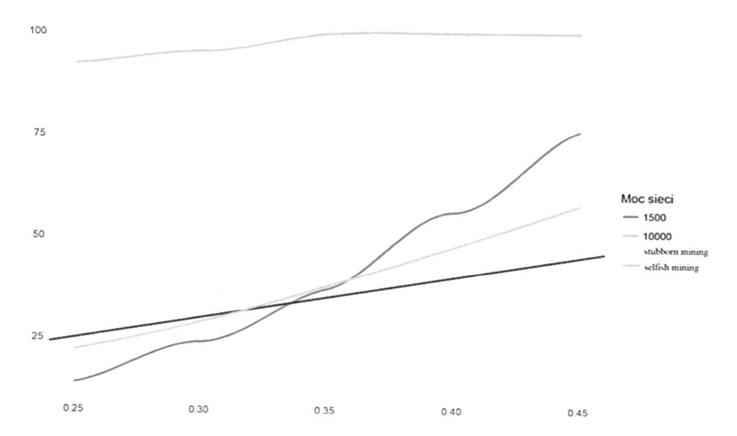

Fig. 7. Comparison of the results obtained in this work for the best representative data for j = 3, P = 1500, $\theta = 0.9$ and the best results obtained in work for P = 10,000, $\theta = 0.9$, $\alpha = 0.9$, j = 4 compared to selfish extraction [1], optimal selfish extraction [6] and persistent extraction [5] for $\gamma = 1$. The black reference line is the reference income line resulting from the honest observance of the cryptocurrency protocol

4 Summary

The paper presents vector of attack on the mechanism of achieving a proof of work consensus, such as the application of selfish mining strategy. The aim of the work was to analyze to what extent these risks may affect the fairness of the cryptocurrency extraction process. This goal was achieved by performing simulations using appropriate models. As a result of the conducted research, it was found that both the underlying mechanisms of cryptocurrencies, such as blockchain or consensus algorithms are not 100% resistant to the malicious activity of cybercriminals, which is also acknowledged by their creators, also indicating the possibility of attacks: attack 51% [7], nothing at stake [10], long range attack, or deanonymization of cryptocurrency network users on the basis of tracking their transactions [12, 14, 15]. Particularly noteworthy, however, are threats that are not widely considered, and were the subject of research, which include attack such as use of malicious extraction strategies.

Compared to the previously obtained results [1, 5, 6], the results obtained in this work, for a representative group of results, cannot be taken directly due to the inability to trace the γ parameter. This is due to the fact that the information about taking up work was not included in the implementation of the tracking model on a block from a particular node, so you can not verify how the value of parameter γ of the fair part of the network, which started working on the block, changed coming from the mine. For this reason, it was decided to compare the best results obtained for values: P = 1500 hash due to the smallest amount of s branching that arose during the simulation and $\theta = 0.9$ due to obtaining the best results at the given power (Fig. 7). Additionally, the results for P = 10000, $\theta = 0.9$, $\alpha = 0.9$, j = 4 were summarized, because they were the best results, without taking into account the fact of branching, in order to show how changes in the number of arising branches influence the algorithm's operation. The results presented for the selfish and persistent extraction algorithm were presented for $\gamma = 1$, for which the results obtained by them were the best. It can be noticed that the algorithm presented in this work in comparison to the self-extracting algorithm [1] works better for the value of $\alpha > 0.37$ and as the relative amount of blocks extracted by the dishonest mine approaches the results achieved by [5, 6], which were presented as one course marked as “stubborn mining”, because their results in the presented works did not diverge from each other. What is visible, however, is the significant influence exerted by the formation of numerous branches on the income of dishonest miner, which is disproportionately larger than in the situation when they do not exist.

References

1. Eyal, I., Sirer, E.G.: Majority is not enough: Bitcoin mining is vulnerable. Commun. ACM **61**(7), 95–102 (2018)
2. Zheng, Z., Xie, S., Dai, H.N., Wang, H.: Blockchain challenges and opportunities: a survey. Work Pap.–2016 (2016)
3. Kozlowski, P., Kedziora, M., Marianski, A., Jozwiak, I.: Implementation of Research on the Feasibility of a Sybil Attacks in Blockchain Networks, ICSAI 2019 – in review (2019)
4. Bitcoin Cash. Bitcoin Cash. https://www.bitcoincash.org/
5. Nayak, K., et al.: Stubborn mining: generalizing selfish mining and combining with an eclipse attack. In: 2016 IEEE European Symposium on Security and Privacy (EuroS P), pp. 305–320 (2016). https://doi.org/10.1109/eurosp.2016.32
6. Sapirshtein, A., Sompolinsky, Y., Zohar, A.: Optimal selfish mining strategies in Bitcoin. CoRR abs/1507.06183 (2015). arXiv:1507.06183
7. Nakamoto, S.: Bitcoin: A peer-to-peer electronic cash system (2008). http://bitcoin.org/bitcoin.pdf
8. Fischer, M.J., Lynch, N.A., Paterson, M.S.: Impossibility of distributed consensus with one faulty process. J. ACM **32**(2), 374–382 (1985). https://doi.org/10.1145/3149.214121. http://doi.acm.org/10.1145/3149.214121, ISSN 0004-5411
9. Lamport, L., Shostak, R., Pease, M.: The Byzantine generals problem. ACM Trans. Program. Lang. Syst. **4**(3), 382–401 (1982). https://doi.org/10.1145/357172.357176. http://doi.acm.org/10.1145/357172.357176, ISSN: 0164-0925
10. Li, W., et al.: Securing proof-of-stake blockchain protocols. In: Garcia-Alfaro, J., et al. (eds.) Data Privacy Management, Cryptocurrencies and Blockchain Technology, pp. 297–315. Springer, Cham (2017). ISBN: 978-3-319-67816-0
11. Olfati-Saber, R., Fax, J.A., Murray, R.M.: Consensus and cooperation in networked multi-agent systems. Proc. IEEE **95**(1), 215–233 (2007). https://doi.org/10.1109/jproc.2006.887293, ISSN: 0018-9219
12. Quesnelle, J.: On the linkability of Zcash transactions. In: CoRR abs/1712.01210 (2017). arXiv:1712.01210
13. Zheng, Z., et al.: Blockchain challenges and opportunities: a survey, December 2017
14. Fanti, G., Viswanath, P.: Deanonymization in the Bitcoin P2P network. In: Guyon, I., et al. (eds.) Advances in Neural Information Processing Systems 30, pp. 1364–1373. Curran Associates, Inc. (2017). http://papersnips.cc/paper/6735-deanonymization-in-the-bitcoin-p2p-network.pdf
15. Reid, F., Harrigan, M.: An analysis of anonymity in the Bitcoin system. ArXiv e-prints, July 2011. arXiv:1107.4524 [physics.soc-ph]

Cloud Computing and Web Performance

Cooperation of Neuro-Fuzzy and Standard Cloud Web Brokers

Krzysztof Zatwarnicki(✉) and Anna Zatwarnicka(✉)

Department of Electroengineering, Automatic Control and Computer Science, Opole University of Technology, ul. Prószkowska 76, 45-758 Opole, Poland
k.zatwarnicki@gmail.com, anna.zatwarnicka@gmail.com

Abstract. Nowadays we cannot imagine the world without the World Wide Web which connects us with the surrounding world. The best way to maintain the availability and high quality of the Web services is to host Web systems in the cloud computing environment. The cloud-based Web systems use in their design numerous Web servers as well as appropriate strategies and mechanisms for the distribution of HTTP requests. In the article we discuss intelligent and non-intelligent distribution strategies and present our Fuzzy-Neural Request Distribution strategy. We also attempt to answer the question whether the cooperation of the non-intelligent and intelligent HTTP request distribution strategies eliminates the shortcomings of these strategies and increases the quality of service in the Web cloud system. We describe modern solutions, present the test-bed and the results of conducted experiments. In the end we discuss the results and present final conclusions.

Keywords: Web cloud system · Cloud computing · Web systems simulation · Fuzzy-neural network · Fuzzy-neural modeling · HTTP-based request distribution

1 Introduction

In the modern world, the digital revolution is taking place. There are many IT systems that improve various aspects of human life. Using Web services, we can perform many of our daily tasks, such as: shopping, borrowing books, learning, watching TV and keeping in contact with colleagues and friends from around the world. The very existence of Web systems providing useful services does not give users much. For users, Web systems become useful when they are available and when they offer content and services in a timely manner. This time should be acceptable to users. According to the latest studies, the so-called user's patience time does not exceed 2 s.

Due to the specifics of network traffic, providing an acceptable time of access to the Web content is not trivial. Web traffic is characterized by self-similarity and burstiness [3, 5, 15]. When certain content is more popular for users then the traffic increases. High server loads caused by a sudden increase in the inflow of requests are very difficult to predict. Classical Web systems do not cope with them, or they cope in a way that does not satisfy users.

L. Borzemski et al. (Eds.): ISAT 2019, AISC 1050, pp. 243–254, 2020.
https://doi.org/10.1007/978-3-030-30440-9_23

The solution to the problem is the use of the cloud computing. A cloud platform is a properly prepared technological infrastructure containing servers, storages, network and Internet connections along with the appropriate dedicated software. Thanks to its flexible construction and processing capabilities, the cloud provides extensive access to resources and more efficient service. Many companies transfer their services to the cloud in order not to maintain an aging and thus less efficient infrastructure. According to the forecasts, this trend will continue and by 2022 as much as 90% of institutions and enterprises will have purchased services in the cloud computing systems [4].

We can distinguish three types of cloud computing Models [4]:

- Infrastructure as a Service (IaaS) – hardware is provided by an external provider and managed for a customer. The consumer has control over deployed applications as well as operating systems and storage;
- Platform as a Service (PaaS) – customer hosts applications and makes them available to end users over the Internet. The consumer does not control the structure of the lower cloud infrastructure level;
- Software as a Service (SaaS) – consumer is able to use and manage the provider's applications running on a cloud infrastructure but not the cloud infrastructure and software itself.

The cloud computing platform can contain many Web and database servers providing exactly the same content for the end users. Thanks to this, Web services using cloud computing can service a large number of HTTP requests simultaneously. Distribution of requests, and in particular making decisions to redirect requests to specific parts of the cloud, can be taken on several different levels of the network infrastructure. This reduces the load of Web brokers responsible for distribution.

It is very important to choose the appropriate strategy for distribution of HTTP requests among individual elements of the infrastructure in the Web cloud. Strategies properly tailored for working conditions significantly improve processing and accelerate the operation of the Web cloud. A lot of work has been done to adapt existing HTTP request distribution strategies to work in cloud infrastructure and also many innovative mechanisms have been developed. Among HTTP request distribution strategies we can distinguish non-intelligent and intelligent ones. Among intelligent strategies special attention should be paid to self-learning mechanisms using artificial intelligence. Intelligent strategies are the only ones able to cope with the burstiness of network and Web traffic and the sudden increase in the load of web cloud services and infrastructure.

We have decided to investigate the cooperation of Web brokers using intelligent and non-intelligent requests distribution strategies. The aim of the research was to examine whether the intelligent and non-intelligent strategies cooperating with each other could achieve the synergy effect and thereby overcome their shortcomings. In the experiments, as the non-intelligent strategies we took the strategies which were used in Amazon Web Services (AWS) - the most popular cloud platform [4]. We compared those algorithms with the intelligent Fuzzy-Neural Request Distribution (FNRD) strategy we had developed. We examined how those strategies cooperated in different configurations and how the cooperation affected the process of servicing HTTP requests under conditions of high load.

The rest of the paper is structured as follows: Sect. 2 contains related work, Sect. 3 describes request distribution strategies to be examined. In Sect. 4 the test-bed, results of experiments and discussion are presented. Lastly, Sect. 5 concludes the paper.

2 Related Work

In recent years many organizations and enterprises have been putting more and more emphasis on sharing their resources and services via the WWW. The slogan “Web first” is still valid and shows the direction of development of information systems. Currently, we cannot imagine the world without the Web services which connect us with the surrounding world. The evolving Internet of Things segment further strengthens the need for an efficient, effective and secure delivery of content and services.

In order to ensure high efficiency and stability of Web services, cloud Web systems that use groups of cooperating Web servers are created. These systems are able to handle the very large load generated by the inflow of a significant number of HTTP requests during a short time. An important issue is to ensure an even load of infrastructure elements of cloud systems. Load balancing is one of the main challenges that the cloud computing faces.

Used for years in cloud Web systems, request distribution strategies greatly contribute to improving the service of HTTP requests. Simple strategies have been popular from the beginning of cluster-based Web systems and they are still subject to various improvements and modifications. Based on the popular Round-Robin algorithm, a slightly more intelligent version was created by Xu Zongyu and Wang Xingxuan. The transformed version is called Modified Round-Robin (MRR) and takes into account the Web server load [16].

Modern Web systems usually have a more complex structure than a one-layer architecture where distribution decisions are taken in a single point. Often two- and multi-layer architectures are used, where distribution decisions are taken in a cascade. An interesting example of a two-layer Web system was proposed by Ramana and Ponavaikko. It is called Global Dispatcher-Based Load Balancing (GDLB) [13]. The GDLP system uses the DNS servers to distribute HTTP requests among server rooms and Web switches to distribute requests in server rooms.

Babu and Samuel indicate in [14] that static load balancing strategies work well only when there is a small variance in the inflow of requests. The authors notice that the dynamic nature of cloud computing environment will need dynamic strategies for effective and efficient load distribution in the cloud. There are many strategies following this trend. Numerous studies are being carried out on the application of artificial intelligence mechanisms in request distribution strategies.

The Artificial Neuro-Fuzzy Inference System mechanism (ANFIS) with Sugeno-Type Fuzzy Inference was used for efficient load balancing, where the decision was made on the basis of different parameters of the servers [7].

Some interesting HTTP request distribution strategies use the natural phenomena-based strategies. An example of such an approach is the strategy called Artificial Bee Colony (ABC) in which a decision mechanism imitates the behavior of a bee colony [14]. Other heuristic intelligent approaches also offer interesting possibilities.

The Particle Swarm Optimization (PSO) strategy is based on heuristic algorithms and is designed to schedule applications to individual components of the cloud [11]. The PSO strategy takes into account both the computational costs of application operation and the cost of data transmission.

Unfortunately, there is not much research on intelligent HTTP request distribution strategies designed for the two- or multi-layer Web cloud. Our recent research [1, 17–19] fits well into the trend of using intelligent strategies for the distribution of HTTP requests. As the Web cloud usually has a multi-layered architecture, we decided to investigate how effectively - in various combinations – non-intelligent request distribution strategies and our intelligent FNRD mechanism cooperated.

3 Two-Layer Web Cloud System

A typical Web cloud system is composed of server rooms named regions, placed in different locations all around the world. Regions are divided into smaller independent parts called zones [4]. They contain all of the infrastructure elements like Web, database and application servers designed for the service of HTTP requests. Brokers are devices responsible for distribution of HTTP requests. They can be placed both in the region and in the zones. If a broker is placed in the region and redirects requests to brokers placed in the zones we call this architecture a two-layer architecture (Fig. 1).

The Web broker receives incoming HTTP requests. Using a distribution mechanism, it redirects the request to a chosen Web server or a zone. We will call the element (a Web server or a zone) servicing HTTP requests an executor. The broker uses the distribution algorithm to choose the executor to service the request. The Web broker can use a simple distribution algorithm like Round-Robin or a more sophisticated intelligent request distribution strategy like the aforementioned PSO, GDLB and FNRD.

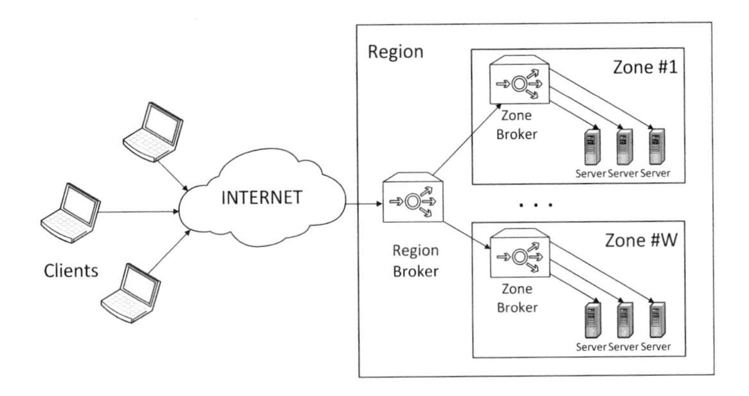

Fig. 1. Web cloud systems with two layer architecture

In the experiments we compared strategies used by the Amazon Web Services, namely:

- Round-Robin (RR) – redirects incoming HTTP requests to subsequent executors,
- Last Loaded (LL) – redirects a request to the server with the smallest number of currently serviced requests,
- Path-based routing (P) – redirects requests based on their URI according to the table where a given pattern of URI is assigned to one of the Web servers.

Additionally, a Web broker distributing requests according to the Fuzzy-Neural Request Distribution strategy (FNRD broker) was used in the experiments. The construction of the broker is complex. According to the FNRD strategy the broker chooses the executor offering the shortest response time.

Most of the Web brokers can work both as the zone broker and as the region broker. The region brokers sometimes are more efficient to handle a high flow of HTTP requests incoming to the cloud.

3.1 FNRD Web Broker

The FNRD broker forwards incoming HTTP requests to the executor with the shortest estimated response time. In our considerations we assume that every executor can provide exactly the same content. The process of calculating response time is not trivial and requires a complex design of the broker. The broker contains the following modules: a decision module, a redirection module, a measurement module and executor models (Fig. 2a).

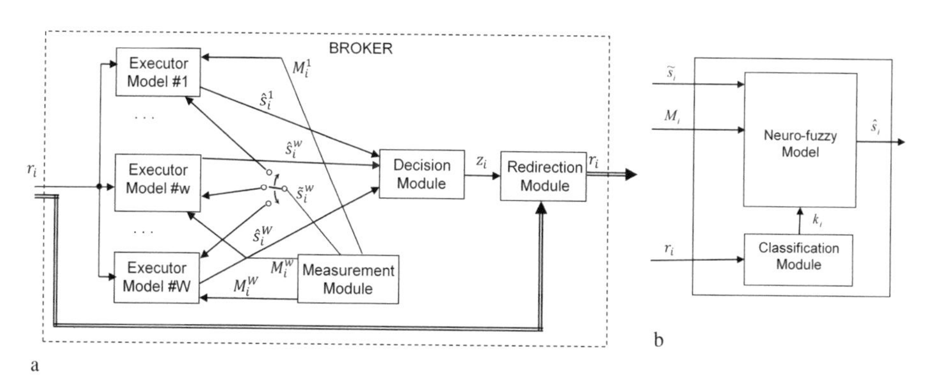

Fig. 2. FN broker design: (a) overall view, (b) Executor model.

The decision module chooses the executor for an incoming HTTP request r_i, where $i = 1, \ldots, I$ is the index of the request. The decision is taken according to the following formula:

$$z_i = min\{\hat{s}_i^w : w \in \{1, 2, \ldots, W\}\} \quad (1)$$

where z_i is the decision taken for i-th request, w is the index of the executor and there are W executors in the system, $\hat{s}_i^w$ is the estimated response time for i-th request and w-th executor.

The redirection module sends physically the request through the LAN network to the z_i-th executor. The measurement module collects and delivers to the executor model modules information about the measured response time $\tilde{s}_i^w$ and the load of executors $M_i^w = \left[e_i^w, f_i^w\right]$, where e_i^w is the number of requests being serviced by the w-th executor concurrently and f_i^w is the number of dynamic requests (the content of a response for this kind of the request is created during the service in the Web server).

Each of the executor models in the broker is assigned to one executor in the cloud system. The module estimates the response time $\hat{s}_i^w$ taking into account the load M_i^w of the w-th executor, the request r_i and the information of previous response times $\tilde{s}_{i-o}^w$. The executor module is a composed of classification module and a neuro-fuzzy module (Fig. 2b).

Because the executor module estimates the response times for only one executor in the system, the index of the executor will be omitted in the formulas provided below.

The classification module determines the class k_i, $k_i \in \{1, \ldots, K\}$, to which the request belongs. Requests belonging to the same class have similar response times. Requests connected with fetching static objects (html, pictures, style files etc.) are classified on the basis of the size of an object. The dynamic requests connected with objects created dynamically are assigned to the classes separately for each object, which means that a different class is created for each dynamic object.

The neuro-fuzzy model module estimates the response time by means of a neuro-fuzzy network. For each class k a separate neuro-fuzzy network is created. The parameters of the model are as follows: $Z_i = [Z_{1i}, \ldots, Z_{ki}, \ldots, Z_{Ki}]$, where $Z_{ki} = [C_{ki}, D_{ki}, S_{ki}]$ contains parameters of input fuzzy set functions $C_{ki} = [c_{1ki}, \ldots, c_{lki}, \ldots, c_{L)ki}]$, $D_{ki} = [d_{1ki}, \ldots, d_{mki}, \ldots, d_{Mki}]$ and output fuzzy set functions $S_{ki} = [s_{1ki}, \ldots, s_{jki}, \ldots, s_{Jki}]$. Input fuzzy set functions $\mu_{F_{el}}(e_i)$, $\mu_{F_{fm}}(f_i)$, $l = 1, \ldots, L, m = 1, \ldots, M$, are triangular (Fig. 3b), while the output fuzzy sets functions $\mu_{S_j}(s)$, $j = 1, \ldots, J$ are singletons (Fig. 3c). The input parameters e_i and f_i in the input fuzzy set functions are the elements of the load M_i of the executor.

After conducting the preliminary experiments the number of input fuzzy sets was set to $L = M = 10$ while the number of output fuzzy sets is equal to $J = L \cdot M$. After the phases of fuzzification and inference the estimated response time is calculated in a defuzzification process according to the formula $\hat{s}_i = \sum_{j=1}^{J} s_{jki}\mu_{R_j}(e_i, f_i)$, where $\mu_{R_j}(e_i, f_i) = \mu_{F_{el}}(e_i) \cdot \mu_{F_{fm}}(f_i)$. The adaptation phase is conducted after finishing the service in the executor and receiving the measured response time $\tilde{s}_i$. The following parameters $S_{k(i+1)}, C_{k(i+1)}, D_{k(i+1)}$ are calculated with the Back Propagation Method: $s_{jk(i+1)} = s_{jki} + \eta_s \cdot (\tilde{s}_i - \hat{s}_i) \cdot \mu_{R_j}(e_i, f_i)$,

$$c_{\varphi k(i+1)} = c_{\varphi ki} + \eta_c(\tilde{s}_i - \hat{s}_i)\sum_{m=1}^{M}\left(\mu_{F_{fm}}(f_i)\sum_{l=1}^{L}\left(s_{((m-1)\cdot L+l)ki}\partial\mu_{F_{el}}(e_i)/\partial c_{\varphi ki}\right)\right),$$
$$d_{\gamma k(i+1)} = d_{\gamma ki} + \eta_d(\tilde{s}_i - \hat{s}_i)\sum_{l=1}^{L}\left(\mu_{F_{el}}(e_i)\sum_{m=1}^{M}\left(s_{((l-1)\cdot M+m)ki}\partial\mu_{F_{fm}}(f_i)/\partial d_{\gamma ki}\right)\right),$$

where η_s, η_c, η_d are adaptation ratios, $\varphi = 1, \ldots, L-1$, $\gamma = 1, \ldots, M-1$ [18].

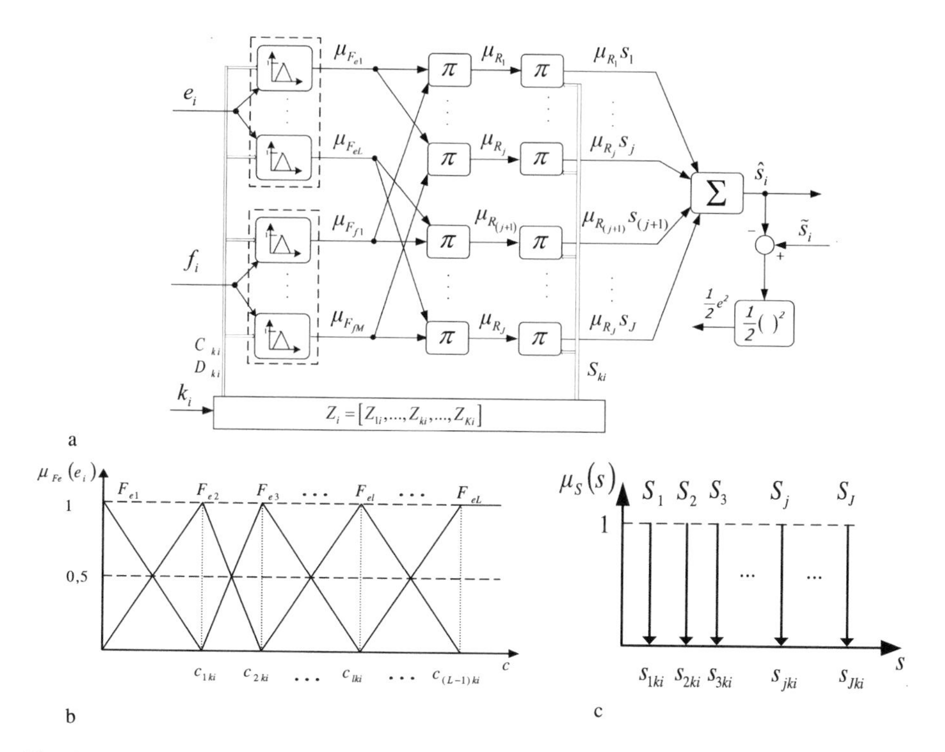

Fig. 3. Neuro-fuzzy model: (a) overall view, (b) input fuzzy sets functions, (c) output fuzzy sets functions.

4 Simulation Experiments

The main aim we wanted to achieve in our research was to examine if neuro-fuzzy Web brokers could cooperate with standard non-intelligent Web brokers and make the whole cloud system attain good results. To conduct the experiments a simulation approach was chosen. The simulation program was written with a discrete event simulator OMNET++ [10] and contained many modules imitating the behavior of real elements in the system, e.g.: clients sending HTTP requests, region and zone brokers, Web servers and zone database servers. Figure 4 presents the model of the simulator.

The request generator in the simulation contained a number of clients behaving similarly to Web browsers. Each of the clients behaved similarly to a typical Web browser and sent the first request regarding a given Web page, and after downloading the first element of the page (HTML content) the client opened up to 6 concurrent TCP connections to download the rest of the resources nested in the page. The number of Web pages downloaded by one user was modelled with use of the Inverse Gaussian distribution ($\mu = 3.86, \lambda = 9.46$) (the user simulated the behavior of a human). After finishing the session each user was deleted and a new user was created to keep a constant number of users in the simulation. The user think time (the time to open a next page) was modelled on the Pareto distribution ($\alpha = 1.4, k = 1$) [2]. Clients downloaded

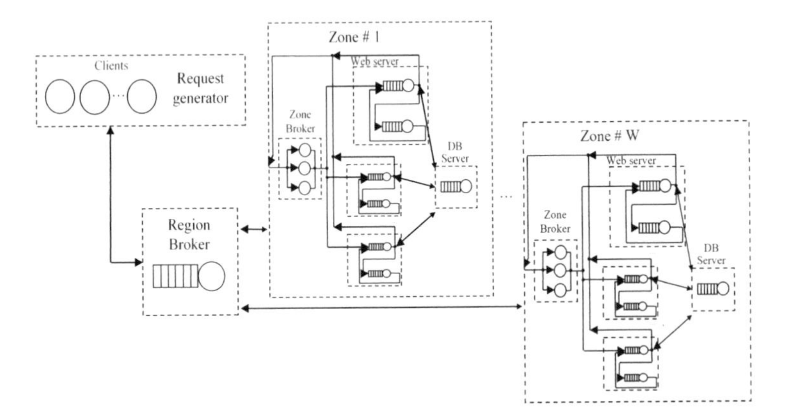

Fig. 4. Simulation model.

simulated Web pages, the size and character of which reflected the real and complex Web service of Opole University of Technology [8]. The website used WordPress, the most popular CMS system in the world which is used on more than 25% of the world's websites [9].

Each availability zone in the experiment contained one Web broker, a number of Web servers and one database server. The times of servicing HTTP requests in the simulated Web server were similar to the times gained in the real Web server containing Intel Core i7 7800X CPU, a Samsung SSD 850 EVO driver and 32 GB of RAM, and running Wordpress service. The processor was modelled as a single queue, and the SSD driver likewise. The RAM memory acted as the cache memory for the file system.

Both region and zone brokers were modelled in the same way, which let us distribute HTTP requests according to the following strategies: Round-Robin (RR), Last Load (LL), Partitioning (P) and Fuzzy-Neural Request Distribution (marked shortly FN in some of the further descriptions). In the simulator the broker contained one single queue. The service times of making a decision for subsequent algorithms were measured for Intel Xeon E5-2640 v3 processor. The obtained service times were as follows: FNRD 0.2061 μs, LL 0.0103 μs, RR 0.00625 μs, P 0.0101 μs. In our experiments we used a modification of the P algorithm, making it behave similarly to the LARD algorithm [12]. According to the LARD algorithm during the first service, a given type of request is forwarded to the server with the last load. During the next service, the request is redirected to the same server as long as the server is not overloaded, and the number of requests serviced concurrently by the server is not much higher than on other servers. If the load of the server is too high, the request is redirected to the server with the last load.

In the experiments the region brokers running different request distribution algorithms sent requests to zone brokers forming together a two-layer decision system. Because the time to make a decision in the FNRD strategy is much longer than for other strategies, the FNRD region broker distributing all requests in the system can potentially become a "bottleneck" in the system. In the experiments we used non-intelligent strategies in the region broker and the FNRD strategy in the zone broker. The list of the strategies used is as follows: LL (region) and FNRD (zone) (marked as LLFN), RR and FNRD (RRFN), P and FNRD (PFN). In addition, to gain reference results for the purposes of comparison, we performed experiments in which both brokers used the same strategy: FNRD and FNRD (FNFN), LL and LL (LLLL), RR and RR (RRRR), P and P (PP).

Subsequent experiments were run for an increasing number of clients (from 100 to 2500) sending HTTP requests. During each experiment about 30 million of HTTP requests were served and for 20 million mean response time was measured.

The experiments were conducted for four different configurations of regions and zones. However, in every experiment exactly 12 Web servers were used. The following configurations were used:

- two zones and six Web servers in each zone (2×6),
- three zones and four Web servers in each zone (3×4),
- four zones and three Web servers in each zone (4×3),
- six zones and two Web servers in each zone (6×2).

Results of the experiments are presented on Fig. 5, where plots from (a) to (d) present the results of the experiments for configurations: 2×6, 3×4, 4×3, 6×2, and Fig. 5(e) contains the best results obtained for all of the four configurations.

As it can be noticed the best results (the shortest mean response times) were obtained for both brokers running the intelligent FNRD strategy. Taking into account only configurations where the region broker is non-intelligent the best results are gained for the LLFN strategy and in all four configurations the results are better than for the LLLL strategy. The distance between those two strategies is almost the same in all configurations of the region. Also quite good results (better than LLLL) were obtained for the PFN strategy but only for two configurations (2×6 and 3×4), where the number of zones was small.

As it was mentioned before, in the simulation we took into account the service time of making a distribution decision. Despite the fact that the service time for the FNRD is almost two orders of magnitude lower than for other strategies, the region broker was not a bottleneck for the system in the experiments.

It should be stressed that despite the fact that the LLLL strategy is simple and very effective (especially in practical applications) the results for that strategy are not the best. The results obtained for LLFN are significantly better than for LLLL, and as we can see on Fig. 5(e) the results for LLFN are very stable even when the size of zones changes (with the same number of Web servers in the region). It can be noticed that the LL region broker cooperates well with the intelligent FNRD zone broker.

The results obtained for the FNFN configuration of two intelligent and adaptive brokers are surprisingly outstanding. In many cases the HTTP response time was twice as low for FNFN as for other strategies, especially for LLFN and LLLL. As we can guess the brokers learned the mutual behavior of each other in an adaptation process and began to cooperate in some way. It is worth noticing that when the number of zones increases the results for FNFN become better (Fig. 5(e)). Perhaps the region broker works better when the number of zones is greater and it can control the request flow more strictly.

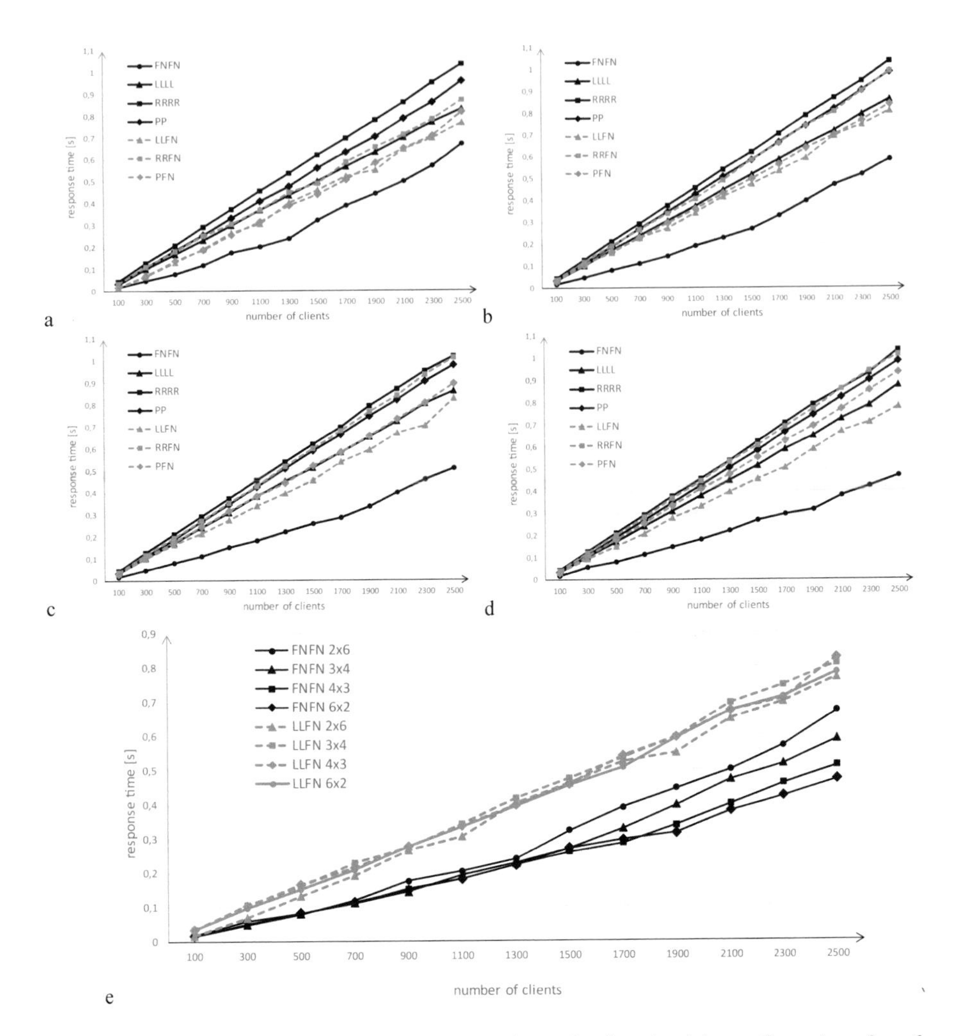

Fig. 5. Response time in load function (number of clients); (a) configuration 2 × 6; (b) configuration 3 × 4; (c) configuration 4 × 3; (d) configuration 6 × 2; (e) comparison of the best results.

Evaluating the results, the intelligent cloud Web brokers can cooperate well with non-intelligent brokers gaining quite good results. However, the results obtained for cooperating intelligent brokers are very good. We have also shown that the FNRD broker can distribute HTTP requests effectively without becoming a "bottleneck" in the cloud system.

5 Summary

In this article we tried to figure out if an adaptive intelligent neuro-fuzzy Web broker working in Web cloud-based systems could cooperate well with standard non-intelligent brokers and make the whole system work effectively. For the purposes of comparison we chose three non-intelligent strategies, which were most often used in practice in Web brokers and we prepared the simulation program to conduct the research.

The simulation application was able to imitate well the behavior of Web clients, as well as the work of Web brokers and both the Web and the database servers. The results of conducted experiments revealed that the system in which an intelligent Web broker worked always got better results than the same system running two brokers with the same non-intelligent strategy.

Good results were obtained for the Web cloud containing a region broker with the Last Load strategy and a zone broker with the adaptive neuro-fuzzy strategy. This confirms that intelligent brokers can work with non-intelligent brokers. However, the best results were gained for two brokers running the same neuro-fuzzy strategy. In this case the brokers cooperated closely and the obtained response times were almost twice as short as for other strategies. It also occurred that a region broker running an intelligent strategy could effectively distribute HTTP requests and did not become a "bottleneck" in the system.

References

1. Borzemski, L., Zatwarnicki, K., Zatwarnicka, A.: Adaptive and intelligent request distribution for content delivery networks. Cybern. Syst. **38**(8), 837–857 (2007)
2. Cao, J., Cleveland, W.S., Gao, Y., Jeffay, K., Smith, F.D., Weigle, M.C.: Stochastic models for generating synthetic HTTP source traffic. In: Proceedings of Twenty-Third Annual Joint Conference of the IEEE Computer and Communications Societies, INFOCOM 2004, Hong-Kong, pp. 1547–1558 (2004)
3. Crovella, M., Bestavros, A.: Self-similarity in world wide web traffic: evidence and possible causes. IEEE/ACM Trans. Netw. **5**(6), 835–846 (1997)
4. Documentation of Amazon Web Services, How Elastic Load Balancing Works. https://docs.aws.amazon.com/elasticloadbalancing/latest/userguide/how-elastic-load-balancing-works.html. Accessed 23 Feb 2019
5. Domańska, J., Domański, A., Czachórski, T.: The influence of traffic self-similarity on QoS mechanisms. In: Proceedings of SAINT 2005 Workshops, 31 January – 4 February, Trento, Italy (2005)
6. Gartner Press Releases. https://www.gartner.com/en/newsroom/press-releases/2018-09-12-gartner-forecasts-worldwide-public-cloud-revenue-to-grow-17-percent-in-2019. Accessed 09 Mar 2019
7. Seok-Pil, L., Eui-Seok, Nahm, E-S.: Development of an optimal load balancing algorithm based on ANFIS modeling for the clustering web-server. In: Communications in Computer and Information Science, vol. 310, pp. 783–790 (2012)
8. Main page of Opole University of Technology. https://www.po.opole.pl/. Accessed 03 Jan 2019

9. Munford, M.: How WordPress Ate The Internet in 2016… And The World in 2017. https://www.forbes.com/sites/montymunford/2016/12/22/how-wordpress-ate-the-internet-in-2016-and-the-world-in-2017/. Accessed 02 Feb 2019
10. OMNeT++ Discrete Event Simulator. https://www.omnetpp.org/. Accessed 01 Jan 2019
11. Suraj, P., Wu, L., Guru, S.M., Buyya, R.: A particle swarm optimization-based heuristic for scheduling workflow applications in cloud computing environments. In: Proceedings of 2010 24th IEEE International Conference on Advanced Information Networking and Applications, Perth, WA, Australia (2010)
12. Pai, S.V., Mohit, A., Banga, G., Svendsen, M., Druschel, P., Zwaenepoel, W., Nahum, E.: Locality-aware request distribution in cluster-based network servers. In: Proceedings of the 8th International Conference on Architectural Support for Programming Languages and Operating Systems, San Jose, California, USA, ACM SIGOPS Operating Systems Review, vol. 32(5), pp. 205–216 (1998)
13. Ramana, K., Ponnavaikko, M., Subramanyam, A.: A global dispatcher load balancing (GLDB) approach for a web server cluster. In: Kumar, A., Mozar, S. (eds.) International Conference on Communications and Cyber Physical Engineering ICCCE 2018, Hyderabad, India, Lecture Notes in Electrical Engineering, vol. 500, pp. 341–357 (2019)
14. Remesh Babu, K.R., Samuel, P.: Enhanced bee colony algorithm for efficient load balancing and scheduling in cloud. In: Snášel, V., Abraham, A., Krömer, P., Pant, M., Muda, A. (eds.) Innovations in Bio-Inspired Computing and Applications. Advances in Intelligent Systems and Computing, vol. 424, pp. 67–78. Springer, Cham (2016)
15. Suchacka, G., Dembczak, A.: Verification of web traffic burstiness and self-similarity for multiple online stores. In: Advances in Intelligent Systems and Computing, vol. 655, pp. 305–314 (2017)
16. Xu, Z., Wang, X.: A predictive modified round robin scheduling algorithm for web server clusters. In: Proceedings of 34th Chinese Control Conference, IEEE, Hang-Zhou, China (2015)
17. Zatwarnicki, K.: Adaptive control of cluster-based web systems using neuro-fuzzy models. Int. J. Appl. Math. Comput. Sci. **22**(2), 365–377 (2012)
18. Zatwarnicki, K.: Guaranteeing quality of service in globally distributed web system with brokers. In: Jędrzejowicz, P., Nquyen, N.T. (eds.) Proceedings of Computational Collective Intelligence Technologies and Applications: Third International Conference, ICCCI 2011, September 21–23, Gdynia, Poland, vol. 6923, pp. 374–384. Springer-Verlag, Heidelberg (2011)
19. Zatwarnicki, K., Zatwarnicka, A.: Two-layer cloud-based web system. In: Borzemski, L., Świątek, J., Wilimowska, Z. (eds.) Information Systems Architecture and Technology: Proceedings of 39th International Conference on Information Systems Architecture and Technology – ISAT 2018. Advances in Intelligent Systems and Computing, vol. 852, pp. 125–134. Springer, Cham (2019)

Microservice-Based Cloud Application Ported to Unikernels: Performance Comparison of Different Technologies

Janusz Jaworski, Waldemar Karwowski(✉), and Marian Rusek

Faculty of Applied Informatics and Mathematics, Warsaw University of Life Sciences, Nowoursynowska 166, 02–787 Warsaw, Poland
waldemar_karwowski@sggw.pl

Abstract. Microservice architecture is nowadays a popular design pattern for cloud applications. Usually microservices are launched on the servers of a cloud datacenter inside virtualization containers. This provides their isolation and performance comparable to virtual machines. However, security of virtualization containers remains a problem. They share the hosts machine operating kernel and thus their mutual isolations depends on proper implementation of kernel isolation features like cgroups or namespaces. Moreover the usage of kernel security modules like SELinux is hindered by different base Linux distributions used for containers creation. The concept of unikernels provides much better security for microservices. They are launched on a hypervisor instead of a full operating system kernel and contain only the base files needed to run the application. However their performance in realistic environments still remains a question. In this paper we port a microservice-based cloud application from Docker containers to Rumpkernel and OSv unikernels and analyse its performance. It is shown, that the performance of this unikernel-based port can match or exceed the more traditional approaches.

Keywords: Microservice architecture · Cloud computing · Unikernel

1 Introduction

Development of new internet technologies is connected with cloud services model. For information systems that are in the cloud and share public services publicly, security and safety are very important. In case of safety, the proper isolation of users using the application, guaranteeing the availability of the data only to the appropriate persons is crucial. At the same time, the protection of environment in which the application runs is very important. Moreover, a hundred or even million users use the software in parallel, which means that scaling is also necessary. To meet these requirements, new software architectures were created. The most prominent example is the microservice architecture. Nowadays, Docker virtualization containers are the most widely used technology in the cloud. Containers share a common operating system kernel with their host system and can use applications present on the host [1]. Traditional virtual machine image contains own copies of each application needed, which means that disk space used is bigger than for containers. Containers use the host's processes, so the start of newly

L. Borzemski et al. (Eds.): ISAT 2019, AISC 1050, pp. 255–264, 2020.
https://doi.org/10.1007/978-3-030-30440-9_24

added container is almost immediate. Another solution is the unikernels technology designed to make applications be safer and more efficient than traditional, and highly scalable. Unikernels are lighter than containers, they are equally easy to scale, but they are better isolated from the environment and therefore safer. It is therefore important to investigate the practical performance of unikernel-based applications.

The history of the unikernels technology is connected with operating systems architecture research called libOS, which dates back to nineties of the last century. The concept of this architecture is based on the many small libraries that implement certain operating system mechanisms like file system, access control, device drivers etc. However, direct access to hardware and resources has allowed one application to use resources of other applications. To solve this problem, exokernel was designed and implemented [2], whose only task is to make access to specific resources only to eligible applications. Exokernel as opposed to a microkernel or a monolithic kernel does not use high-level abstraction but low-level interfaces. However, this solution has a major disadvantage, because the development of hardware libraries is not up to date and they must be adapted. Unikernels overcome this inconvenience using virtualization machine monitor (a.k.a. hypervisor) as the hardware abstraction layer. A unikernel is small, fast, secure virtual machines that lack operating systems [3] and can be booted up as a standalone virtual machine. Compared to virtual machines, unikernels eliminate unnecessary code, which results in memory savings and shorter boot time. Thanks to hypervisor there are no IPC (Inter Process Communication) and application components (microservices) are isolated. They communicate via Representational State Transfer (REST) interface only.

There are currently several implementations of unikernels. MirageOS uses the OCaml language, with libraries that provide networking, storage and concurrency support [2, 4], application code is compiled into a fully-standalone, specialised unikernel that runs under a Xen or KVM hypervisor. HalVM [5] is close to the MirageOS philosophy, but it is based on the Haskell language. IncludeOS [6] compiles C and C++ applications natively into a standalone application that boots without an operating system. RuntimeJS is an libOS for the cloud that runs JavaScript, it can be bundled up with an application and deployed as a lightweight and immutable VM image for KVM hypervisor. OSv and Rumprun provide compatibility for existing POSIX applications. OSv runs unmodified Linux applications under a Xen, KVM, and VMware. It supports many managed language runtimes including unmodified JVM, Python 2 and 3, Node.JS, Ruby, Erlang as well as languages compiling directly to native machine code. The Rumprun enables running POSIX applications as unikernels for hypervisors such as Xen and KVM. The research on HermiTux unikernel providing binary-compatibility with Linux applications is presented in [7].

There are not much performance tests of unikernel technologies. In [6] CPU-time required to execute 1000 DNS-requests, on IncludeOS and Ubuntu, running on Intel- and AMD systems was measured. On AMD, IncludeOS used 20% fewer CPU-ticks on average total than Ubuntu, on Intel it was only 5.5%. Performance evaluation for MirageOS, OSv and Linux for DNS server was investigated in [8]. Final benchmarks showed that OSv significantly exceeded the performance of Linux, but MirageOS failed. In [9] was performed an evaluation of KVM (Virtual Machines), Docker (Containers), and OSv (Unikernel), when provisioning multiple instances concurrently in an OpenStack cloud platform. Authors concluded that OSv outperforms the other options. The performance of unikernels versus Docker containers in the context of REST services and heavy processing workloads, written in Java, Go, and Python was investigated in [10]. The Go service was about 38% faster when running as a unikernel than as a container, while the Java version was about 16% faster as a unikernel. Python service was much slower according to Go and Java but with 15% improvement as a unikernel than as a container. In [11] performance evaluation of containers (Docker, LXD) and unikernels (Rumprun and OSv) is presented. Nginx HTTP server (version 1.8.0) was used to evaluate the HTTP performance for static content and whole-system virtualization. HTTP performance ranging from 0 to 1000 concurrent connections was measured. As a second benchmark application Redis (version 3.0.1) a key-value store, was used, performance of GET and SET operations ranging from 0 to 1000 concurrent connections was measured. Regarding application throughput, most unikernels performed at least equally well or even better than containers. Redis GET and SET benchmarks performed in [7] for Rumprun, OSv, Linux KVM, Docker and HermiTux confirmed above conclusion.

The tests mentioned above were made for isolated functionalities not for real applications. In this paper we tested performance of sample microservice application ported to OSv and Rumprun unikernels respectively. The rest of this paper is organized as follows: in Sect. 2 the architecture of different versions of a sample microservices-based application is introduced. In Sect. 3 performance experimental results for different technologies are presented. We finish with summary and brief conclusions in Sect. 4.

2 Sample Application

To make performance comparisons between different unikernels technologies a sample Docker example-voting-app[1] microservices-based application was chosen. It is an application that allows conducting a vote for a more popular household pet, the choice is between Dog and Cat. User can vote only for one candidate. It consists of two data containers and three microservices:

Vote—service written in Python using the Flask library2. It provides a Web page with two available options. User can give voice on one of them i.e. set environment variable. The getenv function from the standard Python os module returns the value of the chosen option. In addition, the application generates voters ID placed in the cookie to prevent user to give two valid votes. The HTML code is generated using the JINJA2 engine from the template file that is sent in response to the voter's browser. With this solution, it is possible to dynamically change the contents of the website code, it is not necessary to change the static HTML code. The voting ID and selection are saved in the JSON format and stored in the Redis database.

Worker—this service uses the Command Query Responsibility Separation (CQRS) design pattern. CQRS separates the logic of recording and reading data from each

[1] https://github.com/dockersamples/example-voting-app.

other. In example-voting-app, each unit has its own database respectively to read and write data. Worker is written in the C# language, it reads the data stored by the voting-app service in the Redis database and saved processed data in a relational database PostgreSQL in format readable for result-app. Moreover, worker verifies if the user with the specified identification number has already given a vote. If this happens, it does not add another vote but exchange the previous one. Very important advantage of worker is that if there is no connection to any database it does not stop its action, but periodically checks whether the connection can be created again. Therefore, if the connection to the PostgreSQL database is lost, the votes stored in the Redis database are not lost. After reconnection to the base, the votes may be recorded.

Result—the last element of the example-voting-app application is a service that displays the results of voting in a website. The service is written in the JavaScript language, script is running in the Node.js environment. When started, it connects with PostgreSQL database in which the worker service has recorded results. Service periodically retrieves results from PostgreSQL using polling. Results are displayed in the user's browser both in text and graphic form. Thanks to polling and the Node.js environment, the service updates the customer's website automatically. For the result-app service, the time window is 1000 ms, exactly every one second it executes the query to the database of the latest voting results. Naturally, if the connection will be lost, service displays the results that were available to it before the crash and tries to connect the next call.

Because the application consists of five services implemented in different programming languages, Rumprun and OSv were selected as a possible tools for preparing the unikernel version. Other tools would require code to be rewritten to other programming languages, for example OCaml (MirageOS unikernel).

3 Experimental Results

3.1 Setup

The performance tests described below were conducted on a commodity PC with an ASRock Z77 Extreme4 mainboard; Intel Core i5-3470 (3.2 GHz) processor; 16 GB DDR3 (666 MHz) memory; NetLink BCM57781 Gigabit Ethernet PCIe card; OCZ-VECTOR150 hard drive running Ubuntu 16.04.03 LTS operating system with Linux kernel 4.4.0-92-generic. KVM hypervisor was also installed and all microservices packaged as unikernels were launched on it. Also NetBSD was used inside a KVM virtual machine. It's kernel settings from the sys/conf/param.c file were modified to bring it on pair with Linux configuration. For example the MAXFILES parameter was set to the value of 200000 which is the same as used by the authors of [11] in their Linux kernel configuration (fs.file-max setting).

3.2 Redis

Redis is in memory key-value store, popular in cloud applications. It is also a part of the example-voting-app introduced in Sect. 2. In our Rumprun unikernel tests Redis version 3.0.6 from rumprun-packages repository was used[2]. For OSv unikernel tests Capstana porting tool was used[3], but we slightly modified the source code of OSv Redis package, so the same Redis server and redis.conf versions as in the Ruprun tests were used. Next both unikernels were launched on the KVM hypervisor.

The same Redis source code was used to build service on Ubuntu and NetBSD operating systems. Note that it is different from a regular Redis which utilizes the fork system call for background savings of the database into a /data/dump.rdb file. Unikernels run only a single process and do not allow for fork system call, so this periodic save feature is missing from the Rumpkernel port used in our experiments.

Redis package includes redis-benchmark utility which was used in our tests. It was started for 10 parallel connections and 10 pipeline requests. The results from this and similar measurements for OSv unikernel, Ubuntu running on the bare metal, and Ubuntu and NetBSD running inside virtual machines are expressed in requests per second and presented in Table 1.

Table 1. Results of redis-benchmark testing of the Redis microservice launched on bare metal, Rumprun and OSv unikernels and Ubuntu and NetBSD virtual machines. All numbers are expressed in requests per second.

	Ubuntu	Rumprun	OSv	Ubuntu (KVM)	NetBSD (KVM)
PING_INLINE	632911.38	389105.06	518134.72	483091.78	89126.56
PING_BULK	1098901.12	578034.69	806451.62	714285.69	97943.19
SET	8223.01	3750.52	6579.81	3648.84	2729.33
GET	746268.62	406504.06	649350.62	495049.50	88652.48
INCR	8140.01	3914.81	7007.71	3659.38	2856.82
LPUSH	8293.94	4036.98	6844.63	3638.22	2908.41
LPOP	8152.62	4318.35	7150.01	3702.74	2911.63
SADD	657894.75	387596.91	458715.59	512820.53	81967.21
SPOP	980392.19	483091.78	675675.69	613496.94	87873.46
LRANGE_100	94161.95	73421.44	53418.80	91157.70	37893.14
LRANGE_300	29291.15	21199.92	15401.20	25940.34	14457.13
LRANGE_500	16806.72	14496.96	9737.10	14176.35	10117.36
LRANGE_600	12253.40	10563.01	7275.37	9441.09	7376.26
MSET	7833.31	2917.41	5296.33	3503.73	1903.96

To ease comparison rows were normalized to the largest value in a row and plotted as colour bars in Fig. 1.

[2] https://github.com/rumpkernel/rumprun-packages/tree/master/redis.

[3] https://github.com/cloudius-systems/osv-apps/tree/master/redis-memonly.

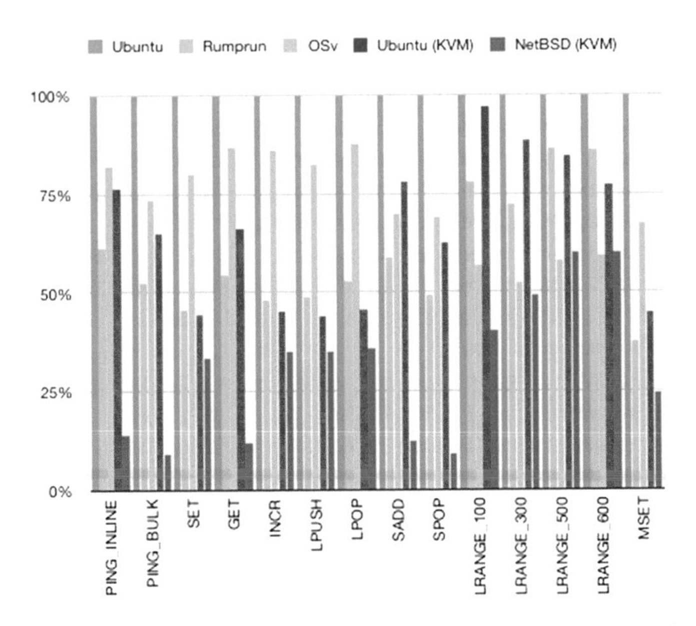

Fig. 1. Bar plot of the results of redis-benchmark normalized to the highest value (100%).

It is seen from inspection of this plot that Ubuntu running on bare metal is always the fastest one (blue bar). However OSv unikernel (orange bar) catches up and in the first 7 tests (up to LPOP) it offers at least 80% of the bare metal performance. Surprisingly it is also faster than Ubuntu running in a KVM virtual machine. The situation changes for the LRANGE tests where Ubuntu on KVM takes the second place but also Rumprun (green bar) starts to offer acceptable performance. In the SADD and SPOP tests OSv performs almost as good as Ubuntu on KVM and outperforms it again in the last MSET test. Note that the Rumprun unikernel is always faster than NetBSD (violet bar) whose drivers it utilizes. The large difference between Linux and NetBSD may come from the fact, that the Linux network stack is heavily optimized and Linux runs better in a KVM environment (NetBSD weren't tested on bare metal). In [11] only GET requests were studied for 100 concurrent users and similar results to the corresponding row from Table 1 were obtained—the performance of Rumprun, OSv and Ubuntu/KVM was at about 70% of Ubuntu performance (these authors did not study NetBSD).

3.3 NodeJS

In that paper [11] performance of a Nginx web server with a static web page running on Rumprun and Ubuntu VM was also studied. However, according to Netcraft May 2019 Web Server Survey[4] Nginx accounts only for 21% of active sites. The competing monolithic web server Apache has 30%, but the most rapidly growing category of web

[4] https://news.netcraft.com/archives/2019/05/10/may-2019-web-server-survey.html.

servers is Other at 35%. These are custom web servers probably run as microservices in the cloud. In this section we analyse the performance of the result microservice from the example-voting-app described in Sect. 2. This microservice is written in NodeJS and coupled to PostgreSQL database running as Docker container on the host. We ported it to Rumprun[5] and OSv[6] unikernels using the procedure outlined in Subsect. 3.2. To test their performance the Apache HTTP server benchmarking tool[7] was used. The results are presented in Table 2 and Fig. 2.

Table 2. Results of Apache HTTP benchmark (ab) testing of the result micoservice launched on bare metal, Rumprun and OSv unikernels as well as Ubuntu and NetBSD virtual machines. Latencies are expressed in miliseconds.

	Ubuntu	Ubuntu (KVM)	Rumprun	OSv	NetBSD (KVM)
REQUESTS/SEC	2238.73	2567.29	4083.43	2494.76	2056.50
KBYTE/SEC	4234.79	4856.28	7724.22	4719.08	3890.07
LATENCY 50%	44.0	38.0	23.0	39.0	46.0
LATENCY 90%	48.0	45.0	28.0	47.0	64.0
LATENCY 100%	152.0	78.0	74.0	83.0	127.0

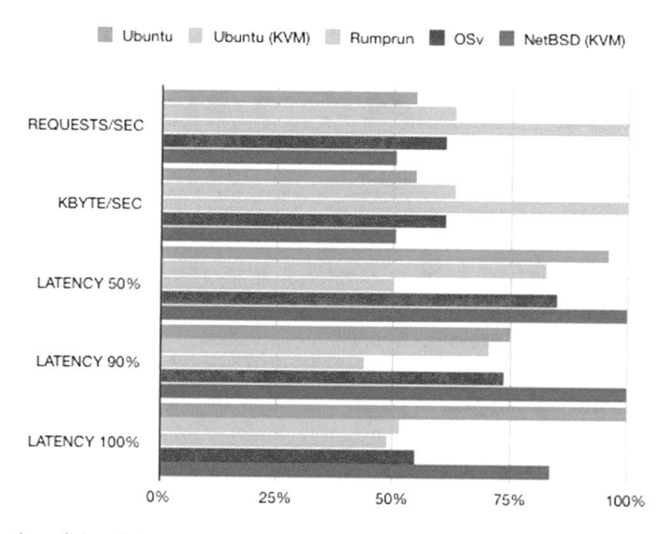

Fig. 2. Bar plot of the ab benchmark of the result micoservice normalized to the highest value.

Surprisingly Rumprun is the clear winner now offering the highest throughput and lowest response times. OSv offers only mediocre performance at the level of 60% but

---

[5] https://github.com/rumpkernel/rumprun-packages/tree/master/nodejs.

[6] https://github.com/cloudius-systems/osv-apps/tree/master/node.

[7] https://httpd.apache.org/docs/2.4/programs/ab.html.

does not fall behind other solutions (this may be due to the fact that the Rumprun repository is at version 4.3 whereas in OSv version 4.1 is used). This is stark contrast with results of the static web server testing from [11] where Rumprun performed only at 38% of the bare metal server, and Ubuntu on KVM at 75% of bare metal Ubuntu (at 100 concurrent users). We double checked our results using weighttp but the conclusions are similar: Rumprun achieved 4092 requests per second and OSv—2482. The ratio is again 60%.

3.4 Python

The last microservice studied was vote coupled to the Redis database. It is written in Python and utilizes the Flask library. We managed to successfully build Rumprun image[8] but OSv creation using Capstana[9] failed. Probably because the python3x app available in the repository is tweaked to filter out static libraries to lower image size. The results of the performance tests with ab are presented in Table 3 and Fig. 3.

Table 3. Results of Apache HTTP benchmark (ab) testing of the vote microservice launched on bare metal, Rumprun unikernel as well as Ubuntu and NetBSD virtual machines. Latencies are expressed in miliseconds.

	Ubuntu	Ubuntu (KVM)	Rumprun	NetBSD (KVM)
REQUESTS/SEC	1040.19	564.52	106.75	139.121
KBYTE/SEC	1505.36	816.42	155.22	1041.65
LATENCY 50%	93.0	171.0	934.0	134.0
LATENCY 90%	103.0	245.0	951.0	152.0
LATENCY 100%	148.0	341.0	1035.0	242.0

This time the Rumprun version performs very poorly, even worse than the NetBSD VM. Poor performance of Python unikernels was noticed by the authors of [10]. They studied a REST microservice written in different programing languages (Go, Java and Python) using an OSv unikernel. For Python Flask RESTfull[10] was used but their application was not coupled to Redis—this may be another reason why our image failed to build properly.

3.5 Scalability

Scalability of microservices is very important for cloud applications. Their launch time should be as low as possible. In Table 4 we have the times measured from image startup to the first result obtained from a microservice it contains. We see that unikernel-based images start up to 3 times faster than corresponding Ubuntu virtual

[8] https://github.com/rumpkernel/rumprun-packages/tree/master/python3.

[9] https://github.com/cloudius-systems/osv-apps/tree/master/python3x.

[10] https://flask-restful.readthedocs.io/en/latest/.

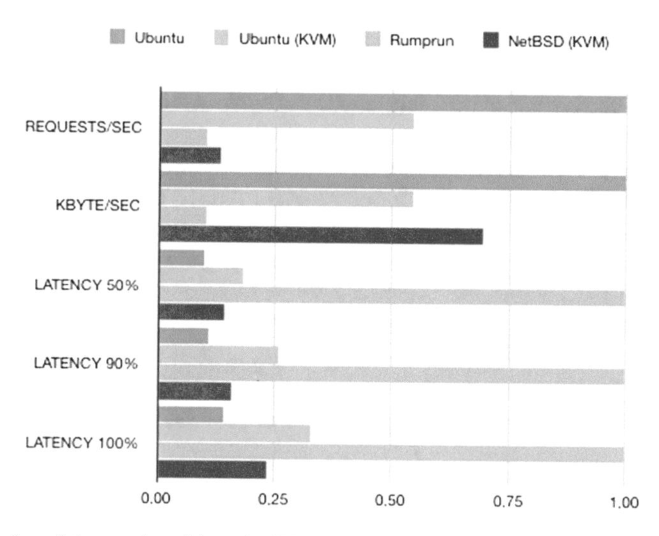

Fig. 3. Bar plot of the results of Apache HTTP benchmark for vote microservice normalized to the highest value (100%).

machine. The NetBSD VM is again an exceptionally poor performer on KVM Linux system. It is interesting to note, that OSv version of result starts faster than a Rumprun version. This is contrary to the common believe that OSv is a "fat" unikernel. For example the authors [11] reported that images generated by this tool are twice as large as Rumprun images.

Table 4. Startup times of different microservice versions measured in seconds.

	Vote	Result
Ubuntu (KVM)	9.12	9.17
Rumprun	2.29	2.39
OSv	–	1.65
NetBSD (KVM)	29.71	31.36

4 Summary

Performance of a relatively simple cloud application ported to unikernel architecture is studied. The application was originally written for Docker virtualization containers and consists of five containers: vote, results, worker, Redis and database. Worker is a trivial one, it only copies votes from Redis to database. The database used is PostgreSQL, there was no reason to port it to unikernels because it is not scalable and runs on a dedicated server. Therefore only vote, results and Redis were ported and studied.

Two unikernel technologies were chosen for our study. Rumprun is an implementation of a rump kernel and it can be used to transform just about any POSIX-compliant program into a working unikernel. The rump kernel project provides the modular drivers from NetBSD in a form that can be used to construct lightweight, special-purpose virtual machines. The second choice is OSv which was originally created by Cloudius Systems as a means of converting almost any application into a functioning unikernel. It can accept just about any program that can run in a single process.

They both provide performance better than monolithic virtual machines and sometimes better than a bare-metal implementation. Thus, they make a good basis for modern cloud-based applications. The only problem we encountered was poor performance of Python, which uses a disproportionate amount of operations that makes it run slower in a Rumpkernel environment. We were not able to port the Python-based vote microservice onto the OSv unikernel. This is because Cloudius Systems has been transformed into a new company called ScyllaDB and they are no centred on servicing OSv. The OSv open source community is now maintaining the software, which is fully operational, but some packages like Python do not receive sufficient support.

References

1. Madhavapeddy, A., Mortier, R., Rotsos, C., Scott, D.J., Singh, B., Gazagnaire, T., Smith, S. A., Hand, S., Crowcroft, J.A.: Unikernels: library operating systems for the cloud. ACM SIGPLAN Not. **48**(4), 461–472 (2013). ASPLOS'13
2. Engler, D.R., Kaashoek, M., O'Toole, J.: Exokernel: an operating system architecture for application-level resource management. ACM SIGOPS Oper. Syst. Rev. **29**(5), 251–266 (1995)
3. Pavlicek, R.C.: Unikernels: Beyond Containers to the Next Generation of Cloud. O'Reilly Media (2017)
4. Madhavapeddy, A., Scott, D.J.: Unikernels: rise of the virtual library operating system. Commun. ACM **51**(1), 61–69 (2014)
5. Galois Inc.: The Haskell Lightweight Virtual Machine (HaLVM) source archive. https://github.com/GaloisInc/HaLVM. Accessed 24 May 2019
6. Bratterud, A., Walla, A., Haugerud H., Engelstad, P.E., Begnum, K.M.: IncludeOS: a minimal, resource efficient unikernel for cloud services. In: 2015 IEEE 7th International Conference on Cloud Computing Technology and Science, pp. 250–257 (2015)
7. Olivier, P., Chiba, D., Lankes, S., Min, C., Ravindran, B.: A binary-compatible unikernel. In: Proceedings of the 15th ACM SIGPLAN/SIGOPS International Conference on Virtual Execution Environments, pp. 59–73 (2019)
8. Briggs, I., Day, M.J., Guo, Y., Marheine, P., Eide, E.: A Performance Evaluation of Unikernels (2014). http://media.taricorp.net/performance-evaluation-unikernels.pdf. Accessed 24 May 2019
9. Xavier, B.G., Ferreto, T., Jersak, L.C.: Time provisioning evaluation of KVM, docker and unikernels in a cloud platform. In: 2016 16th IEEE/ACM International Symposium on Cluster, Cloud and Grid Computing, pp. 277–280 (2016)
10. Goethals, T., Sebrechts, M., Atrey, A., Volckaert, B., Turck, F.D.: Unikernels vs containers: an in-depth benchmarking study in the context of microservice applications. In: 2018 IEEE 8th International Symposium on Cloud and Service Computing, pp. 1–8 (2018).
11. Plauth, M., Feinbube, L., Polze, A.: A performance evaluation of lightweight approaches to virtualization. In: CLOUD COMPUTING 2017: The Eighth International Conference on Cloud Computing, GRIDs, and Virtualization, pp. 34–48 (2017)

A Study on Effectiveness of Processing in Computational Clouds Considering Its Cost

Mariusz Fraś(✉), Jan Kwiatkowski, and Michał Staś

Department of Informatics, Faculty of Computer Science and Management, Wroclaw University of Science and Technology, Wybrzeze Wyspianskiego 27, 50-370 Wroclaw, Poland
{mariusz.fras, jan.kwiatkowski}@pwr.edu.pl, 220954@student.pwr.edu.pl

Abstract. The paper concerns issues related to evaluation of processing in computational clouds. The effectiveness of the computational clouds depends on many factors and different metrics can be used to asses effects of processing. In this work potentiality for evaluation of processing in cloud virtual machines that takes into account also financial cost constraints is investigated. There are considered KVM and Hyper–V based cloud platform configurations. In the paper the APPI index (Application Performance and Price Index) that can be used for selection of recommended configuration considering acceptable price of processing is defined. Presented experiments show the impact of use of the proposed metric on the choice of recommended platform for processing.

Keywords: Cloud computing · Effectiveness of computing · Processing cost · Virtual machine

1 Introduction

Current user requirements related to hardware, hardware availability, and speed of data processing are increasing all the time. Maintaining own dedicated servers requires adequate security and services, and increasing its performance is not possible in a short time due to necessity of replacing their physical elements or adding the new one. It is even not easy when the virtual machines are used.

On the other hand, when cloud computing is used, it is possible dynamically change the available resources allocated to task. It can be done by hand or automatically due to possible auto-scaling property. Using such a solution allows to make quick changes to the virtual server parameters in the environment provided by the service provider. Moreover, additional advantage of cloud computing is delegating the risk related to securing of physical servers.

Currently cloud computing has become very popular and more and more applications are supported by this environment. Considering the data available at Internet [1] up to 57% of applications used by corporations are clouded, and in the case of small enterprises it is 31%. The data for European Union are also similar. In 2018, 26% of enterprises used cloud computing, and 55% of these companies used advanced business applications, for example financial management of customers [2].

L. Borzemski et al. (Eds.): ISAT 2019, AISC 1050, pp. 265–274, 2020.
https://doi.org/10.1007/978-3-030-30440-9_25

Considering above the question is raised, how to choose the most convenient virtual machine configuration, its location and finally the its vendor. For the first view the answer is very simply, that one which offer the shorter response time. As a consequence of it all currently used metrics, even Apdex index [3] used for evaluation of efficiency different parameters related to "execution time".

However, it is not enough to fulfill the user satisfaction, he is interested in the cost of the service as well. Therefore, the paper the new metrics is proposed. It is some extension of Apdex index [4], which additionally to response time takes into consideration the cost of using cloud services.

The paper is organized as follows. Section 2 briefly describes different approaches to cloud efficiency evaluation as well as the new proposed index. The plan of measurement cloud efficiency is presented in the Sect. 3. The next section illustrates the obtained experimental results and analyzing of them. Finally, Sect. 5 outlines the work and discusses ongoing work.

2 Effectiveness of Cloud Processing

The effectiveness of the computational clouds depends on many factors, for example, the way how resources are allocated to the job, types of used virtual machines, localization of computational center, etc. It causes that different metrics for evaluation the effectiveness of computational clouds can be used. Considering the effectiveness of used virtual machines one of the most frequently used measure is percentage usage of CPU or memory. However, it is not enough. For the user much more important metric is response time. Additionally, the effectiveness can change during day. For example, in the paper [5] authors compare the speed of disk reading as a function of time. They noticed the variability of read speed from the disk within 24 h. Moreover, they observed that the speed of the disk on a virtual machine also changes during the week.

These daily changes are very important for the user because it can cause different virtual machine effectiveness during the day. It can be observed the differences between day and night. It means that the metrics should take into consideration changes of effectiveness and can entail higher fees for using the clouds.

Very similar results were obtained by the authors of the paper [6]. They present variability of CPU, RAM and a disk efficiency of virtual machines for 6 different computational clouds. The largest variability coefficient was obtained for the disk and it was about 10%, so it confirms the results from paper [5]. Moreover, in the paper [7], authors present results of experiments related with data transfer (virtual network) speed performed for Amazon, Google and Microsoft clouds. They noticed that data transfer is different for different providers and locations.

Scalability can be defined as the ability of system to keep efficiency at the fixed value by simultaneously increase of the number of used resources and problem size. On the other hand, it can be treated as the ability to perform specific tasks and in-creasing resources depending on needs that are rapidly changing [8]. In this case there are two types of the scalability: vertical and horizontal. The vertical scalability means that allocation of resources increases on a single virtual machine instance, when the horizontal means that we increase the number of virtual machine instances.

In case of cloud computing so called auto-scaling service is available. It changes allocation of resources in automatic way during job execution depending on the current loading [9], horizontal as well as vertical scaling is possible. For example, if it will be determined resources overloading, auto-scaling in automatic way increase the number of processors allocated to a virtual machine or add the new instance of the virtual machine. In this case mostly use metrics for taking decision is percentage of used resources, mainly CPU by the virtual machine. When it is larger than 80%, the number of allocated resources increased automatically. The main disadvantage of this solution is that, the change of the virtual machine efficiency will be not in real visible, but the price of the single virtual machine will be higher.

In the paper [10], the results of experiments performed at local server and at cloud were compared. As a benchmark, the implementation of algorithm that calculate prime numbers was used. The average response time was measured, and as expected the local server responded faster. The results of similar experiments have been presented in the paper [11]. In the research a multithread algorithm of Salesman was used, its execution time was measured on various configurations at local server and at Azure computing cloud. The obtained results show for example, that the time of task performed on the local server with four virtual processes is shorter than with eight in the cloud. Therefore, considering results presented in [10, 11] we can conclude that the local server respond faster to tasks comparing with a virtual machine in the cloud.

In the paper [12] the effectiveness of a web application using the Eucalyptus system was tested. It is used to create private clouds. During experiments the matrix multiplication algorithm implemented in Java EE was used. The tests were carried out using different configurations of virtual machines, and applying load balancing. The virtual machine was assessed on the basis of the number of queries per minute. In case of a single instance, a virtual machine grew as resources increased.

Concluding above brief presentation of different efficiency metrics, it can be stated, that the following metrics are mainly used:

- percentage of CPU resource usage, RAM memory,
- average time of performing a specific task,
- supported number of queries per second.

Unfortunately, these metrics ambiguously represent customer needs. A lot of these problems can be solved when the Apdex (Application Performance Index) index is used. This index takes into account user satisfaction of serving its request by means of response time, and variance for this satisfaction.

Even when using of the Apdex index solves some above-mentioned problems still one problem is skipped. The user is interested not only in the satisfied response time but also wants to pay for it as less as possible. It means that there is a need to consider the cost of using cloud.

Taking above into consideration, the new metric, which is based on Apdex, has been proposed. The metric called APPI index (Application Performance and Price Index) is defined by Eq. (1):

$$\text{APPI} = \frac{1}{N}\sum\nolimits_{j=1}^{N}\left(Sat_j + \frac{Tol_j}{2}\right)\cdot \min\left(\frac{P_{AC}}{P_{VM}}, 1\right) \quad (1)$$

where:

- N – number of performed requests for service (processing),
- j – index of j-th request,
- Sat – satisfaction of serving given request defined as follows: $Sat = 1$ if $t_r < t_s$, and $Sat = 0$ otherwise, where: t_r – response time of request, t_s – time that satisfies client (assumed value),
- Tol – tolerance for given request defined as follows: $Tol = 1$ if $t_r > t_s$ and $t_r < (t_s + t_t)$, and $Tol = 0$ otherwise, where: t_t – the tolerated time value to exceed the satisfaction time (assumed value),
- P_{VM} – virtual machine price – it is a cost per 1 h of using a virtual machine instance according to the cloud price list,
- P_{AC} – acceptable price – it is a cost for 1 h of using the virtual machine which customer wants to spend.

The expected response time for tasks usually does not depend on the application. It results from standards and demands of users. The accepted price depends on the cloud client and it can change depending on its financial capabilities. During carried out experiments the various values of the accepted price were taken into account.

The measure assumes the value of 1, when all response times for requests have a value less than the time of satisfaction and the price is possible to be higher than the price for a virtual machine. However, it takes the value 0, when all response times for requests are greater than the sum of satisfaction and tolerance time. The rating decreases depending on the price ratio. If the acceptable price is less than the price for the virtual machine, then the rating decreases proportionally. These restrictions have been introduced to make the pattern of values from 0 to 1.

3 Experimental Study

The main goal of performed experiments was to characterize effectiveness of processing in two considered cloud computing solutions, for selected environment configurations and processing conditions, and to investigate how considering the financial cost of processing can impact the choice of given configuration as the recommended one. The last was performed with use of proposed APPI index.

As the considered cloud configurations are based on two types of virtual machines, KVM and Hyper–V in the first step these two solutions were tested in pure, isolated environment build with 4-core local computer. Next, for the same testing conditions the measurements were performed in real clouds, which currently are getting more of a market, namely Google Clouds and Microsoft Azure.

The experiments were performed with the following settings:

- for configurations build with 1, 2, and 4 virtual machines,
- for 3 different loads – the load task executed by 50, 100, and 200 users,

- for several acceptable price per hour,
- there were performed series of measurements during day (from 10:00 to 12:00 of local time) and night (from 22:00 to 24:00 of local time),
- every measurement value was calculated from 10 probe.

The evaluation of effectiveness with use of proposed APPI index was performed for the index parameters selected according to work [13], i.e.:

- the satisfaction value equal 1.5 s.
- the tolerance value equal 0.5 s.

The assumption for the experiment was that the measurements should be performed for processing of not a specific type. That's why the processing task was the developed own Java based application task with regard to specified above satisfaction and tolerance values. The task was run as a service which consists of the following operations:

- downloading static web page content of size 1,9 MB,
- resolving travelling salesman problem for 10 nodes (cities), 200 iterations, population size 50, and mutation 0,01,
- encryption and decryption of 10,000 bytes message using the AES algorithm with 256 bit encryption key length,
- performing a sorting algorithm for a 100,000 set,
- reading 100 objects from the database in JSON format (data size 44000 bytes),
- adding a new user to the database with password encryption,
- adding a new object to the database.

After the measurements the evaluation Apdex index and APPI index were calculated for all tested configurations. These indexes can be considered to choose the recommended environment for individual processing needs. The APPI index was evaluated for 4 different acceptable price C from 0.08 $/h (US dollars per hour) to 1,4 $/h.

The local testing environment was built on Intel i5, 8 GB RAM, 4-core based PC, with Ubuntu 16.04 LTS operating system. There were tested 3 KVM and 3 Hyper–V configurations, each composed of 2 virtual machines with the parameters presented in the Table 1.

Table 1. Local virtual machine configurations

Configuration name	No. of virtual CPUs	RAM size [GB]
CPU-1	1	1
CPU-2	2	2
CPU-4	4	3

For tests in real clouds there were selected virtual machines located in US (precisely in Virginia) and EU (precisely Holland). The tested configuration has parameters presented in the Tables 2 and 3.

Table 2. Google Cloud virtual machines

Configuration name	Vendor name	No. of CPUs	Location	Price [$/h]
GC-EU CPU-1	n1-standard	1	EU	0.0346
GC-EU CPU-2	n2-standard	2	EU	0.072
GC-EU CPU-4	n4-standard	4	EU	0.144
GC-US CPU-1	n1-standard	1	US	0.038
GC-US CPU-2	n2-standard	2	US	0.076
GC-US CPU-4	n4-standard	4	US	0.154

Table 3. Azure virtual machines

Configuration name	Vendor name	No. of CPUs	Location	Price [$/h]
Az-EU CPU-1	DS1 v2	1	EU	0.068
Az-EU CPU-2	DS2 v2	2	EU	0.136
Az-EU CPU-4	DS4 v2	4	EU	0.272
Az-US CPU-1	DS1 v2	1	US	0.07
Az-US CPU-2	DS2 v2	2	US	0.14
Az-US CPU-4	DS4 v2	4	US	0.279

From both vendors there were selected standard configuration in view of its widespread use and moderate cost.

4 Measurements and Analysis

The first experiment was performed to compare processing in virtual machines based on two solutions used in later tested real clouds,. in pure, isolated local environment (local computer), i.e. KVM and Hyper–V based machines. The Fig. 1 presents the evaluation of effectiveness of processing with use of Apdex index, for the configurations CPU–1, CPU–2, and CPU–4, for load L generated by 50, 100, and 200 users.

The measurements were performed with and without additional constant load. The results of comparison are similar in both cases apart one fact visible in the Fig. 1. Hyper–V platform is slightly less effective (however very close to KVM) apart from two cases. With additional constant load its effectiveness more decreases on configuration with one CPU for all loads, but works clearly better for bigger load when more CPUs are available (L = 200, CPU–4 configuration).

The comparison of processing in real clouds was performed without taking into account the financial cost of processing first. Using Apdex index there were not observed significant differences between measurements during the day and the night, therefore the results have been aggregated. In the Fig. 2 is presented evaluation of effectiveness of processing for the configurations located in EU and US for CPU–1, CPU–2, and CPU–4 configurations, for load L = 50, 100, and 200.

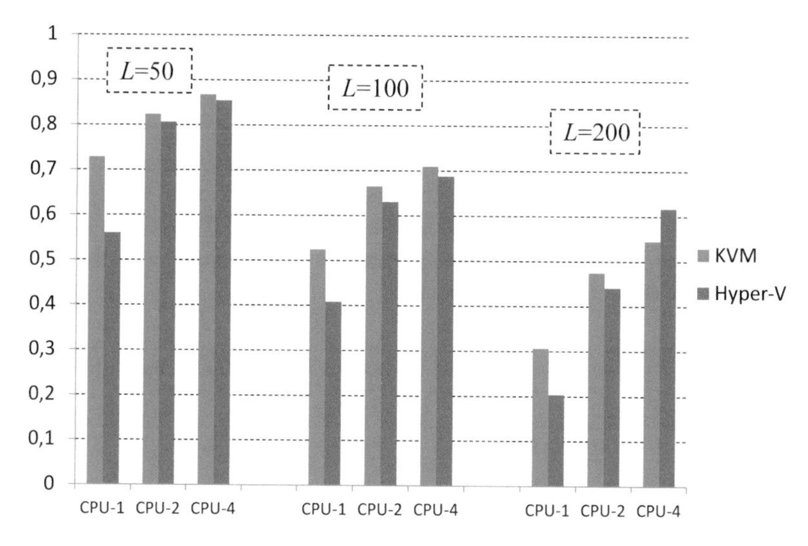

Fig. 1. The evaluation of effectiveness of processing with use of Apdex index for KVM and Hyper–V virtual machines in local environment (with additional constant load) for configurations CPU–1, CPU–2 and CPU-4, and for load L = 50, 100, and 200.

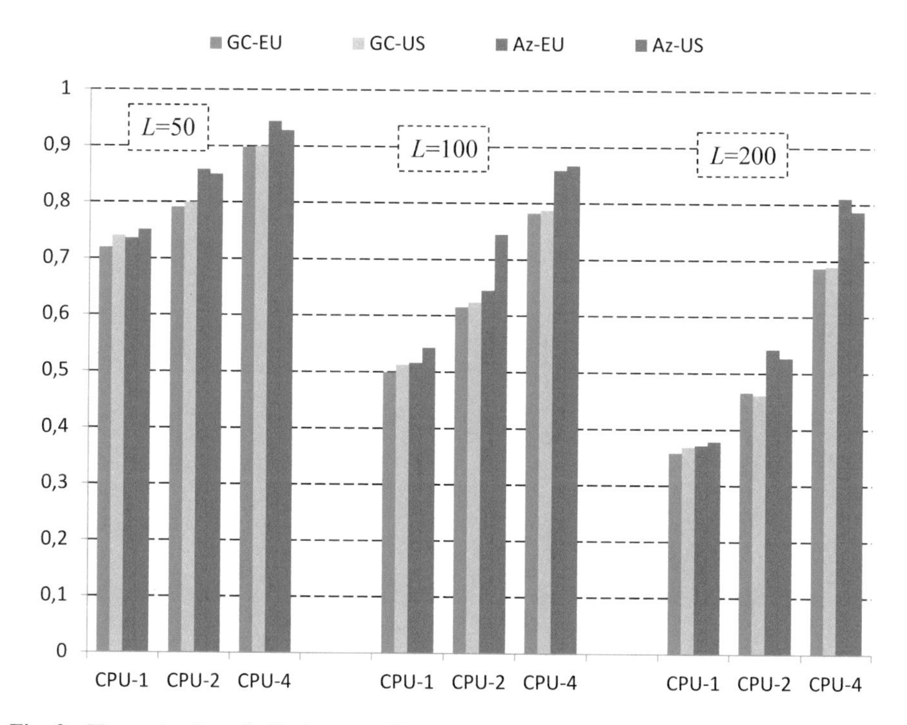

Fig. 2. The evaluation of effectiveness of processing with use of Apdex index for Google Cloud and Azure, in EU and US, for configurations CPU–1, CPU–2 and CPU–4, and for load L = 50, 100, and 200.

The differences between location in EU and US are small (however machines in US are usually slightly faster). In contrast to local measurements, machines in Azure are a little more efficient. This is probably caused by using more powerful equipment. The difference is bigger for bigger load and when more CPUs are used, what matches local measurements. The only discrepancy is for Az-EU CPU-2 configuration. It seems that during this test something happened in Azure cloud and processing performance decreased globally in EU location. Detailed results show that it happened for daily test. The standard deviation of measurements presented in the Table 4 show that results for Google Cloud are more stable in general.

Table 4. Standard deviation for aggregated measurements in tested clouds, in EU and US.

Cloud	GC-EU	GC-US	Az-EU	Az-US
Standard deviation	0.024	0.025	0.035	0.028

Finally, on the basis of collected measurements and the cost of computing for each configuration the effectiveness considering financial cost was evaluated with use of APPI index. The assessment was performed for various acceptable price. The two examples are presented in the Figs. 3 and 4.

In the Fig. 3 are compared all configurations for load $L = 50$ and for acceptable cost of processing $C = 0.14$ $/h. While for no cost limit (using Apdex index) almost in all cases the best solution for processing was Azure, in this case the acceptable price does not affect less expensive configurations CPU–1 and CPU–2, and the choice is the same, but the impact of price for configuration CPU–4 is obvious and the cost limit now show strongly the GC-EU configuration as the best solution.

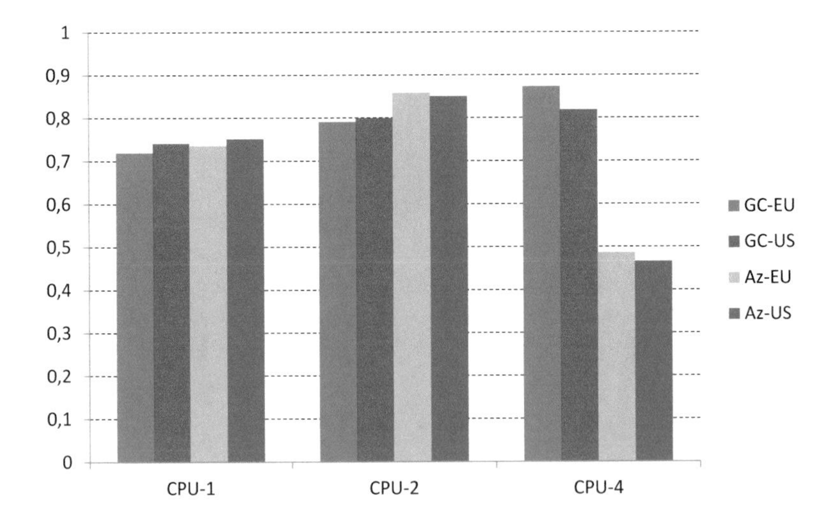

Fig. 3. The evaluation of effectiveness of processing with use of APPI index for Google Cloud and Azure, for configurations CPU–1, CPU–2 and CPU–4, and for load L = 50, and acceptable price $C = 0.14$ $/h

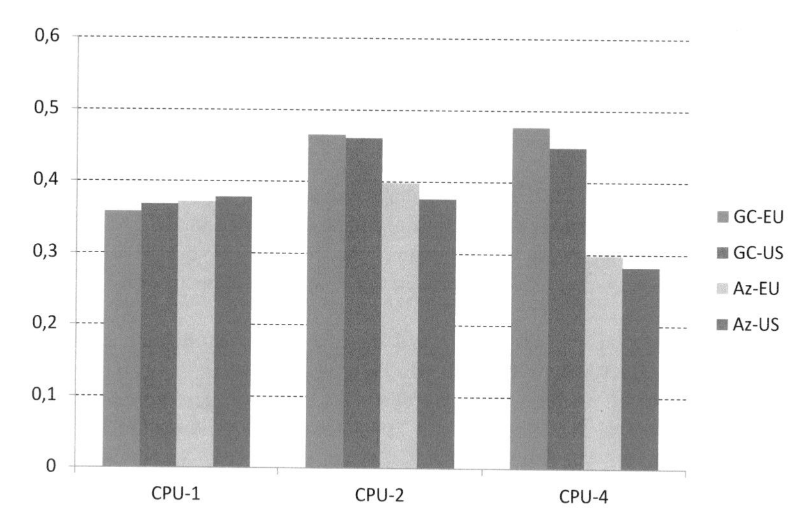

Fig. 4. The evaluation of effectiveness of processing with use of APPI index for Google Cloud and Azure, for configurations CPU–1, CPU–2 and CPU–4, and for load L = 200, and acceptable price $C = 0.10$ $/h

In the Fig. 4 are compared configurations for bigger load $L = 200$ and for acceptable cost of processing $C = 0.10$ $/h. The lower acceptable price now points to Google Cloud as recommended solution as well for CPU–4 as for CPU–2 configurations.

Another conclusion derived from the Figs. 3 and 4 is that new APPI index decreases evaluation of effectiveness when the cost of computing goes beyond the acceptable price significantly. It raises some doubts if this impact is not too strong.

5 Final Remarks

The paper focuses on issues related to evaluation of processing environment for cloud computing. The goal of presented study was to investigate potentiality for assessment of processing in selected computational clouds (i.e. virtual machines used in considered clouds) taking into consideration also financial cost constraints.

In the paper the APPI index (Application Performance and Price Index) is proposed. This index can be used for selection of recommended cloud platform configuration considering not only response time or speed of computing, but also acceptable price of processing. Considered cases showed how for various load and demanded resources the choice of "best" platform can change.

The proposed APPI index in its preliminary form is not flexible. It rigidly treats overrun of accepted price of processing and decrease evaluation of processing environment heavily. It seems that it should be tuned more flexible. Considering some more advanced balancing between financial cost and speed of processing (response time) could be more interesting for potential cloud service customer. This is worth to study in future work.

References

1. Cloud Computing Trends: 2017 State of the Cloud Survey. https://www.rightscale.com/blog/cloud-industry-insights/cloud-computing-trends-2017-state-cloud-survey. Accessed 21 May 2019
2. Kaminska, M., Smihily, M.: Cloud computing - statistics on the use by enterprises 2018. https://ec.europa.eu/eurostat/statistics-explained/index.php/. Accessed 21 May 2019
3. Sevcik, P.: Apdex interprets app measurements. Network World 2005. https://www.networkworld.com/article/2322637/apdex-interprets-app-measurements.html. Accessed 21 May 2019
4. Staś, M.: Performance evaluation of virtual machines in the computing clouds. Master's Thesis, Wroclaw University of Science and Technology (2019)
5. Leitner, P., Cito, J.: Patterns in the chaos – a study of performance variation and predictability in public IaaS clouds. ACM Trans. Internet Tech. **6**(3), 15 (2016)
6. Shankar, S., Acken, J.M., Sehgal, N.K.: Measuring performance variability in the clouds. IETE Tech. Rev. **35**(6), 1–5 (2017)
7. Popescu, D.A., Zilberman, N., Moore, A.W.: Characterizing the impact of network latency on cloud-based applications performance. Technical report number 914, University of Cambridge - Computer Laboratory. https://www.cl.cam.ac.uk/techreports/UCAM-CL-TR-914.pdf. Accessed 21 May 2019
8. Dhall, C.: Scalability Patterns - Best Practices for Designing High Volume Websites. 1st edn, Apress, Berlin (2018)
9. Chen, T., Bahsoon, R.: Toward a smarter cloud: self-aware autoscaling of cloud configurations and resources. Computer **48**(9), 93–96 (2015)
10. Aminm, F., Khan, M.: Web server performance evaluation in cloud computing and local environment. Master's Thesis, School of Computing Blekinge Institute of Technology (2012)
11. Fraczek, J., Zajac, Ł.: Data processing performance analysis in Windows Azure cloud. Studia Informatica **34**(2A), 97–112 (2013)
12. Habrat, K., Ładniak, M., Onderka, Z.: Efficiency analysis of web application based on cloud system. Studia Informatica **35**(3), 17–28 (2014)
13. Everts, T.: Time Is Money - The Business Value of Web Performance, 1st edn. O'Reilly Media Inc., Sebastopol (2016)

WebAssembly – Hope for Fast Acceleration of Web Applications Using JavaScript

Krystian Fras and Ziemowit Nowak(✉)

Faculty of Computer Science and Management, Wrocław University of Science and Technology, Wybrzeże Wyspiańskiego 27, 50-370 Wrocław, Poland
ziemowit.nowak@pwr.edu.pl

Abstract. One of the most difficult experiences of users of modern web applications is their performance, which is too slow. Those responsible for the development of technology and applications put a lot of effort to meet the requirements of users. It turns out that their actions do not always bring the expected results, as the hindrance is on the client's side – the monopoly of JavaScript language.

The authors of this study reviewed issues in the context of the performance of front-end applications (Single Page Applications). They presented the metrics for comparing application performance, discussed the issue of the multi-threaded character of web applications and the principles of Progressive Web Applications.

The key issue raised by the authors is the revolutionary functionality introduced recently in leading internet browsers. Currently, browsers can receive and execute low-level machine code. WebAssembly (WASM), which is the name of the new binary instructions format, allows applications to run at a speed comparable to native (desktop) solutions. Low-level application code is the result of compiling programs written in C, C++ or Rust and is fed into web browsers by web servers.

The authors listed the possibilities offered by the WebAssembly and explained how it cooperates with JavaScript. They described two cases of the application of WebAssembly in business, and characterized its limitations. Finally, they looked at the Yew framework, which is the most advanced one among those using the WebAssembly code.

Keywords: Single Page Application · Web Worker · Progressive Web Application · Framework · Yew

1 Introduction

1.1 JavaScript

JavaScript is a relatively young programming language. May 1995 is generally considered its creation date, when a Netscape programmer developed its current premises in just 10 days [1]. Initially, it was called Mocha, later LightScript, and finally JavaScript. It was first used for simple tasks performed in the browser (opening menus, alerts). Over time, it became the basis for more complex operations and larger systems. The development of the Gmail application by Google in 2004 is often considered a turning point. It was unusual at the time to be able to update the content of the website,

L. Borzemski et al. (Eds.): ISAT 2019, AISC 1050, pp. 275–284, 2020.
https://doi.org/10.1007/978-3-030-30440-9_26

without the need to refresh the browser tab simultaneously. It showed that using AJAX technology (Asynchronous JavaScript and XML) was sensible and paved the path for the currently popular Single Page Applications (SPA).

The dynamically developing language needed standardization in order to maintain support for web browsers developed by many independent companies. Netscape began the standardization process with the Ecma International association. Since then, several versions of the standard named EcmaScript have been created. The one that lasted the longest (7 years) was EcmaScript 5, created in 2009.

Since the emergence of EcmaScript 6 (ES6) version, the standardization process has changed and now a new standard is published every year. This helps to avoid complex changes compared to previous versions and ensures easier implementation for web browser developers.

The reason for the dynamic development of JavaScript is the continuous development of front-end applications. Essentially, the very term "web application" reflects the changes in the perception of the field. Formerly, most web systems acted as simple websites. All complex operations were performed on the server's side. Advanced client applications, on the other hand, were mostly native programs installed directly on the platform.

The continuous development of JavaScript allowed, first of all, to create the technology of Single Page Application (SPA) [2, 3], i.e. web systems in which a large part of the operations is shifted onto the client in the client-server architecture. In such systems, the page is not reloaded as the front-end application has internal routing enabling the change of page elements using JavaScript. SPA has changed the perception of web browsers. It is now the launch environment for an advanced application that can often work even in the offline mode (Progressive Web Applications) [4].

The low performance of front-end applications is the major obstacle to the development of web systems. The developers of applications are familiar with the problem and various solutions are suggested [5]. One of them is the introduction of multithreading into JavaScript, another one – the aforementioned progressiveness of the application with the intensive use of buffering. Still, the technology that seems the most promising to the programming community is the introduction of the option to handle the new format of binary instructions into browsers – WebAssembly.

1.2 Contribution

The authors of this study reviewed the problem area in the context of front-end application's (SPA) performance. They presented the metrics for comparing application performance, discussed the issue of multi-threaded web applications and the principles of functioning of progressive applications (PWA).

The main issue they discussed is the use of WebAssembly in web applications – a new format of binary instructions that allows applications to run at a speed similar to native (desktop) solutions. The authors presented the potential of the WebAssembly code and how it works with JavaScript. They described two cases of the application of WebAssembly in business, and discussed its limitations. They also looked at the Yew framework, which is the most advanced one among those using the WebAssembly language.

2 Measuring SPA Performance

Performance in the context of front-end applications refers to two measures [6]:

- *web performance* – time in which the application will be delivered and displayed,
- *runtime performance* – time in which the application responds to an interaction with a user.

The first measure is primarily related to the way the browser downloads resources – the TCP/IP model and the HTTP protocol. It is an extensive subject, which, however, is more relevant to the web system architecture itself than to the application performance analysis. The second measure refers to the time in which the application processes the information. *Runtime performance* determines the response time to an interaction with the user, for example a change of page after clicking the button.

Runtime performance is a complex measure influenced by many factors such as: algorithms applied, asynchronous operations, memory and CPU load, resource downloading, as well as optimization of the code processed by the JavaScript interpreter.

Both *web performance* as well as *runtime performance* are measures which allow for the "technical" expression of performance. The very purpose of this process, on the other hand, is primarily to improve user experience.

Another, frequently encountered metric is *time-to-interactive* [7]. It is a measure that determines the time in which the application will be delivered, displayed and will start to respond to user events. It can be considered as the sum of earlier, technical measures, but only at the first launch of the site. An ideal example is an online store. Let us assume that the user will go the following path:

- opening the product page,
- indicating the amount of products,
- moving to the basket.

A frequent case is a relatively fast display of the page with incomplete content (e.g. displaying the loaders) and no response to the interaction with the user. It is only the moment when the content is full and the page responds to user actions that marks *time-to-interactive*. This period in which the application was downloaded but does not respond to user events is often caused by the build-up of JavaScript functions on the call stack and is referred to as *ui-freeze*.

3 Multithreading

In order to implement multithreading in web applications to the HTML5 standard, *Web Workers* were introduced. They provide an API that enables running JavaScript in a separate physical thread. *Web Workers* mainly allow for the relief of the main thread. The perfect application of *Web Workers* is performing complex arithmetic calculations and thus eliminate *ui-freeze* that can occur when processing the same data in the main thread.

When creating new *Workers* it should be borne in mind that they burden the user's system. For this reason, their number should not theoretically exceed the number of available processor threads.

Detailed analysis of the impact of the number *Workers* on application performance is presented in paper [8]. According to the authors, the current implementation method of *Web Workers* does not allow for estimating their optimal number that developers should create in the application. This is due to many factors, the most important of which are: system architecture, number of threads, type and version of the browser. However, the authors managed to show a positive influence of *Workers* on the application. Moreover, increasing their number over the number of processor threads did not affect the performance of the application negatively. In their opinion, it is safe to create at least one *Web Worker* for complex arithmetic calculations performed in parallel to the main thread.

The registration of a *Worker* involves creating its object by means of providing the requested JavaScript code. Then communication with the *Worker's* thread via the *postmessage* and *onmessage* method is possible. Methods are available in the main thread through the *Worker* object, and natively in the *Worker* via the browser's API.

Additionally *Workers* have access to all elements of the EcmaScript standard supported by a given browser. The main limitation of *Web Workers* is, first of all, no access to the Document Object Model (DOM) tree and the *window* object. When *Web Workers* are used communication becomes a problematic issue. The main thread communicates with *Web Worker* by sending and receiving messages along with the value. The value can be sent in the following way:

- *Duplication* – operation available in many browsers and used from the beginning of introduction of *Web Workers*. It is expensive, at the same time, with complex data structures and problematic for object types.
- *Transfer* – the method incorporated in Chrome 17, which is based on *Transferable Objects* and enables the transfer of objects between threads. An object sent from the thread simultaneously ceases to be available in it. This means no access to the value sent. This type of transmission is characterized by higher efficiency.

Communication with *Web Workers* is difficult and takes place through a complicated, native JavaScript interface. As a result, there are many libraries that are native API wrappers and enabling easier communication. One of them is the *Comlink* library implemented by Google for ease of *Web Workers'* use and optimizing data transfer.

4 Progressive Web Applications

Progressive applications are those which follow the rules listed below [4].

- Progressivity – working on any browser, in any environment and for each user.
- Responsiveness – adaptation to each device: mobile, tablet, desktop.
- Independence from connection – offline and online operation.
- App-like operation – the user's experience is similar to the native platform (e.g. desktop and mobile applications).

- Data consistency – the application version is always updated as a result of the refreshing of the so-called *Service Worker* [9].
- Safety – TLS standard.
- Detection – due to the W3C manifest and *Service Worker* applications are treated as native.
- Installability – the ability to create a shortcut to the application and stop it, for example, on the device's desktop.
- Linkability – the ability to share the application via a traditional URL link.

From the above rules it results that progressive applications are primarily to imitate native applications while maintaining the structure (and implementation) of the web application.

Progressive applications consist of two elements: the *manifest* file and *Service Worker*. The manifest contains information on the elements necessary to install the application and display it in the system menu. It includes, among others, the logo, application name, and colors.

Service Worker allows to run the application offline and to install it. To maintain compatibility with new versions, when running the application, a query is performed to check if there is a newer version of the application files and the *Service Worker* itself. When an update is necessary, a new *Service Worker* is installed, but the activation is suspended until the last tab of the browser with the application is closed. In this way it is possible to maintain compatibility for the earlier version of the application.

Service Workers allow greater control over what and how it is saved and delivered from the cache memory (RAM and disk) [9]. Unlike traditional cache memory, they allow applications to be started offline easily. The publication [9] presents the results of performance analysis of "Google I/O web app" application. The application was enriched with analytical tools that monitored the performance on end-user devices. The research showed that the use of the Service Worker resulted in a double reduction in page loading time, which translates into better user experience. Additionally, further analysis of the collected data showed that nearly 5% of users make use of the opportunity to display applications offline.

5 Web Assembly

As mentioned before, JavaScript has been the only language natively supported by web browsers for many years. There are many languages transpiled into JavaScript (e.g. TypeScript). For years, even languages such as C or C++ have received compilers that generate a subset of the JavaScript language called asm.js [10, 11]. The increasing JavaScript load on virtual machines resulted in a number of improvements in the engines of web browsers. Currently, JavaScript code is optimized more efficiently than it used to be. Still, there are problems that are related to the very nature of the language – the use in web applications was not its original intended use.

Engineers from companies: Google, Mozilla, Microsoft and Apple have jointly developed a low-level binary code called WebAssembly (WASM). The author's design decisions are described in the publication titled "Bringing the Web up to Speed with

WebAssembly" [12]. It is a joint work of the companies responsible for the four major Internet browsers: Chrome, Firefox, Safari, Edge.

WebAssembly consists of two elements: text files with the.wat extension in the format easy to understand by humans, and binary code with the .wasm extension, which is delivered to the browser. The language is, by assumption, an abstraction over the physical platform of the device, which is to become independent of a particular programming language. WASM is to be compiled from popular languages such as Kotlin or Rust [13]. The basic premises of WebAssembly are primarily:

- safety of performance,
- speed of execution,
- independence from language, equipment and platform,
- simplicity of cooperation with the web platform,
- ease and speed of decoding,
- ease of validation and compilation,
- ease of generation,
- streamability and concurrency.

WASM is the first language developed by the four major browser developers and an independent community, designed, by definition, as a low-level language, efficient on any platform. According to the authors of the publication [12], it is the first industrial language modeled from scratch with the preservation of correct, formal semantics. Moreover, WASM is designed with the assumption of concurrent activity, which is uncommon practice in JavaScript where it is used reluctantly. WebAssembly, by definition, is intended to be interpretable through JavaScript interpreters inside web browsers – without the need to create new engines.

5.1 Applications

Figma. In November 2017, the official, full WASM support was announced in the four major browsers (listed above) [14]. One of the first companies to use WASM was Figma [15]. Figma is a popular web tool used to create website designs and mobile applications. The main performance challenge of the system is the loading time of files, parsing and rendering of the DOM tree.

Figma is an extensive web system and downloads a lot of files to create a frontend application. The creators managed to speed up *time-to-interactive* three times as a result of WASM use. The files were still to be downloaded via JavaScript code due to language limitations. WASM files themselves, on the other hand, had a considerable advantage over the former JavaScript counterparts, as WASM was designed with the smallest size of files and the speed of parsing by the JavaScript interpreter in the browser in mind. Additionally, Figma formerly relied on the C++ code from which asm.js was generated.

Now asm.js is a subset of the WebAssembly language, which is more similar to C++. For this reason, the C++ compiled code had to emulate 64-bit Integer values. In addition, the C++ code is optimized by LLVM (*Low Level Virtual Machine*) [16] even before it is recompiled to WASM. It results in the ease of interpretation by the browser.

JavaScript code requires a multi-layer optimization and parsing process (in the browser) before it can be interpreted. The WASM code is compiled to binary code, and the interpretation follows with the omission of the optimization and parsing stage.

eBay. In May 2019, eBay described the latest business use of the WebAssembly on its website [17]. According to the authors of the publication, the use of WASM in the e-commerce industry is still largely experimental and it is difficult to find a good business reason to use it. WebAssembly is supported by few browsers, and performance gains are not always sufficient reasons for its use.

In 2017 eBay introduced a barcode scanner to its web application corresponding to the scanner available in the mobile application. Research showed that in 80% of cases users stopped using the scanner because the *runtime performance* of the tool was extremely low. Users assumed that the application stopped responding to interaction and turned it off.

Differences in the performance of browser engines affect the different execution times of algorithms. For this reason, eBay chose three independent algorithms:

- ZBar library [18] and WebAssembly implementation,
- own eBay solution using WebAssembly,
- own eBay solution using JavaScript.

WASM files were then loaded into the application via JavaScript code (the so-called *glue code* – the only way to load the WebAssembly in the browser to date). All three implementations were running in separate threads through *Web Worker*. The control over *Workers* was accomplished by means of using the main thread. Each of the algorithms processed the barcode independently to get a quick result. Then the first algorithm that designated the bar code numbers returned the result via *Web Worker* and the main thread interrupted the remaining operations and removed all three *Workers*.

EBay then created a simple A/B test. Some users of the platform saw the version of the application with a field to enter the barcode manually, and some with the barcode scanner. The test involved checking the percentage of users who finish the process of entering the code positively in both cases (i.e. continue until the end of the process). The manual entry of the code scored below 40% while the code scanner was just under 70%. Moreover, it was examined which code recognition algorithm completed the process first and proved to be the most efficient. The implementation based on ZBar was in 53% the fastest algorithm. Additionally, in 86% of cases the WASM implementation proved to be the most efficient solution.

5.2 Limitations

WebAssembly has many limitations. One of the most serious ones is the lack of access to the DOM tree of the application. To overcome it, WASM communication with JavaScript is required. Fortunately, the developers are working on removing this limitation.

The current approach assumes the use of WASM mainly for processing complex logic, and JavaScript for manipulating the DOM tree (application presentation layer). Both languages complement each other quite well. JavaScript allows for relatively simple integration with WASM – it supports the import of .wasm modules, and bundles (e.g. the *webpack*) in the latest versions support the use of WebAssembly.

Currently, the compilation to WASM is most fully supported by the C, C++ and Rust languages, whose common feature is the *manual* memory management. WebAssembly still does not have a garbage collector which is to be added in the forthcoming versions and will facilitate the compilations of languages such as Java or C#. Compilers for these languages are already available, but they work in a rather limited form.

5.3 Frameworks

The most advanced and mature (from the business perspective) front-end based framework supporting the compilation to WASM is Yew [19]. This is a framework for Rust, with architecture modeled on ReactJS and Elm. The Rust files are compiled to WebAssembly or asm.js.

At the core of the framework there lies the multithreading of the implemented systems, as well as the component architecture. Each component may create its own process called an agent (or actor) and communicate with it. Agents are separate processes that run in parallel and are managed by a scheduler. Unfortunately, the framework is still in the testing phase, and the implementation of applications with its use is problematic due to frequent API changes, lack of additional libraries and illegible documentation.

The author of the Yew framework made a study of the application performance written in WASM and based on Yew [20]. The author implemented the TodoMVC application. It is a well-known system that has been implemented in over 16 popular frontend frameworks to compare them. The application is open-source and is used to present the use of popular frameworks in manufacturing conditions. This is to help programmers looking for the right technology for their own applications. The author compared the performance of the TodoMVC application implemented using the Yew framework with the performance of applications implemented using other popular frontend frameworks. The TodoMVC application using the Yew framework ranked first. In other words, the investment made in WASM offers an advantage over the competition.

6 Conclusions

JavaScript language is the foundation of modern web applications operating on the client's side. Yet, both users and application developers are increasingly aware of the limitations of using the language. The basic issue is application performance. To solve this problem, various improvements are described in this publication. Among them, WebAssembly is certainly revolutionary due to its detachment from the JavaScript language. Why is it so? Because only compiled .wasm files (containing compact binary code) are sent over the network. Because the analysis and conversion of JavaScript code in the Abstract Syntax Tree (AST) is bypassed. Because the code is not optimized using compilers that consume processor resources – the binary code is optimized and ready for immediate operation in every browser engine.

WASM has a lot of limitations. It does not use the Garbage Collector known from JavaScript. It cannot call complex JavaScript elements (it cannot mutate the DOM). If it needs to communicate with the outside world using the API, it needs to do so through JavaScript. It does not support source maps, so it is not possible to debug the code. It is single-threaded, but it is only a matter of time when it becomes multi-threaded.

Currently, game developers benefit from WASM the most. Also developers of libraries that use the processor intensively should start using it as soon as possible. It is almost certain that, as the technology develops, an ever-growing body of web programmers will use WASM. The biggest winner in this revolution will be the user who will benefit from the hitherto unknown speed of web applications, but also less power consumption of their device – after all, the code will be compiled outside the device.

References

1. Rauschmayer, A.: Speaking JavaScript: An In-Depth Guide for Programmers, 1st edn. O'Reilly Media, Sebastopol (2014). http://speakingjs.com/es5/ch04.html
2. Karabin, D., Nowak, Z.: AngularJS vs. Ember.js – performance analysis frameworks for SPA web applications [in Polish]. In: Kosiuczenko, et al. (eds.) From Processes to Software: Research and Practice, pp. 137–152. Polish Information Processing Society, Warsaw (2015)
3. Stępniak, W., Nowak, Z.: Performance analysis of SPA web systems. In: Borzemski, L., Grzech, A., Świątek, J., Wilimowska, Z. (eds.) Information Systems Architecture and Technology: Proceedings of 37th International Conference on Information Systems Architecture and Technology – ISAT 2016 – Part I. Advances in Intelligent Systems and Computing, vol. 521, pp. 235–247. Springer, Cham (2017)
4. Osmani, A.: Getting started with Progressive Web Apps (2015). https://addyosmani.com/blog/getting-started-with-progressive-web-apps/. Accessed 15 Jul 2019
5. Chęć, D., Nowak, Z.: The performance analysis of web applications based on virtual DOM and reactive user interfaces. In: Kosiuczenko, P., Zieliński, Z. (eds.) Engineering Software Systems: Research and Praxis. Advances in Intelligent Systems and Computing, pp. 119–134. Springer, Cham (2019)
6. Barker, T.: Pro JavaScript Performance: Monitoring and Visualization. Apress, New York (2012)
7. Osmani, A.: The Cost Of JavaScript In 2018 (2018). https://medium.com/@addyosmani/the-cost-of-javascript-in-2018-7d8950fbb5d4. Accessed 15 Jul 2019
8. Pajuelo, A., Verdú, J.: Performance scalability analysis of javascript applications with web workers. IEEE Comput. Archit. Lett. **15**(2), 105–108 (2016)
9. Walton, P.: Measuring the Real-world Performance Impact of Service Workers (2016). https://developers.google.com/web/showcase/2016/service-worker-perf. Accessed 15 Jul 2019
10. Herman, D., Wagner, L., Zakai, A.: asm.js Working Draft (2014). http://asmjs.org/spec/latest/. Accessed 15 Jul 2019
11. Van Es, N., Nicolay, J., Stievenart, Q., D'Hondt, T., De Roover, C.: A performant scheme interpreter in asm.js. In: Proceedings 31st Annual ACM Symposium on Applied Computing, SAC 2016, pp. 1944–1951. ACM (2016)

12. Haas, A., Rossberg, A., Schuff, D., Holman, M., Gohman, D., Wagner, L., Zakai, A., Bastien, J.: Bringing the web up to speed with webassembly. In: Proceedings 38th ACM SIGPLAN Conference on Programming Language Design and Implementation, PLDI 2017, pp. 185–200. ACM (2017)
13. Belyakova, J.: Language support for generic programming in object-oriented languages: peculiarities, drawbacks, ways of improvement. In: Castor, F., Liu, Y. (eds.) Programming Languages. SBLP 2016. Lecture Notes in Computer Science. Springer, Cham (2016)
14. McConnell, J.: WebAssembly support now shipping in all major browsers (2017). https://blog.mozilla.org/blog/2017/11/13/webassembly-in-browsers/. Accessed 15 Jul 2019
15. Wallace, E.: WebAssembly cut Figma's load time by 3× (2017). https://www.figma.com/blog/webassembly-cut-figmas-load-time-by-3x/. Accessed 15 Jul 2019
16. Lattner, C., Adve, V.: LLVM: a compilation framework for lifelong program analysis & transformation. In: Proceedings International Symposium on Code Generation and Optimization: Feedback-Directed and Runtime Optimization, CGO 2004. IEEE Computer Society, Washington (2004)
17. Jha, P., Padmanabhan, S.: WebAssembly at eBay: A Real-World Use Case (2019). https://www.ebayinc.com/stories/blogs/tech/webassembly-at-ebay-a-real-world-use-case. Accessed 15 Jul 2019
18. Brown, J.: ZBar bar code reader (2010). http://zbar.sourceforge.net/. Accessed 15 Jul 2019
19. Kolodin, D.: Yew Framework (2018). https://github.com/DenisKolodin/yew. Accessed 15 Jul 2019
20. Kolodin, D.: TodoMVC Benchmark (2018). https://github.com/DenisKolodin/todomvc-perf-comparison. Accessed 15 Jul 2019

Measured vs. Perceived Web Performance

Leszek Borzemski(✉) and Maja Kędras

Faculty of Computer Science and Management, Department of Computer Science, Wrocław University of Science and Technology, 50-370 Wrocław, Poland
leszek.borzemski@pwr.edu.pl

Abstract. We considered the measured vs. perceived web performance case study. The work aimed to find a relationship between the automatically measured times of loading elements on web pages with zoological articles and the feelings of users regarding these loading times. We found that the following automatically measured performance metrics, translate the best into the perceived performance evaluated by the users: (i) *page_loading*, measured by Selenium; (ii) *first_meaningful_paint*, measured by PageSpeed Insights; (iii) *domInteractive* measured by Browsertime; and (iv) *contentEventLoad*, measured by Browsertime. In research carried out with the use of Selenium; however, there is a problem of changes in the interface that make the measurement unable to be performed correctly. A possible solution to this problem could be to change the used CSS selectors from selectors picked from the website itself to selectors defined based on the analysis of the selected HTML code. Another solution would be to use tools that allow, for example, to detect all graphic files on the website and design navigation based on them.

Keywords: Web performance · Measured performance · Perceived performance · Human-Computer Interface

1 Introduction

More and more areas of life are moving to the Internet. A typical example of such a trend is e-commerce. The technical preparation of such digital transformation is particularly important. This work includes, among other things, the development of the website, desktop and mobile user interface, preparing resources available, and optimization of website performance. The web designers and administrators can use different ways to optimize the web pages, following performance experts and rules-of-thumb.

While monitoring the performance of a website, we can apply different performance measures. However, the set of possible choices is vast. Our choice may depend on many factors, including the measurement method (active/passive), observation location (client/server/in-the-middle), the methodology of building individual performance measures (W3C, own) and tools implementing these methodologies. In each of the choices, we base our performance evaluation on the measured value of the selected performance index. We can then talk about *measured performance*.

L. Borzemski et al. (Eds.): ISAT 2019, AISC 1050, pp. 285–301, 2020.
https://doi.org/10.1007/978-3-030-30440-9_27

However, the question is still open. Which choice is the best one? Which performance index we should take into account for a particular tool? Why different tools, when measuring the same defined performance index, present different results? Why do we need to use all tools as suggested by some experts?

We need to take into account how the customer, now a digital customer, experience the transformation. Therefore, we also consider the second type of performance that we meet daily - the so-called *perceived performance*, which is related to how the end-user on his/her device notices the operation of the system.

Google shows that 53% of mobile website users resign from reading them when the web page loading time is more than 3 s [1]. Google's representatives, referring to issues related to user experience, argue that profit-oriented websites should try to reduce further user waiting time - so that it does not exceed two seconds [2]. We can see here that the discussion is about the measured performance. The system can measure the time spent in the realization of a particular system function. However, it is difficult to imagine the situation in which the user with the stopwatch in his/her hand measures time while accessing an e-commerce website.

From the other side, for most websites, however, such speed remains unattainable, especially if we take into account the fact that the size of websites is steadily increasing, with more and more resources including these requiring user feedback, such as information on the use of cookies or the proposal to subscribe to the newsletter.

Research [3] shows that the average loading time of a commercial website in 2018 was 8.66 s. This result exceeds both values suggested by Google, only confirming that these are standards that often go beyond the impossible.

Despite technical difficulties, striving to shorten the page loading time remains an important strategy related to the profit generated by the website [4].

In this paper, we study the relationship between the measured and perceived performance for users of a set of websites to determine which measured performance indices best match perceived performance. Four websites related to the sale of pet products were selected for the study. This choice was due to the interests of one of the authors, shared with people invited to the survey. The research on measured performance was carried out using the same computer equipped with chosen web performance measurement tools. The group of respondents who took part in the survey evaluating the perceived performance of individual pages consisted mainly of people aged 18 to 30, regularly using computers or mobile devices. At least half of the respondents had a technical education.

The paper is organized as follows. The second section provides an overview of related work. The third section includes an overview of the tools used to test the performance of websites and assess their usefulness in conducting academic research. The fourth section describes the research carried out with the three tools described in the second section. The fifth section contains a description of the survey carried out. The sixth section discusses the results of the tests described in the third and fourth chapters. The last section presents conclusions and possible future research.

2 Related Work

There have been many efforts to web performance auditing theory and practices that changed over the years from early web existence until now. Note that most of these efforts focus on either measured performance or perceived performance. There has been surprisingly little work on the relationship between both kinds of performance.

2.1 Auditing Approaches

The complete performance auditing requires the inspection of the whole website, which means to access different subpages related to particular user activities [5]—using such an approach, we can detect in the system the states that are performance bottlenecks, and correct separation of resources in cases where, for example, a dispersed application is involved. One of the ways to use such an audit is to rely on user logos. However, there is the problem of unambiguous identification of a set of queries made by one person. A possible solution to the problem of user identification is the use of cookies, user sessions (in the case where the application is logging in) or the browser fingerprinting method [6]. A useful feature of such an audit model is the ability to detect a part of the website that practically does not reach users, which may allow optimizing the management of resource consumption. It is also worth noting that such a model will change with the addition of a new subpage or the creation of another connection between already existing subpages, and its generation will be possible only after collecting the appropriate amount of data on user behavior.

In often-used evaluations [7], we examine website behavior with increasing load, taking the response time as one of the performance indexes, and testing the system's performance while overloading. Performance indexes can be evaluated in the long-term, which may lead to the detection of irregularities, such as memory leaks and errors in the database. There are three ways to run such tests - simulating virtual users, by testing usage scenarios or by testing specific website objects. The first approach is based on generating the traffic of different users and examining their impact on the system. The randomness of their behavior may, however, make the most critical points of the application unnoticed. The second approach assumes some orderly behavior, but its disadvantage is the need to define how potential users would use the system. The third approach requires breaking down the entire system into individual elements, such as links and buttons, and testing each one separately, for example, by recursively passing through the whole website.

Paper [8] studies web performance degradation due to website complexity. Another set of sizes used to test performance can be DNS server response times, TCP connection establishment time, and HTTP object load time [9]. A different approach will be to measure page loading times by changing the loading order, in particular, the components that depend on each other [10]. Another aspect that is considered in the performance study is the location of resources loaded by the page, where one of the approaches is to check how many resources are from another server [11]. The next set of compared performance indexes may be the times characteristic for the HTTP client behavior to get the web page - DNS, DNS2SYN, CONNECT, ACK2GET, FIRST BYTE, and LEFT BYTES [12, 13]. Web performance is also auditing to show the impact of the HTTP/2 protocol [14]. Another performance evaluation concern the use of the browser's cache by [15, 16].

When considering different performance auditing, we can not forget about the recommended specification to measure the performance of resource downloading, developed by the World Wide Web Consortium (W3C) [17]. This specification describes not only under which attribute names should the measurement results be saved, but also how the performance testing process should proceed. W3C has published a description of the attributes and the process of loading the whole webpage.

The difference in the performance measures can be due to the type of resource loaded, in particular, image files, video files and text files [18], which are resources directly received by users, in contrast to, for example, javascript files. The interface, content quality, as well as the appropriate technical preparation, will also have an impact on the website performance [19].

There are also proposals in the literature to investigate the impact of incorrectly loaded resources, such as, for example, graphics files, on the quality of the website [20]. The cases of loading only part of files, prolonged loading of files, and not loading files at all can be studied. These considerations can be done with CSS files, whose incorrect or incomplete loading changes the interface of the site, which is visible to the user. Another type of incorrectness that may affect page performance is to execute requests in order that the resource required for the next user action has not yet been executed, which either slows down the page loading by increasing the request processing time that has not been able to be successful or makes the resource will never be downloaded [21].

More and more users use websites via mobile devices. Mobile usage of websites makes a new aspect that has not been considered too much; that is the energy consumption of web page loading on smartphones [22]. This is important because mobile devices, unlike computers, are not connected to a power source during their typical use. This case may also be relevant for laptops that are designed to operate without an external power source. In the case of mobile devices, it is also necessary to appropriately adapt the website and its performance to various types of available Internet protocols and networks [23].

The audits discussed in this section allow determining numerical indicators of the web page performance. Usually, these are the times that elapse from the start of the observed system operation to the occurrence of a certain event, which we consider to be important for measured performance. It can be, such as the total page load time and time to-be-interactive.

In contrast, the perceived performance is to evaluate by methods and tools that are not necessarily accurate and reliable. The most commonly used method is to make a survey on this topic for a given set of webpages among a group of users. But there are many open questions before us. What websites do we need to take into account to get the "ultimate" and the proven result? What group of users do we need to define? Does a Top 100 websites research is enough?

2.2 Web Performance Measurement Tools

There are many tools to evaluate the measured web performance. In this section, we present only some of them. Therefore, this presentation should not be considered as a comprehensive report on these systems

mPulse. mPulse [24] is a SOASTA's software; now, the Akamai's product to collect data about the performance of a web application from the level of the browser to monitor its behavior in a real-time. It records information about the user, such as its geographical location and data about the device it uses. Also, it saves bandwidth data, page load time, and conversion rate. mPulse enables calculating the correlation for particular performance indicators, based on which the solutions that may increase the performance of the system can be proposed. Another metric offered by this tool is to identify those parts of the system that have the most significant impact on the conversion and duration of the user session. Also, it is possible to predict the level of user involvement based on the performance of the application, which allows deciding on the steps to improve its operation. Another feature of mPulse is the ability to present data in the form of various types of charts. Thanks to this, the program is particularly well suited to the business world, enabling to explain to non-technical people the motivations behind the decisions related to the development of specific software. An essential drawback in the academic/educational use of this tool is the fact that it is payable. mPulse has a repository of open and free software [25], but these are only overlays and extensions to paid software, utterly useless without a basic program.

GTmetrix. GTmetrix [26] is a software written by GT.net. It was developed for customers interested in purchasing hosting websites to test their performance. It allows collecting statistics such as page load time, page size, and the number of requests. It uses page design rules that have been set by Google and Yahoo. For this, we can compare the collected statistics with the results of other users who use this tool.

The measurements that can be done using this software allows comparing the website's response to requests sent from different locations of the world. Another application of this tool is the simulation of devices, including mobile devices, used by potential users. Thanks to this, it is possible, among other things, to test the compatibility of the layout of pages adapted to mobile devices with the screen resolution settings. GTmetrix allows conducting studies with two different browsers (Mozilla Firefox and Google Chrome), simulating the use by the user of a plug-in blocking part of the page content (Adblock), and simulating a weak internet connection, including 2G mobile network.

Like mPulse, GTmetrix is equipped with the option of generating charts showing the collected information. It can be accessed by the API, which facilitates the automatic maintenance of statistics. GTmetrix provides an incomplete version of its platform for free. It allows making reports using seven servers with different geographical locations. However, automatic monitoring of the website status is possible only from a server in Vancouver, Canada. An additional drawback of the free version is the limitation of the number of measurements taken to one test per day (while in the paid version it is possible to perform tests every hour). Moreover, we can monitor only three web pages at the same time. The free version of GTmatrix allows performing twenty queries a day, using the API.

Lighthouse. Lighthouse [27] is a tool developed by Google to measure the performance of websites. It is an open software helping the improvement of the performance of websites by measuring it and then generating hints about elements and solutions of the system that may improve its performance. Four versions have been developed:

- as one of the default tools of Chrome DevTools, which allows intuitive studies of web pages, especially those that require authentication by the user;
- as a plugin for Google Chrome, offering the same functions as the tool in Chrome DevTools, but in the form of a plugin. (The developers of this plugin point out that in the case when no arguments argue for using a plugin instead of a built-in tool, it is recommended to use the built-in tool);
- as the command-line tool, which allows full automation of tests using shell scripts;
- as a Node module to use the Lighthouse tool in a continuous integration system.

PageSpeed Insights. PageSpeed Insights [28] is a tool written by Google for testing the performance of websites. It is based, among other things, on the previously described Lighthouse tool. The system can perform two types of measurements - under controlled conditions, allowing easy search for elements of the website that are characterized by poor performance and by simulating real user traffic. The first of these approaches allows for the use of more metrics and is resistant to data interference caused by the unpredictability of users, but the latter approach, although more limited, enables finding parts of the system that generate problems for users. Also, data about the behavior of users on websites is collected by the Google Chrome web browser, which makes them very high quality and there are enough of them to make the research similar to live tests with real people. PageSpeed Insights offers the generation of basic charts presenting data collected during research, but it is more distinguished by providing APIs as a package for popular programming languages, such as Python, JavaScript, Java or PHP.

YSlow. YSlow [29] is a tool written by Yahoo! to study the performance of websites by a previously defined set of design rules. It takes the elements from the page (pictures, scripts, styles, etc.) and then checks their size and whether they have been compressed. It also studies web page headings. In summary, based on the data collected, it assesses the performance of the website. Various indicators are examined, each of which has a weight attached. The highest importance is to avoiding empty src and href tags, and the lowest - to ETags and the appearance of low-resolution page icons, that can be stored in the cache memory. Other indicators are, for example, CDN using, reducing the number of DNS queries, or not scaling images in HTML. YSlow has no data presentation function for collected data. It was written for Web 1.0, so it does not work well with the solutions currently used in web applications.

Selenium. Selenium [30] allows automatization of user actions and is intended to write end-to-end tests, i.e., those that test the entire system from the user's point of view. The advantage is that regardless of the reason, an error related to displaying data will be detected. The disadvantage is that Selenium will not indicate in which component or connection between which components an error appeared. The software can also be used to test the site's performance, using the ability to measure the time it takes to execute subsequent commands in the user's behavior scenario. In contrast to the previously described tools, Selenium permits to check the performance of specific solutions of the website, in particular, the speed of execution of different user activity. The disadvantage, however, is that it does not provide any functions related to the analysis of the data received.

Browsertime. Browsertime is the tool which has been developed based on the Selenium [31]. It launches the selected web page and collects information about it, such as:

- times on the navigation on the page, such as redirection times, connection establishment or termination of the connection;
- user times, i.e., times how page queries are performed;
- first paint, the time from sending a web page request to loading the first element on the page;
- RUM Speed Index [32] (Real User Monitoring Speed Index), i.e., the time of user interaction with page elements.

The web page performance research is supported by the generation of a HAR file that represents all requests to the web site and the responses. The JSON file is created with metrics collected using javascript. The tool is open and free, and thanks to returning data in JSON format and support from the console, it is suitable for automated tests. Unfortunately, just as Selenium does not have functions related to the development and analysis of data, it requires the use of additional, external tools.

Sitespeed.io. Sitespeed.io is a set of Open Source tools that makes it easy to monitor and measure the performance of a website [33]. It is based on the Browsertime. It is significantly different from its original. The introduced innovations include, among others, the default generation of the report in the form of an HTML file. It makes it much easier for people who are not familiar with technology to analyze results, but it makes it difficult to carry the tests automatically. Sitespeed.io uses the Coach [34] tool, which is described by its authors as a modern version of YSlow tool. Its most important function is to generate tips on how to improve the performance of a website. Guidelines are presented for both HTTP/1.1 and HTTP/2 protocols. Other tools that use sitespeed.io are PageXray [35], which is used to convert HAR files to JSON format, and Throttle [36], which is responsible for simulating free network connections, which allows for more detailed tests.

3 Case Study

3.1 Websites Under Evaluation

The case research was carried out on four Polish on-line pet shops dedicated to the sale of accessories for animals, namely:

1. https://www.mobilkarm.pl/
2. https://kakadu.pl/
3. https://www.bitiba.pl/
4. https://www.keko.pl/

We selected these websites as every second Google's search results for the query "zoological," bypassing websites that are not on-line shops, but the promotion of stationary stores only. One of the positioning criteria applied by Google search is website performance. Therefore, it is expected that speeder pages will get a higher position in the list of search results, while the slower ones will have a lower position. It is worth noting, however, that the impact of other positioning criteria used by Google may disrupt the tendency described above, and thus, from the user's perspective, the dependence of the position on the performance of the website may be irrelevant.

Initially, the pet shop https://fera.pl/ was also tested. However, it turned out that it changes the interface so often that performing tests in Selenium, imitating user traffic, was prevented due to the need to update the tests. The problem of the changing interface also appeared in the case of the rest of the sites studied; however, it occurred with much less severity. Therefore, it did not affect the ability to perform tests, but for this resulted in some periods without measurements, related to the need to update the tests.

Mobilkarm. This page has a simple interface. At the top of the page, there is a basic menu, while below a set of selected products currently covered by the promotion is displayed. The right part of the page has an additional menu divided into categories of products. Unfortunately, the page is not suited to the most popular screen resolutions, so to make the menu fully visible, it is necessary to move the screen view. In addition to the individual categories, the side menu also has a reference to the basket, which, however, remains inactive until the first product is added.

There are no windows and messages on Mobilkarm that are gaining popularity on many websites (in particular, there is no communication about the use of cookies here).

It is worth paying attention to the error in the interface, which consists in updating the reference to the basket located in the side menu only after adding two separate products or after refreshing the page. Note that there is a separate link that allows going to the subpage with the contents of the basket, although the responsible element of the interface is not yet active. From the waterfall chart, we see that 88 requests are made to page downloading, mainly to download graphics files but considering the size of downloaded files, the most data is downloaded as javascript files. From this chart, we can also find an error – a request for a resource that is not on the server.

Kakadu. The other side investigated was the Kakadu site. It contains not only an online store but also a lot of information about pets. Most of the links on the home page redirect to various types of animal articles, which means that the users interested in shopping must first find the appropriate link in the top menu. Our research was carried out on the home page because it is directed to it by Google search after searching for a pet store. 275 requests were made to load this page. This is the page that performs the most requests from all studied. Most requests were made for image files.

Bitiba. Bitiba website stands out graphically against the background of other analyzed pages, using intense colors, especially yellow and red. Similarly to the other websites considered, Bitiba is equipped with an upper menu - however, unlike Mobilkarm and Kakadu, it allows a direct transition to the user's category of interest. Bitiba is also the only website on which the side menu is located on the left. The other elements of the interface that should be mentioned are a banner informing about current promotions and a link to the basket. 87 requests were submitted to load the page.

Keko. Here also the menu at the top of the page is developed - it allows for more convenient navigation on the page. Right below the menu, there are links to information about additional offers and services provided by the shop, such as delivery or loyalty program. Also, some of the space was allocated for the promotion of selected products. At the top of the page, there is a link to the subpage with the view of the basket, active even when the basket is empty. Keko is downloaded via 120 requests. - most requests are made to get graphics files.

Mobilkarm has the fewest number of HTTP requests for given asset category, apart from the requests for image files, where Bitib has the least (Table 1). Kakadu has a higher number of requests for most categories. This is not true in case of requests for HTML files, where Keko has a higher number or in the case of requests for other resources, where Bitiba has the most. If we compare the percentage of requests, we can see that most of them are to get image files. Considering the size of the resources required, Mobilkarm has the smallest size in three categories (CSS scripts, images, other). However, the size of the HTML page is not the smallest - Bitiba is the lightest. In the case of javascript files, the lightest is Kakadu (Table 2). Images are the most popular resource type on these websites.

Table 1. Number of HTTP requests per page load/percentage share [%]

Resource	Mobilkarm	Kakadu	Bitiba	Keko
HTML/text	7/7.95	8/2.91	9/10.34	10/8.33
Javascript	13/14.77	29/10.55	14/16.09	23/19.17
CSS	3/3.41	23/8.36	8/9.20	4/3.33
Image	63/71.59	203/73.82	42/48.28	73/60.83
Other	2/2.27	12/4.36	14/16.09	10/8.33
Total	88	275	87	120

Table 2. Weight of resource [kB]/percentage share [%]

Resource	Mobilkarm	Kakadu	Bitiba	Keko
HTML/text	61.60/8.44	40.40/1.71	32.50/3.19	71.70/5.27
Javascript	353.50/48.44	247.10/10.43	309.00/30.33	550.00/40.39
CSS	46.65/6.39	58.60/2.47	69.60/6.83	75.00/5.51
Image	244.35/33.48	1700.00/71.78	505.50/49.61	550.00/40.39
Other	23.70/3.25	322.40/13.61	102.30/10.04	115.00/8.45
Total	729.8	2368.50	1018.90	1361.70

Note that, besides Mobilkarm, all websites used the HTTP/2 protocol - this may justify why the designers of Mobilkarm used the small number of requests. However, even in the case of Kakadu, with so many requests, it is less number than 660 requests that have been found for some website in the study [37] for the top 100 attractive e-commerce websites according to Alexa. The average number of requests performed per load was 192, with 60% of the sites performing less than 200 requests, and the fewest number of requests was 45. According to HTTP Archive, the average page size as of January 2018 was 3.54 MB [37]. This means that the sizes of tested websites are comfortably below the average size. Summing up, tested websites have been quite well designed, taking into account the number of HTTP requests per page load and page total size. Of course, we realize that the websites examined (probably) were much less complex than those from the Top 100.

3.2 Measured Performance Experiment

We used three tools to measure the performance: PageSpeed Insights, Browsertime, and Selenium. The individual scripts regarding particular tools were collected into one Python script, to run a single measurement experiment. Experiments were repeated in a loop along the required time. Four minutes gap was defined between successive experiments. This gap was in order to finish the browser work correctly and avoid possible complications, which might occur after incorrect termination of the Selenium test. The results of the tests were saved to JSON files. Despite the common approach to the performance audit, the structures and the contents of these files were different. Differences between the results related to the implementation of these tools could also be expected.

Moreover, Selenium due to the structural differences between tested websites required the definition of individual scenarios for every website.

The results were collected for 1213 experiments, from 2019-06-05 07:49:15 to 2019-06-12 21:21:53. Besides, due to changes in the interface of the site, in the case of experiments conducted with the use of Selenium, there are missing periods of study due to the need to update the scenario definition.

3.3 Perceived Performance Survey

The survey was carried out regarding the perceived speed of page loading. It consisted of the following questions:

1. From what device was the survey carried out?
2. Go to https://www.mobilkarm.pl/ page and rate how quickly it is loaded in the browser.
3. Find any dry dog food at https://www.mobilkarm.pl page and rate the speed of going through the next subpages.
4. Go to https://kakadu.pl/ page and rate how quickly it is loaded in the browser.
5. Find any dry dog food at https://kakadu.pl/ and rate the speed of passing through subsequent subpages.
6. Go to https://www.bitiba.pl/ page and rate how fast it is loaded in the browser.
7. Find any dry dog food at https://www.bitiba.pl/ and rate the speed of passing through subsequent subpages.
8. Go to https://www.keko.pl/ page and rate how fast it is loaded in the browser.
9. Find any dry dog food at https://keko.pl/ and rate the speed of passing through subsequent subpages.
10. Which page did you load the fastest?

11. Which page did you load the slowest?
12. Which site loaded the next subpages the fastest?
13. Which site loaded the next subpages the slowest? In which store would you like to shop?
14. Why?

The first question had two possible answers: "computer" or "mobile device." Questions from the second to the ninth included a choice between 1 and 10, where 1 meant very slow page loading, and 10 very fast. Questions from ten to thirteen inclusive had four possible answers – "Mobilkarm," "Kakadu," "Bitiba," and "Keko." All the questions described earlier were answered once. The last open question was to justify the choice of the page on which the surveyed users would most preferably do shopping.

3.4 Results

Measured Performance

We present the measured performance results obtained using Selenium. A wide range of measured metrics has been defined for this tool, namely:

- *page_loading* - the time from loading the test to loading the main page,
- *cookies* - the time of closing the message about the use of cookies by the website,
- *dropdown_menu_opening* - the time of expanding the slider menu after its launch,
- *category_chose* - the time of loading the category counted from clicking on the link opening it,
- *subcategory_chose* – the time of loading a sub-category counted from the time of clicking on the link that opens it on the category page,
- *subsubcategory_chose* – the time of loading the next subcategory counted from the time of clicking on the link opening it on the page of the previous subcategory,
- *item_details* - the time of opening the detailed product view counted from the moment you click on the link in the sub-category or category view,
- *cart_adding* - the time of adding an item to the cart counted from the moment you click on the button adding the product to the response of the interface informing about the success,
- *cart* – the loading time of basket contents counted from the moment of clicking on the link or button leading to the basket until the elements showing its content are loaded.

We could measure them for individual websites, as shown in Table 3.

We show only selected examples of results, including the metric *page_loading* metric. Selenium measurements of the *page-loading* are shown in Fig. 1 with one percent of the highest results removed. Histograms of this metric are shown in Fig. 2.

The histogram for Mobilkarm has two vertices. This suggests two different populations under observation, which is justified by the fact that the script collecting the measurements of this website should be updated during the experiments because of website interface changes. We also noticed high instability of this page.

Table 3. Performance metrics measured for tested websites

Metric	Mobilkarm	Kakadu	Bitiba	Keko
page_loading	+	+	+	+
cookies				+
dropdown_menu_opening		+		
category_chose	+	+	+	+
subcategory_chose		+	+	+
subsubcategory_chose			+	
item_details	+	+	+	+
cart_adding	+	+	+	+
cart	+	+	+	+

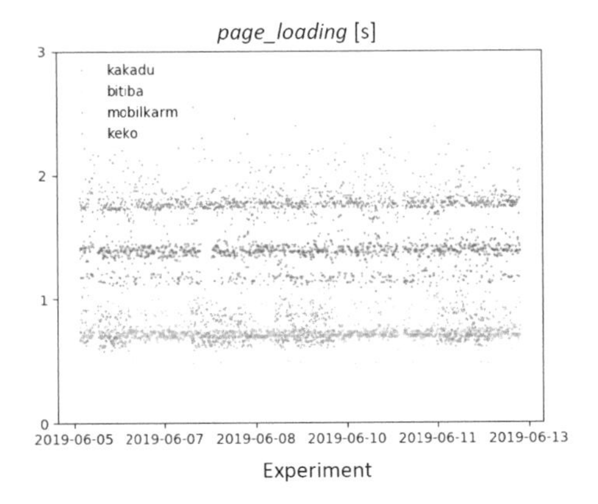

Fig. 1. *page-loading* vs. experiments

In further analysis, the page_loading, item_details, and cart_times metrics collected with the Selenium tool were compared using the Mann-Whitney-Wilcoxon test. It was decided to choose this method because based on histograms, it was assessed that the data considered did not have a normal distribution. The p-value was 0.05. The test showed that the differences between the considered data sets are statistically significant so that they can be analyzed separately

Perceived Performance

A total of 152 responses were collected in the survey. The group of respondents consisted mainly of people aged 18 to 30, regularly using a computer or mobile devices. At least half of the respondents had a technical education. The surveys took place within a period of measured performance experiment without any synchronization with those measurements or between the surveyors. The survey result is shown in Fig. 3. Table 4 shows basic statistical characteristics.

Analyzing the survey data, it can be seen, the fastest in the context of loading the main page, taking into account the median, is Bitiba. This follows the measurements of domInteractive and domComplete measured by the Browsertime, where Bitiba was

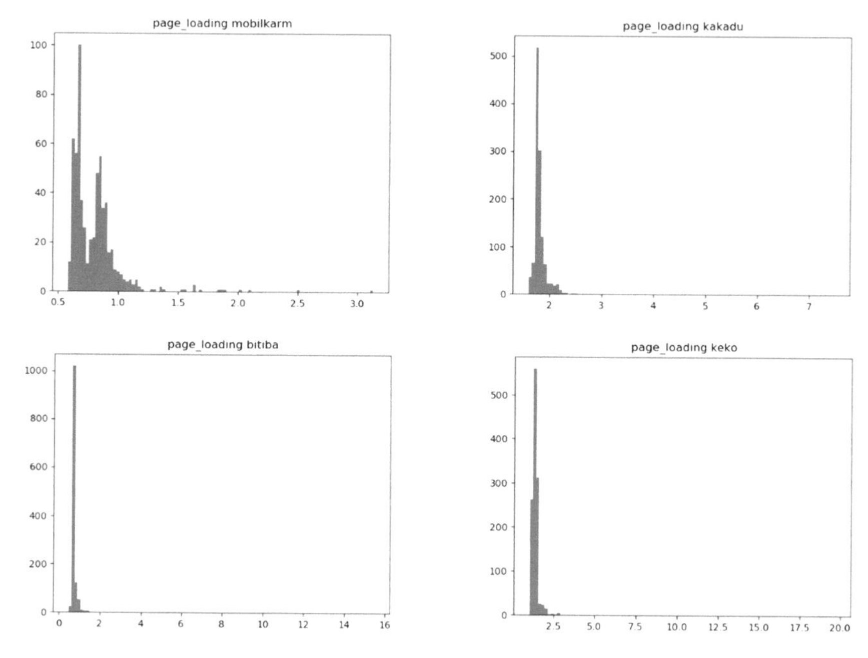

Fig. 2. *page_loading* histograms

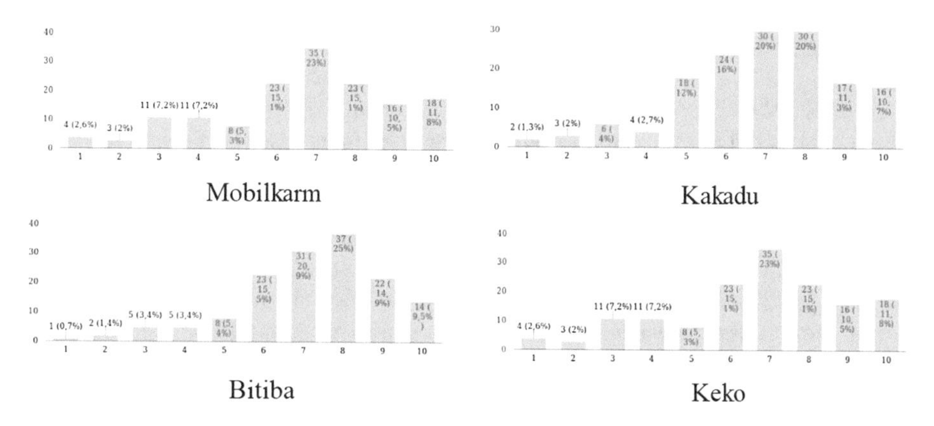

Fig. 3. Perceived performance in the scale 1-10

also the best. Considering the third quantile, Keko performs best - this website was never considered as the fastest one during previous measurements of measured performance. Based on the collected data, unfortunately, the slowest website can not be distinguished. It can be noted, however, that the Mobilkarm site is characterized by the highest average difference between actual response and median response, which indicates its instability - this was also demonstrated while measured performance experiments.

Table 4. xxx

	Mobilkarm	Kakadu	Bitiba	Keko
First quantile	6	6	6	6
Median	7	7	8	7
Third quantile	8	8	8	8.5
The average difference between actual response and median response	1.8	1.7	1.6	1.5

4 Conclusion

We considered the measured vs. perceived web performance case study. The work aimed to find a relationship between the automatically measured times of loading elements on web pages with zoological articles and the feelings of users regarding these loading times.

Based on the conducted research, we can say that the following automatically measured performance metrics, translate the best into the perceived performance evaluated by the users:

- *page_loading*, measured by Selenium;
- *first_meaningful_paint*, measured by PageSpeed Insights;
- *domInteractive* measured by Browsertime;
- *contentEventLoad*, measured by Browsertime.

In research carried out with the use of Selenium; however, there is a problem of changes in the interface that make the measurement unable to be performed correctly. A possible solution to this problem could be to change the used CSS selectors from selectors picked from the website itself to selectors defined based on the analysis of the selected HTML code. Another solution would be to use tools that allow, for example, to detect all graphic files on the website and design navigation based on them.

PageSpeed Insights experiments showed that the *first_meaningful_paint* time measured, indicating the moment when the elements essential for the user appeared on the page, justifies the user's feelings. The *first_cpu_idle* and *interactive* metrics, which inform about when the page starts responding to user actions, will be a bit worse for this tool. When referring to these metrics, one should take into account their dependence depending on whether the user interacted with the site. Both of these metrics work well when trying to evaluate the fastest page, but they do not give results consistent with the slowest page, and therefore the least efficient.

Considering the results from the Browsertime tool, the only metrics that have reference to the results of the survey are *domInteractive* metric and having almost identical values *contentEvenTLoad* metric. The analogous *interactive* metric measured with the Page-Speed Insights experienced worse results.

Our research has shown that in addition to determining the right metric of performance, which we may measure automatically, and which may determine the quality of the website matching the performance perception by the user, it is crucial to use the right tool. We defined the metrics, choose some popular tools, and performed experiments on the chosen set of websites following particular setups.

Faster websites are perceived to be more exciting and may have higher conversion rates. In studying measured vs. perceived performance, we should take into account some solutions and actions that may boost perceived performance via some technological tricks. These "magical" mind hacks do not change actual time, but only its perceived view. The problem is also that perceived time is relative to many things, including our age, our enjoyment, job engagement, task completion status, our health, and many many other things. Our decision could be irrational, as well. These problems were not under consideration in our study, and they are challenges for the next research.

However, our questions set in the introduction section are, unfortunately, still open.

Whether increasing the number of pages studied, a greater diversity of tools will allow providing general answers to these questions? Further research is needed to answer them. The "light in the tunnel" probably may give knowledge learned from the Big Data analysis.

References

1. New Industry Benchmarks for Mobile Page Speed - Think With Google. https://www.thinkwithgoogle.com/marketing-resources/data-measurement/mobile-page-speed-new-industry-benchmarks/. Accessed 28 May 2019
2. How Fast Should Your Web Page Load and How To Speed Up Your Website. https://www.hobo-web.co.uk/your-website-design-should-load-in-4-seconds/. Accessed 19 Jun 2019
3. Average Page Load Times for 2018 - How does yours compare? - MachMetrics Speed Blog. https://www.machmetrics.com/speed-blog/average-page-load-times-websites-2018/. Accessed 19 Jun 2019
4. Szalek, K., Borzemski, L.: Conversion rate gain with web performance optimization. A case study. In: Information Systems Architecture and Technology: Proceedings of 39th International Conference on Information Systems Architecture and Technology – ISAT 2018, AISC, vol. 852, pp. 312–323. Springer
5. Politi, R., Sereno, M., Sereno, M., Ruffo, G., Schifanella, R.: WALTy: a user behavior tailored tool for evaluating web application performance. In: Third IEEE International Symposium on Network Computing and Applications, 2004 (NCA 2004), Proceedings, pp. 77–86. IEEE (2004)
6. Eckersley, P.: How unique is your web browser? In: International Symposium on Privacy Enhancing Technologies Symposium, pp. 1–18. Springer (2010)
7. Zhu, K., Fu, J., Li, Y.: Research the performance testing and performance improvement strategy in web application. In: 2010 2nd International Conference on Education Technology and Computer, pp. V2–328. IEEE (2010)
8. Butkiewicz, M., Madhyastha, H.V., Sekar, V.: Understanding website complexity: measurements, metrics, and implications. In: Proceedings of the 2011 ACM SIGCOMM Internet Measurement Conference, pp. 313–328. ACM (2011)
9. Liandin, Z.: WebProphet: automating performance prediction for web services. In: NSDI, vol. 10, pp. 143–158 (2010)
10. Wangandin, X.S.: Demystifying page load performance with WProf. In: Presented as part of the 10th USENIX Symposium on Networked Systems Design and Implementation (NSDI 13), pp. 473–485 (2013)

11. Ihm, S., Pai, V.S.: Towards understanding modern web traffic. In: Proceedings of the 2011 ACM SIGCOMM Conference on Internet Measurement Conference, pp. 295–312. ACM (2011)
12. Borzemski, L., Nowak, Z.: An empirical study of web quality: measuring the web from Wroclaw University of Technology campus. In: Engineering Advanced Web Applications 4th International Workshop on Web-Oriented Software Technologies/4th International Conference on Web Engineering, Munich, Germany, 28–30 2004, pp. 307–320 (2004)
13. Borzemski, L.: Testing, measuring, and diagnosing web sites from the users' per- spective. Int. J. Enterp. Inf. Syst. (IJEIS) **2**(1), 54–66 (2006)
14. Prokopiuk, J., Nowak, Z.: The influence of HTTP/2 on user-perceived web application performance. Studia Informatica **38**(3), 73–88 (2017)
15. Gaspard, C., Goldberg, S., Itani, W., Bertino, E., Nita-Rotaru, C.: SINE: cache-friendly integrity for the web. In: 2009 5th IEEE Workshop on Secure Network Protocols, pp. 7–12. IEEE (2009)
16. Cao, P., Zhang, J., Beach, K.: Active cache: caching dynamic contents on the web. In: Proceedings of the IFIP International Conference on Distributed Systems Platforms and Open Distributed Processing, pp. 373–388. Springer-Verlag (2009)
17. Resource Timing Level 2, W3C Working Draft 26 June 2019. https://www.w3.org/TR/resource-timing-2/. Accessed 06 Jun 2019
18. Egger, S., Hossfeld, T., Schatz, R., Fiedler, M.: Waiting times in quality of experience for web-based services. In: Fourth International Workshop on Quality of Multimedia Experience, pp. 86–96. IEEE (2012)
19. Aladwani, A.M., Palvia, P.C.: Developing and validating an instrument for measuring user-perceived web quality. Inform. Manage. **39**(6), 467–476 (2002)
20. Guse, D., Schuck, S., Hohlfeld, O., Raake, A., Möller, S.: Subjective quality of webpage loading: the impact of delayed and missing elements on quality ratings and task completion time. In: 2015 Seventh International Workshop on Quality of Multimedia Experience (QoMEX), pp. 1–6. IEEE (2015)
21. Petrov, B., Vechev, M., Sridharan, M., Dolby, J.: Race detection for web applications. In: ACM SIGPLAN Notices, vol. 47, no. 6, pp. 251–262. ACM (2012)
22. Bui, D.H., Liu, Y., Kim, H., Shin, I., Zhao, F.: Rethinking energy-performance trade-off in mobile web page loading. In: Proceedings of the 21st Annual International Conference on Mobile Computing and Networking, pp. 14–26. ACM (2015)
23. Wang, Z., Lin, F.X., Zhong, L., Chishtie, M.: Why are web browsers slow on smartphones?. In: Proceedings of the 12th Workshop on Mobile Computing Systems and Application, pp. 91–96. ACM (2011)
24. mPulse—Real-Time Performance Management and RUM—Akamai. https://www.akamai.com/us/en/products/performance/mpulse-real-user-monitoring.jsp. Accessed 18 Feb 2019
25. Akamai GitHub. https://github.com/akamai. Accessed 18 Feb 2019
26. Features—GTmetrix. https://gtmetrix.com/features.html. Accessed Feb 2019
27. Lighthouse—Tools for Web Developer. https://developers.google.com/web/tools/lighthouse/#get-started. Accessed 18 May 2019
28. About PageSpeed Insights—PageSpeed Insights—Google Develope. https://developers.google.com/speed/docs/insights/v5/about\#score. Accessed 18 Feb 2019
29. YSlow - Official Open Source Project Website. http://yslow.org/. Accessed Feb 2019
30. Selenium - Web Browser Automation. https://www.seleniumhq.org/. Accessed 18 May 2019
31. Introduction to Browsertime. https://www.sitespeed.io/documentation/browsertime/introduction/. Accessed 18 May 2019
32. Welcome to the wonderful world of Web Performance. https://www.sitespeed.io/. Accessed 18 May 2019

33. Coach Introduction. https://www.sitespeed.io/documentation/coach/introduction/. Accessed 18 May 2019
34. Use PageXray to convert HAR files to a more readable format. https://www.sitespeed.io/documentation/pagexray/. Accessed 18 May 2019
35. Throttle - Simulate slow network connection. https://www.sitespeed.io/documentation/throttle/. Accessed 18 May 2019
36. Web Performance of the World's Top 100 E-Commerce Sites in 2018. https://royal.pingdom.com/web-performance-top-100-e-commerce-sites-in-2018/#section3. Accessed 22 Jun 2019

Resource Management and Performance Evaluation

Resource Management for SD-WANs

Dariusz Gąsior(✉)

Faculty of Computer Science and Management,
Wroclaw University of Science and Technology, Wroclaw, Poland
dariusz.gasior@pwr.edu.pl

Abstract. In this paper, resource management in SD-WANs is introduced as a game between networks. The appropriate formulation is given and the properties of the problem are indicated. Furthermore, the solution algorithm finding the Pareto-optimal equilibrium is presented. The algorithm was evaluated with simulation experiments for small size networks.

Keywords: Software defined networking · Software defined wide are networks · Game theory · Quality of service · Utility

1 Introduction

Recently, software defined network has become one of the fastest growing network concepts. The main idea of this paradigm is to decouple the control plane from the data plane [1,2]. While the simple packet forwarding tasks are executed in simplified network devices, all the necessary calculations are made by a central entity called the SDN controller.

The SDN controller interacts with network devices (data plane), applications and other SDN controllers with suitable interfaces, which are depicted in Fig. 1.

The Northbound interface enables applications to utilize SDN functionality. The Southbound interface enables physical devices to operate according to the solution obtained by the SDN controllers. Both of them are already well-defined. However, the so-called East/Westbound interfaces which define the interaction between SDN controllers of different networks are still under consideration [3].

Many works have been devoted to solving emerging challenges in the local networks. More details in this topic one may be found in surveys like [4]. Now, the leading topic is to extend this paradigm to the wide area networks (SD-WAN) [5].

This paper is devoted to the capacity allocation problem in the SD-WANs consisting of the set of interconnected local software defined networks.

Some approaches to the possible architectures of the cooperation between SDN controllers have been proposed.

The hierarchical orchestration model for the software defined networks is given in [6]. This approach enables scalability and security for the SDN orchestration. The proposed architecture has been evaluated with the simulation experiments.

L. Borzemski et al. (Eds.): ISAT 2019, AISC 1050, pp. 305–315, 2020.
https://doi.org/10.1007/978-3-030-30440-9_28

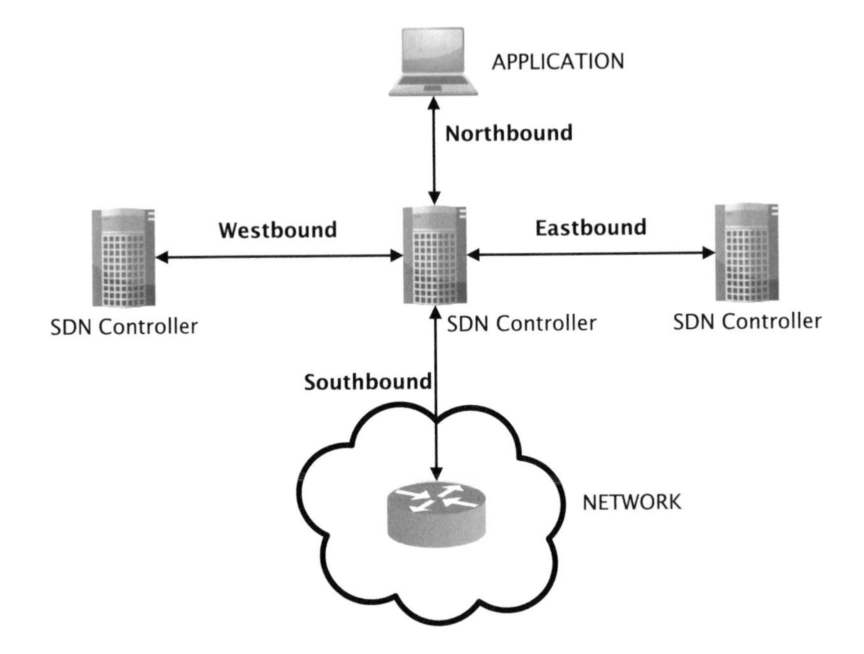

Fig. 1. The schema of the SDN interfaces.

In [7] the author propose an East-/Westbound communication protocol and interface for the SDN controllers, so the software defined networks may be easily scaled and applied in large networks like Internet. The introduced protocol, called INT, is used to exchange information about networks capabilities and to setup the path through the whole SD-WAN. The operating of the proposed architecture was formally verified with the Petri networks.

In [8] the authors propose the distributed multi-domain SDN control plane called DISCO which enables the delivery of end-to-end network services. The presented DISCO controller is responsible not only for intra-domain functionalities but also for the inter-domain cooperation. The latter goal is achieved by using the elaborated communication bus which enables interchanging network information. The messages are sent and received by the special agents. Some of the agents are responsible for making appropriate reservation on the path while others monitor network status. The performance of this architecture was experimentally examined.

All the proposed approaches give the technical solutions of the realization concept of software defined networks in wide are networks but do not provide any solution how to determine the adequate configuration of the network.

In [9] the authors formulate the resource allocation problem as integer linear programming to determine routing for the transmissions with given constant rates. However, this approach bases on the centralized computation for the whole SD-WAN and does not take into account QoE.

To the best of our knowledge, there is still a lack of QoE and QoS capacity allocation algorithms for the structures without central coordination.

Our proposition allows to minimize the necessity of SDN controllers interaction while still offers a very good performance of resource allocation and fulfilling the QoS requirements.

The QoS requirements are crucial for contemporary applications. Similarly, the key is the quality perceived by the users. We call it Quality of Experience (QoE). In this paper, we adapt the model proposed in [12] which allows to describe user's QoE in relation to the QoS parameters using the utility function [13,14]. This approach is the most adequate for the problem formulated in this paper. The solution of the considered problem directly affects only the transmission rates and the utility depends only on this QoS parameter. That is because it is believed that for the capacity allocation other QoS parameters (e.g. delays, loss rate) either depends on the allocated transmission rates or their values are guaranteed with different mechanisms (like routing) [15]. Furthermore, a utility-based approach to the QoE has also the straight economic interpretation which enables possible financial settlements between the networks constituting an SD-WAN. However, the way how the appropriate functions describing perceived quality are obtained is out of the scope of this paper. More details on the approaches to the QoE estimation one may find e.g. in [10] and [11]. We only assume that these functions are given in advance.

In this paper, we propose to apply a game-theoretic approach to solve the resource allocation problem in the software defined wide area networks with quality of experience and quality of service requirements. We introduce the capacity allocation algorithm which finds a Pareto-optimal solution of the introduced game.

2 Mathematical Model and Problem Formulation

The software defined wide area network may be treated as a set of D interconnected networks. Each of them is managed by the independent SDN controller. We assume the network flow model to represent the system under consideration. Each network consists of nodes and links. Links represent connections between nodes inside a network as well as between networks. There are L links in total in the whole wide area network. Each link l is characterized by its capacity C_l. This parameter reflects the maximal amount of data which may be sent between nodes in the unit time. The flow between a pair of nodes represents the data transmission. We assume that there are R transmissions and their routes are given with binary variable a_{rl} which indicates if lth link is used for rth transmission.

The capacity allocation x_{rl} is made on each link l traversed by the rth flow. The capacity allocations are determined in each network independently by its SDN controller. The rth transmission's rate $\overline{x}_r$ results from the capacity allocations made on all links along its route. The rate must be greater than minimal acceptable value $x_{r,\min}$ due to the QoS constraints. As it was aforementioned we limit the QoS parameters under consideration to the transmission rate, while it is believed to be the most crucial one [15]. The fulfilling other QoS requirements may be reached with other network mechanisms (e.g. appropriate routing algorithms, scheduling algorithms on nodes, etc.).

Obviously, the allocations must not exceed links' capacities. We also assume that there are utility functions $f(\overline{x}_r; w_r)$ defined for every transmission which reflects the quality of experience perceived when the transmission rate is $\overline{x}_r$. In the most common approach, it is assumed that it is iso-elastic function and its parameter α corresponds to the particular shape of the QoE dependence. The aim of each SDN controller is to maximize the total quality of experience from all transmissions traversing the corresponding network. We introduce also β_r parameter which denotes how the perceived quality translates to the potential income for each network. The notation is summarized in Table 1. The exemplary architecture of such a system is depicted in Fig. 2.

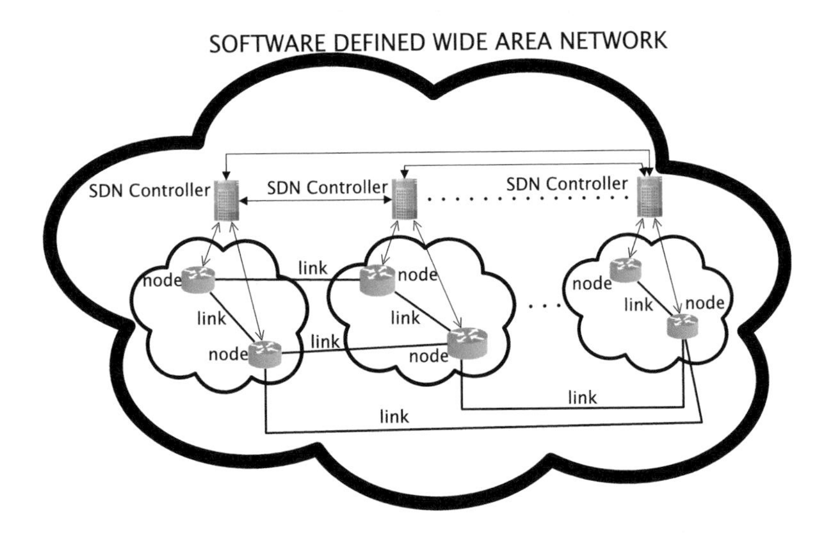

Fig. 2. The exemplary architecture of SD-WAN.

The problem under consideration may be treated as a game [16]. The networks (and associated SDN controllers) are the players (so, there are D players). The feasible capacity allocation for transmissions traversing particular network $\mathbf{x}_d$ is a strategy. The capacity allocations for all networks constitute the strategy profile $\mathbf{x}$. The objectives $Q_d(\mathbf{x}_d, \mathbf{x}_{-d})$ are the players' payoffs. We refer to this game as the Capacity Allocation Game (CAG).

Formally, each player d (network) solves the following optimization problem:

Given: $R, L, a_{rl}, e_{ld}^{(link)}, e_{rd}^{(req)}, C_l, w_r, \beta_r, \alpha$

Find:

$$\mathbf{x}_d^* = \arg\max_{\mathbf{x}_d} Q_d(\mathbf{x}_d, \mathbf{x}_{-d}) \tag{1}$$

Table 1. Notation

L	Number of links
R	Number of flows (transmissions)
D	Number of networks
C_l	lth link's capacity ($l = 1, 2, \ldots, L$)
$\mathbf{a} = [a_{rl}]_{r=1,2,\ldots,R; l=1,2,\ldots,L}$	Routing matrix: $a_{rl} = 1$, rth flow traverse lth link $a_{rl} = 0$, otherwise
$e_{ld}^{(link)}$	Variable indicating if lth link belongs to the dth network
$e_{rd}^{(req)}$	Variable indicating if rth transmission is performed by any link in dth network $e_{rd}^{(req)} = \begin{cases} 1 & \text{gdy } \sum_{l=1}^{L} e_{ld}^{(link)} a_{rl} \geq 1 \\ 0 & \text{otherwise} \end{cases}$
w_r	rth flow priority parameter
$x_{r,\min}$	rth flow minimal transmission rate (QoS parameter)
$x_{rl} \geq 0$	lth link's capacity allocation for rth flow
$\hat{\mathbf{x}}_l$	lth link's capacity allocation vector: $\hat{\mathbf{x}}_l = [x_{rl}]_{r=1,2,\ldots,R}$
$\overline{x}_r \geq 0$	rth flow transmission rate: $\overline{x}_r = \min_{l:a_{rl}=1} x_{rl}$
$\mathbf{x}_d$	The allocation matrix for dth network, $\mathbf{x}_d = [\hat{\mathbf{x}}_l]_{l:e_{ld}^{(link)}=1}$, $\mathbf{x} = [\mathbf{x}_d]_{d=1,2,\ldots,D}$
$\mathbf{x}_{-d}$	The allocation matrix $\mathbf{x}$ without dth component, it is assumed: $\mathbf{x} = [\mathbf{x}_d, \mathbf{x}_{-d}]$
$f(\overline{x}_r; w_r) = w_r \varphi(\overline{x}_r)$	rth flow utility function $\varphi(\overline{x}_r) = \begin{cases} \frac{\overline{x}_r^{(1-\alpha)}}{(1-\alpha)} & \alpha \geq 0 \ \wedge \ \alpha \neq 1 \\ \ln \overline{x}_r & \alpha = 1 \end{cases}$
$\beta_r \geq 0$	Given coefficient for rth flow e.g. $\beta_r = 1$ or $\beta_r = (\sum_{d=1}^{D} e_{rd}^{(req)})^{-1}$
$Q_d(\mathbf{x}_d, \mathbf{x}_{-d})$	The objective of the dth network (payoff): $Q_d(\mathbf{x}_d, \mathbf{x}_{-d}) = \sum_{r=1}^{R} e_{rd}^{(req)} \beta_r f(\overline{x}_r; w_r) =$ $= \sum_{r=1}^{R} e_{rd}^{(req)} \beta_r w_r \varphi(\min_{l:a_{rl}=1} x_{rl})$

such that:

$$\forall_{l:e_{ld}^{(link)}=1} \quad \sum_{r=1}^{R} a_{rl} x_{rl} \leq C_l$$

$$\forall_{r=1,2,\ldots,R} \quad x_{rl} \geq x_{r,\min}$$

It is assumed that at least one feasible solution exists, i.e.: $\forall_{l=1,2,\ldots,L} \sum_{r=1}^{R} a_{rl}$ $x_{r,\min} \leq C_l$, i.e. there is an independent mechanism of the admission control. From the game theory perspective, the solution to such a problem is a game equilibrium.

3 Capacity Allocation Algorithm

Since the formulated problem is the game, the equilibria are perceived as the solutions. One possibility is to find Nash equilibrium [17]. However, in this paper, we focus on the Pareto-optimal equilibrium [18] since we find it more suitable. The weak Pareto-optimal solution is the strategy profile that cannot be changed in such a way that every player's payoff is increased. The strong Pareto-optimal solution is the one that cannot be changed in such a way that at least one player's payoff is increased while no other payoff is decreased. The resource allocation algorithm presented as Algorithm 1 provable stops at strong Pareto-optimal equilibrium.

The idea of the proposed solution method is as follows. We assume that networks determine their final allocation in the sequence, which does not have to be known in advance. Moreover, it may be chosen at random. The only direct communication between networks lays in indicating the next network which has to determine its allocation. Thus, the needs for the East/Westbound interface are limited. Once the network is designated to calculate its allocation, it solves optimization problem finding rates for all transmissions that traverse this network and their rates have not been computed yet.

Algorithm 1. Pareto Optimal Multi-network Utility-based Resource Allocation Algorithm (POMUR)

1: Initialize variables, i.e.: $\forall_{d\in\{1,2,\ldots,D\}} \quad \overline{R}_d = \{r \in \{1,2,\ldots,R\} : e_{rd}^{(req)} = 1 \wedge \sum_{q=1}^{d-1} e_{rq}^{(req)} = 0\}$ and $\forall_{l\in\{1,2,\ldots,L\}} \tilde{C}_l \leftarrow C_l - \sum_{r\in\{1,2,\ldots,R\}} a_{rl}x_{r,\min}$

2: **for** $d \in \{1,2,\ldots,D\}$ **do**

3: Find optimal allocation $\mathbf{y}_d = [\mathbf{x}_l]_{l:e_{ld}^{(link)}}$ for dth network, i.e. solve the following problem:

Given: $\overline{R}_d, L, a_{rl}, e_{ld}^{(link)}, e_{rd}^{(req)}, \tilde{C}_l, w_r, \beta_r, \alpha$

Find:

$$\mathbf{y}_d^* = \arg\max_{\mathbf{y}_d} \sum_{r\in\overline{R}_d} \beta_r f(\overline{x}_r; w_r) \tag{2}$$

such that:

$$\forall_{l\in\{1,2,\ldots,L\}} \quad \sum_{r\in\overline{R}_d} a_{rl}x_{rl} \le \tilde{C}_l + \sum_{r\in\overline{R}_d} a_{rl}x_{r,\min} \quad \wedge \quad \forall_{r\in\overline{R}_d} \quad x_{rl} \ge x_{r,\min}$$
$$\forall_{r\in\overline{R}_d}\forall_{l_1,l_2\in\{1,2,\ldots,L\}} \quad a_{rl_2}x_{rl_2} = a_{rl_1}x_{rl_1}$$

4: $\forall_{l\in\{1,2,\ldots,L\}}\forall_{r\in\overline{R}_d}$ Update: $x_{rl} \leftarrow min_{j:e_{rd}^{(req)}a_{rj}=1} x_{rj}$

5: $\forall_{l\in\{1,2,\ldots,L\}}$ update: $\tilde{C}_l \leftarrow C_l - \sum_{r\in\overline{R}_d} a_{rl}x_{rl}$.

6: **end for**

7: Return $\mathbf{x}$.

The following properties of the Algorithm 1 occur.

Theorem 1. *Algorithm 1 (POMUR) finds strong Pareto-optimal strategy profile of CAG for any enumeration of networks.*

Proof. (Theorem 1 *(proof by induction)).*

1. One cannot increase the payoff for first network $d = 1$ since the allocation was found with solving the optimization problem for this network.
2. Induction assumption: We cannot increase any of $d-1$ first networks, while other networks' payoffs do not decrease.
3. Induction hypothesis: We cannot increase any of d first networks, while other networks' payoffs do not decrease.

According to the induction assumption, the only possibility of increasing any payoff while no other payoff is decreased is to increase the payoff of the d network. If one change allocation for any transmission that traverses any of $d-1$ first networks, at least one of their payoffs will decrease. Thus, one may increase dth network's payoff only be changing allocation for transmissions which does not traverse any previous network (i.e. those which belongs to R_d). But according to step 3 of Algorithm 1 the obtained allocation is optimal for those transmissions. So, we cannot increase dth network's payoff.

From the Theorem 1 we conclude the following remark.

Remark 1. Algorithm 1 (POMUR) may find up to $D!$ Pareto-optimal strategy profiles. (For any enumeration of networks, the solution may be different).

On the other hand, one may easily check the following remark.

Remark 2. Strategy profiles found by Algorithm 1 (POMUR) may not constitute a Nash Equilibrium.

4 Simulation

Some preliminary simulation experiments for network structures corresponding to the small real-life backbone networks [19] have been performed. The network topologies were generated with the method described in [20]. The main objective of the simulation was to pre-evaluate the proposed algorithm. The algorithm was implemented using Python 2.7. The experiments were conducted using the computer with Intel Core M 1.1 processor and 8 GB RAM.

The simulations was run for $\beta_r = 1$ and the following problem parameters:

- wide area network's parameters:
  - $D \in \{2, 3, 4, 5\}$ - number of networks, (each network consist of at least one link),
  - $N \in \{7, 8, 9, 10\}$ - number of nodes,
  - $\rho \in \{0.4, 0.5\}$ - links' density: $L = \frac{1}{2}\rho N(N-1)$,
  - $C_l \sim U(1000, 10000)$ - capacity of lth link [Mbps].

- transmissions' parameters:
  - $R \sim U(\frac{1}{2}\underline{\theta}N(N-1); \frac{1}{2}\overline{\theta}N(N-1))$ - number of transmissions, $\underline{\theta} = 0.3, \overline{\theta} = 0.7$,
  - for each transmission r the source and origin nodes are randomly chosen (i, j),
  - a_{rl} - calculated using shortest path algorithm (from origin node to destination node),
  - $w_r \sim U(1, 10)$
  - $\alpha = 0.5$

The uniform distribution on the interval $[c, d)$ is denoted by $U(c, d)$. The simulation was run three times for each set of parameter values (N, D, ρ).

The following quality criterion was assumed: $\gamma = \frac{SW(\mathbf{x}^{POMUR})}{\max_{\mathbf{x} \in \hat{D}_{\mathbf{x}}} SW(\mathbf{x})}$, where $\mathbf{x}^{POMUR}$ is the solution found by the proposed algorithm, SW is a social welfare [21], i.e.: $SW(\mathbf{x}) = \sum_d Q_d(\mathbf{x}_d, \mathbf{x}_{-d})$ and $\hat{D}_{\mathbf{x}}$ is the set of feasible allocations. The optimal value of the social welfare function $SW(\mathbf{x})$ was found with primal-dual projected gradient method [22].

The statistical results of the simulation experiments are given in Table 2. The representative charts of the average value of quality criterion for a different number of network nodes and a different number of network domains are given in Fig. 3 (for $\rho = 0.4$) and Fig. 4 (for $\rho = 0.5$). As it may be noticed, the proposed algorithm in some cases found not only the Pareto-optimal solution but also global optimal allocation. For the worst case, the obtained result was only 20% smaller than globally optimal, while on average the deterioration was no more than 4%. However, it is hard to find strict dependence between the quality of the proposed algorithm and the number of networks or their size.

Table 2. Simulation results

Statistics for γ	POMUR
average	0.96
median	0.97
variance	0.02
min	0.8
max	1.0

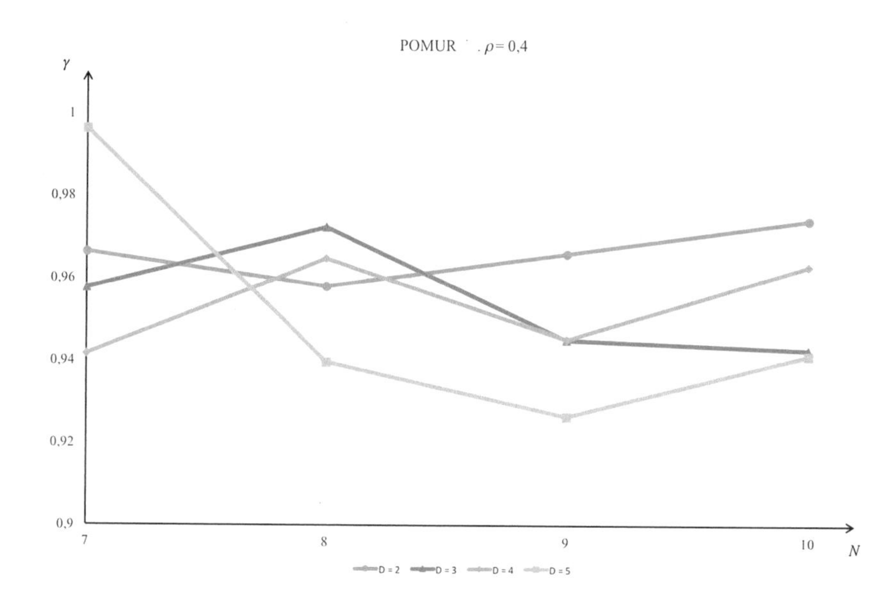

Fig. 3. The average value of quality criterion depending on number of nodes for different number of network domains for $\rho = 0.4$.

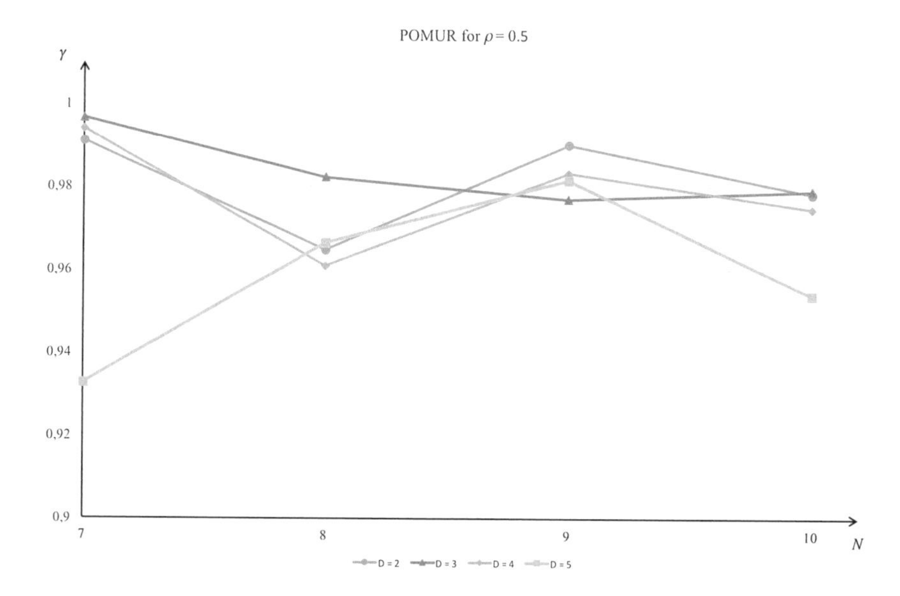

Fig. 4. The avarage value of quality criterion depending on number of nodes for different number of network domains for $\rho = 0.5$.

5 Final Remarks

In this paper, we considered the capacity allocation problem with the quality of experience and quality of service requirements for software defined wide area networks. The problem was formulated in terms of the game theory and the algorithm which finds strong Pareto-optimal equilibrium was presented. Introduced solution method allows SDN controllers to act independently, so almost no communication and no negotiations are required. The initial experiments indicate that the proposed approach is undoubtedly a very promising one. However, the experiments for larger networks are needed, so the more practical conclusions can be made. The future works may also include the application of the presented approach to the software defined wide area networks with virtualization as an extension of the ideas presented in [23] and [24].

References

1. Shin, M.-K., Nam, K.-H., Kim H.-J., Software-defined networking (SDN): a reference architecture and open APIs. In: Proceedings of 2012 IEEE International Conference on ICT Convergence (ICTC) (2012)
2. Nunes, B.A.A., Mendonca, M., Nguyen, X.N., Obraczka, K., Turletti, T.: A survey of software-defined networking: past, present, and future of programmable networks. IEEE Commun. Surv. Tutor. **16**(3), 1617–1634 (2014)
3. Hoang, D.B., Minh, P.: On software-defined networking and the design of SDN controllers. In: 2015 6th International Conference on the Network of the Future (NOF). IEEE (2015)
4. Karakus, M., Arjan, D.: Quality of Service (QoS) in Software Defined Networking (SDN): a survey. J. Netw. Comput. Appl. **80**, 200–218 (2016)
5. Jain, S., Kumar, A., Mandal, S., Ong, J., Poutievski, L., Singh, A., Zolla, J.: B4: experience with a globally-deployed software defined WAN. ACM SIGCOMM Comput. Commun. Rev. **43**(4), 3–14 (2013)
6. Vilalta, R., Mayoral, A., Munoz, R., Casellas, R., Martinez, R.: Hierarchical SDN orchestration for multi-technology multi-domain networks with hierarchical ABNO. In: Proceedings of 2015 European Conference on Optical Communication (ECOC) (2015)
7. Helebrandt, P.: Architecture for core networks utilizing software defined networking. Inf. Sci. Technol. **8**(2), 56–61 (2016)
8. Phemius, K., Bouet, M., Leguay, J.: Disco: distributed multi-domain SDN controllers. In: Proceedings of 2014 IEEE Network Operations and Management Symposium (NOMS) (2014)
9. Fajjari, N.A., Kouicem, D.E.: A novel SDN scheme for QoS path allocation in wide area networks. In: IEEE Global Communications Conference, GLOBECOM 2017, Singapore (2017)
10. Alreshoodi, M., Woods, J.: Survey on QoE QoS correlation models for multimedia services. Int. J. Distrib. Parallel Syst. **4**(3), 53 (2013)
11. Stankiewicz, R., Jajszczyk, A.: A survey of QoE assurance in converged networks. Comput. Netw. **55**(7), 1459–1473 (2011)
12. Khan, M.A., Toseef, U.: User utility function as quality of experience (QoE). Proc. ICN **11**, 99–104 (2011)

13. Kelly, F.P., Maulloo, A.K., Tan, D.K.: Rate control for communication networks: shadow prices, proportional fairness and stability. J. Oper. Res. Soc. **49**(3), 237–252 (1998)
14. Gasior, D.: QoS rate allocation in computer networks under uncertainty. Kybernetes **37**(5), 693–712 (2008)
15. Wydrowski, B., Zukerman, M.: QoS in best-effort networks. IEEE Commun. Mag. **40**(12), 44–49 (2012)
16. Nisan, N., Roughgarden, T., Tardos, E., Vazirani, V.V.: Algorithmic Game Theory. Cambridge University Press, Cambridge (2007)
17. Nash, J.: Non-cooperative games. Ann. Math. **54**, 286–295 (1951)
18. Gasior, D., Drwal, M.: Pareto-optimal Nash equilibrium in capacity allocation game for self-managed networks. Comput. Netw. **57**(14), 2817–2832 (2013). Elsevier
19. Reference Networks. http://www.av.it.pt/anp/on/refnet2.html
20. Bu, T., Towsley, D.: On distinguishing between Internet power law topology generators. In: Proceedings of the Twenty-First Annual Joint Conference of the IEEE Computer and Communications Societies, vol. 2, pp. 638–647 (2002)
21. Koutsoupias, E., Papadimitriou, C.: Worst-case equilibria. In: Annual Symposium on Theoretical Aspects of Computer Science, pp. 404–413. Springer, Heidelberg (1999)
22. Vandenberghe, L.: Convex Optimization. Cambridge University Press, Cambridge (2004)
23. Gasior, D.: Capacity allocation in multilevel virtual networks under uncertainty. In: Proceedings of 15th IEEE International Telecommunications Network Strategy and Planning Symposium (NETWORKS) (2012)
24. Gasior, D.: Game-theoretical approach to capacity allocation in self-managed virtual networks. In: Proceedings of 36th International Conference on Information Systems Architecture and Technology - Part II, Springer International Publishing, pp. 155–164 (2016)

Quality of Data Transmission in the Access Points of PWR-WiFi Wireless Network

Łukasz Guresz and Anna Kamińska-Chuchmała(✉)

Faculty of Computer Science and Management, Wrocław University of Science and Technology, Wybrzeże Wyspiańskiego 27, 50-370 Wrocław, Poland
anna.kaminska-chuchmala@pwr.edu.pl

Abstract. Wireless networks are becoming more popular and increasingly used. Free Wi-Fi network or HotSpot can be found practically at every step. Moreover, it is also important to provide a good quality of data transmission. The aim of this paper is to use mathematical method to define the quality of data transmission for a specific measurement from an access point, and thus for entire network or selected area. By using this method, the value of quality of data transmission is rapidly obtained. This value is the result, which defines the level of quality of data transmission provided by the network. In addition, the method returns values of factors that determine which parameters had the greatest impact on the result. The data and results were analyzed and the conclusions with future plans for extending the proposed method were presented.

Keywords: Wi-Fi · Wireless network · Wireless Network Efficiency · Quality of data transmission · Access points

1 Introduction

Wireless networks based on the IEEE 802.11 standard are found in everyday use. In many cases, wireless network provides unsatisfactory quality of data transmission for the user, who reflects on his dissatisfaction and possible problems that may occur in data transmission.

Every year more users used the PWR-WiFi network, which requires quality of data transmission on high level. That is the cause of searching the solution of the problem of performance and quality of data transmission of Wi-Fi networks. In reference to the mentioned needs, in this paper research on open access Wi-Fi network are conducted. The academic environment is very specific and diverse, which makes it interesting to study.

The paper is constructed as follows: related works are presented in the next section, after that PWR-WiFi network and collected database are described, subsequently standard analysis of access points and proposed mathematical method defining the quality of data transmission are presented in the next section and finally conclusion and plan of future research in the last section are given.

L. Borzemski et al. (Eds.): ISAT 2019, AISC 1050, pp. 316–327, 2020.
https://doi.org/10.1007/978-3-030-30440-9_29

2 Related Work

Till now there have been many published papers about experiments and conducted analyzes in aspect of wireless network performance and quality of data transmission. Researchers and authors proposed different solutions and approaches to the subject.

In paper [1] collected data from free hotspots and access points of the Wi-Fi networks are analyzed. Authors assessed the network performance and its impact on the quality of data transmission. They evaluate the potential offloading data traffic of Wi-Fi networks.

The estimation method, which is scalable and allows to determine the network performance is presented in [2]. Used technique is suitable for estimating the current backhaul bandwidth of different access points and it allows to estimate the number of access points required to maintain the network performance on a high level.

In another paper [3] a frequency analysis method is presented. This method allows to determine the optimal location of new computer stations and access points in university. Proposed method is based on intelligent sensors of the access points.

A mathematical approach in [4] is proposed to find the minimum number of access points required to offload Wi-Fi network. The authors set the average bandwidth per one user to find the minimum.

Another article [5] presents the problem of open Wi-Fi network performance. Collected data from the access points are analyzed and the performance of the university open network is presented.

In paper [6] authors describe the method of measuring network performance for any wireless network or any access point. The tests are conducted using incorrect packets that test the receiving device. The time, number of test packets and confirmation packets are analyzed and the network performance is evaluated.

The research in [7] mainly focuses on selected algorithms that allows to define network performance. The author assume that achieving high efficiency of the wireless network can be obtained by using an appropriate algorithm.

The author present a mathematical algorithm in [8]. The algorithm's task is to analyze the network and used channels and then to propose new settings for access points in the entire wireless network.

In the article [9] author analyze a wireless network based on the 802.11 standard and 5 GHz band. The measurements of quality of data transmission is analyzed and used to improve network performance.

Paper [10] introduces the results of research on the wireless network performance. The research are conducted by using analytical approach and Sym Teredo tunneling.

Concluding, on basis of the described literature there is wide spectrum of research wireless network performance and quality of data transmission, but in this paper authors present research and mathematical method of analysis, which includes all available and relevant parameters that affect network performance and quality of data transmission.

3 PWR-WiFi Network and Collected Database

Wireless network named PWR-WiFi is located at Wroclaw University of Science and Technology (WUST) in Poland. It is an open university network. The data collected for this paper are obtained from over 300 access points located on the main campus, Geocentrum and branches of the WUST. Monitored access points use two frequencies, 2.4 GHz in IEEE 802.11b/g/n and 5 GHz in IEEE 802.11a/n. Access points are wireless connected to switch, they get IP address from the network and then connect to the Wi-Fi controller, creating a direct tunnel to the controller. All client traffic and control traffic is sent through this created tunnel. Access points are connected to Wi-Fi controller in star topology. The wireless and Ethernet clients are not in the same VLAN. All access points use LWAPP protocol which allows to control multiple wireless access points and provides communication between access points and controllers.

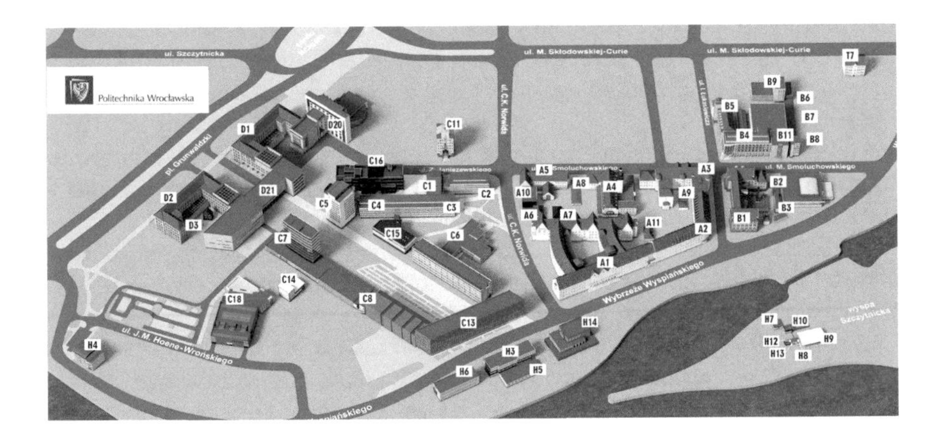

Fig. 1. Considered area in the analysis - main campus of WUST

In this paper research and analysis mainly focus on the area of main campus, which is shown in Fig. 1. The data were acquired from passive experiment, in 2014 measurements were saved from access points every 2 min and data from 2015–2016 were collected every 5 min. From each working access point the value of the transmission parameters were obtained that allows to analyze network performance and quality of data transmission.

3.1 Collected Parameters from Access Points

Many parameters were collected from the access points, but only those essential for the quality of data transmission will be presented and described.

The following important parameters were collected from each access point:

- DateOfMeasurement - date and time of measurement,
- apName - name of access point,

- NoOfUsers - number of users connected to the access point in the last measurement interval,
- TransmittedFragmentCount - this counter shall be incremented for an acknowledged MPDU,
- RetryCount - this counter shall increment when the transmission is successful after one and more retransmissions,
- MultipleRetryCount - this counter shall increment when the transmission is successful after more than one retransmission,
- FrameDuplicateCount - this counter shall increment when a frame is received that the Sequence Control field indicates is a duplicate,
- RTSSuccessCount - this counter shall increment when a CTS is received in response to an RTS,
- RTSFailureCount - this counter shall increment when a CTS is not received in response to an RTS,
- ACKFailureCount - this counter shall increment when an ACK is not received when expected,
- FCSErrorCount - this counter shall increment when an FCS error is detected in a received MPDU,
- TransmittedFrameCount - this counter shall increment for each successfully transmitted MSDU,
- FailedCount - this counter shall increment when an MSDU is not transmitted successfully due to the number of transmit attempts exceeding RetryLimit,
- ChannelUtilization - channel utilization, i.e. the percentage of the communication channel used for data transmission,
- PoorSNRClients - number of users with a low SNR (signal-to-noise ratio) connected to the access point in the last measurement interval.

4 Analysis of PWR-WiFi Network

The analyzed environment is specific and very interesting for research. PWR-WiFi network was mainly used by students and employees of the WUST, therefore the access points were practically not used at night because the buildings were empty at that time. The main network traffic took place from 7:00 am until 9:00 pm from Monday to Sunday, because classes and lectures are being conducted at that time. Mostly classes start at 07:30 am and end around 9:30 pm. Classes do not start at equal hours, usually 15 min after a given hour. Moreover, not every weekend classes are conducted for non-stationary students, as well as for stationary students, some days were free from education. There is a slight increase in the number of users in each starting odd hour when classes begin and end. Students used the network during breaks between classes. All described patterns are identical on each day of classes, both for stationary students and non-stationary students. The main difference between classes on weekdays and weekend is that network traffic in the second case is smaller. Described behavioral patterns are presented in Fig. 2.

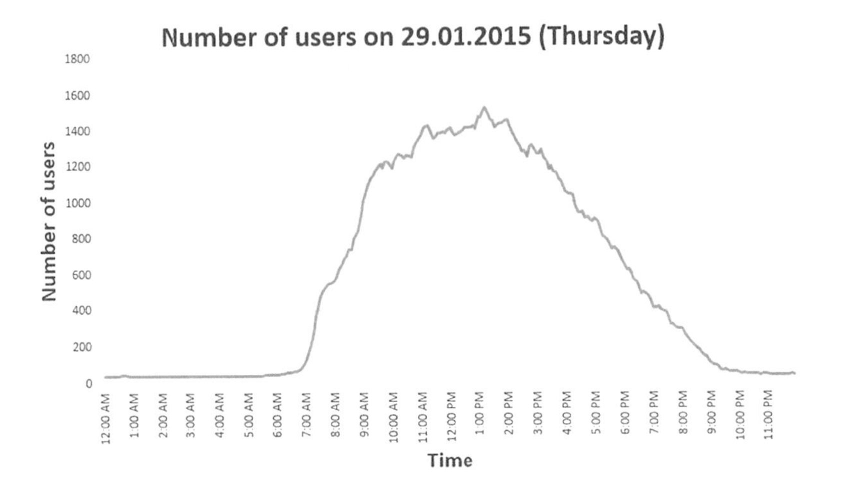

Fig. 2. Number of users in 212 access points on 29.01.2015

Figure 3 presents the day of classes for non-stationary students. The shape of the chart is very similar, the number of users increases and decreases at similar times. It indicates the same behavior people studying at the weekend and during the weekday.

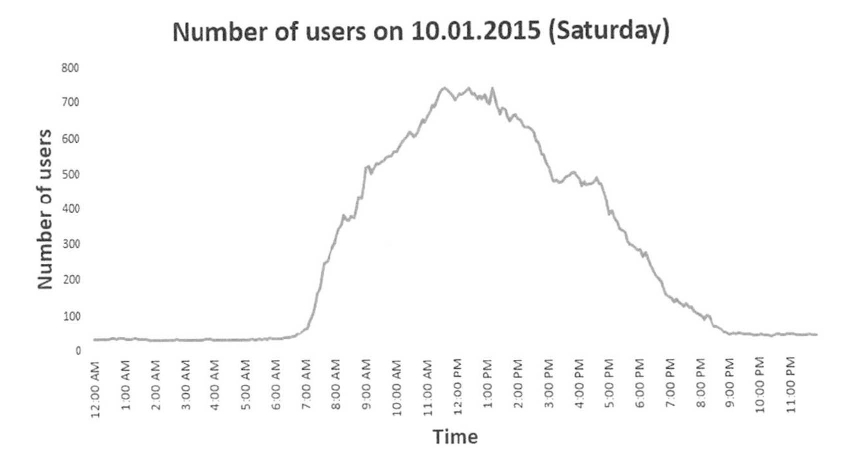

Fig. 3. Number of users in 212 access points on 10.01.2015

Further, many other behavioral patterns can be described by analyzing other parameters. The graph for each of them presents a similar shape and characteristics.

4.1 Mathematical Approach to the Quality of Data Transmission

The proposed approach of data analysis is based on the mathematical nature of the collected data (parameter values). By comparing the values of the appropriately selected parameters, new values are obtained that determine the quality of data transmission. This value can be calculated for one measurement in the access point, for the selected area or for the entire wireless network.

It is important to analyze the parameters and their appropriate selection, as each of them can determine the quality of data transmission in a different way. It is assumed that the obtained value is a percentage unit, the access point can provide a maximum of 100% of the quality of data transmission, when the parameter values are zero.

The next step is to analyze all potentially important parameters that can determine the quality of data transmission, it is necessary to define possible dependences between parameters. For that test one day was selected randomly, then two access points located in the same building and connected to the same Wi-Fi controller were selected. One of these access points provided the quality of data transmission at a good level throughout the day and the second of them definitely had problems providing good quality of data transmission.

After the analysis of these two access points and the collected parameters, eight dependencies were identified (called factors), that together define the quality of data transmission.

These factors are:

- the number of single retransmissions to the number of all transmissions,
- the number of multiple retransmissions to the number of all transmissions,
- the number of unfinished transmissions to the number of all transmissions,
- the number of correct RTS transmissions, number of failed RTS transmissions and number of transmissions without ACK confirmation,
- the number of users with a low SNR to the number of all connected users,
- the number of duplicate frames to the number of all transmitted frames,
- the number of FCS errors to the number of all transmitted MPDUs,
- channel utilization used to calculate data transmission for one user.

It is important to determine for each factor the part it represents in whole quality. The initial value is 100%, which is divided between eight factors. If its value is zero, it means that there were no errors and problems in transmission, and then it does not reduce the result of the overall quality of data transmission, otherwise if the obtained value is greater than 0, then the quality of data transmission is already less than 100%.

Several combinations of the formula have been tested and analyzed to choose the best one and ensure that the appropriate proportions between factors have been selected, which allows to determinate quality of data transmission in the best way. For the tests one random measurement was selected (13.03.2015 - Wednesday). For analysis, the AP-44 access point the measurement at 3:15 PM were selected. The analyzed parameters are presented in Fig. 4.

NoOfUsers	3
TransmittedFragmentCount	5707940 (18101)
MulticastTransmittedFrameCount	135775 (44)
RetryCount	786362 (2774)
MultipleRetryCount	171414 (503)
FrameDuplicateCount	0
RTSSuccessCount	533967 (550)
RTSFailureCount	1797368 (56)
ACKFailureCount	8588631575 (483)
ReceivedFragmentCount	0
MulticastReceivedFrameCount	0
FCSErrorCount	60685218 (9072)
TransmittedFrameCount	5843710 (18145)
WEPUndecryptableCount	39
FailedCount	105831 (92)
RxUtilization	0
TxUtilization	0
ChannelUtilization	3
NumOfClients	3
SNRClients	0

Fig. 4. List of collected parameters of the AP-44 access point from the measurement on 13.03.2015 at 3:15 PM

Figure 4 shows the values of all parameters measured by selected access point. For the parameters considered in the analysis, the difference between this and the previous measurement is calculated in brackets. These values are used in further calculations.

The formula for the quality of data transmission has been presented as follows:

$$QUALITY = 100\% - X \tag{1}$$

The variable X is defined by eight factors. For simplicity, one letter of the alphabet from A to H defines one factor. The value of all factors is subtracted from initial value.

$$A = \frac{\text{the number of single retransmissions}}{\text{the number of all transmissions}} \tag{2}$$

$$B = \frac{\text{the number of multiple retransmissions}}{\text{the number of all transmissions}} \tag{3}$$

$$C = \frac{\text{the number of unfinished transmissions}}{\text{the number of all transmissions}} \tag{4}$$

$$D = \frac{\text{the number of failed RTS transmissions}}{\text{the number of correct RTS transmissions}} + \frac{\text{the number of failed RTS transmissions}}{\text{the number of correct RTS transmissions}} + \frac{\text{the number of failed RTS transmissions}}{\text{the number of correct RTS transmissions}} \tag{5}$$

$$E = \frac{\text{the number of users with a low SNR}}{\text{the number of all connected users}} \tag{6}$$

$$F = \frac{\text{the number of duplicate frames}}{\text{the number of all transmitted frames}} \tag{7}$$

$$G = \frac{\text{the number of FCS error}}{\text{the number of all transmitted MPDUs}} \tag{8}$$

$$H = \frac{\text{the number of all transmissions} \times 1\%}{\text{channel utilization} \times 1\%} \div \text{number of all users} \tag{9}$$

All received results are presented in percentages and they are rounded to the second decimal place except the last factor. The result obtained for the selected measurement is:

$$A = 12,52\% \tag{10}$$

$$B = 2,77\% \tag{11}$$

$$C = 0,51\% \tag{12}$$

$$D = 13,15\% \tag{13}$$

$$E = 0\% \tag{14}$$

$$F = 0\% \tag{15}$$

$$G = 50,12\% \tag{16}$$

$$H \approx 2016 \tag{17}$$

Summing up the first 7 values, the result is 79,07%. After subtracting this value from the assumed maximum value 100% it gives 20,93% quality of data transmission. Such a result is not a bit adequate to how the selected access point worked. In order to achieve the intended result, for each received value it was assigned how important it is for the entire transmission. All eight factors have been multiplied by the values, which add up to 1, i.e. 100%.

In addition, the value of the eighth factor is compared with the average number of transmissions per user to determine if the current value of channel utilization was appropriate. The first attempt to create a QUALITY formula considering the importance of each dependence and factor has the following form:

$$X = 0,1 \times A + 0,1 \times B + 0,3 \times C + 0,2 \times D + 0,05 \times E + 0,1 \times F + 0,1 \times G + 0,05 \times H \tag{18}$$

For the calculated values, the result is:

$$X = 0,099236 \tag{19}$$

$$QUALITY = 1 - 0,099236 = 0,90076443 \approx 90,08\% \tag{20}$$

The obtained value determines that the quality of data transmission in the selected measurement at AP-44 access point was provided at 90,08%.

This results seems to be possible, the value is high and after analyzing all parameters, it can be concluded that the selected access point did not have too many problems with providing high quality of data transmission. The main problem was the parameter defining FCS errors and retransmissions. The other parameters did not indicate major problems with data transmission. That is the reason the result seems to be too low, after analysis it can be concluded that it should be higher.

To improve proposed formula, ratios between factors has to be different. The values were changed until the best formula was found. The most adequate result for what really happened at selected access point was obtained after four iterations. Next iteration did not bring better result.

The final version of QUALITY formula:

$$QUALITY = 1 - X \tag{21}$$

$$X = 0,08 \times A + 0,12 \times B + 0,37 \times C + 0,2 \times D + 0,05 \times E + 0,1 \times F + 0,03 \times G + 0,05 \times H \tag{22}$$

For the calculated values, the result is:

$$X = 0,062558598 \tag{23}$$

$$QUALITY = 1 - 0,062558598 = 0,937441402 \approx 93,74\% \tag{24}$$

The obtained value is higher by 3,66 % points. It can be stated that the AP-44 access point on 13.03.2015 between 3:10 PM and 3:15 PM provided the quality of data transmission at 93,74%.

That is a high result and indicates that users connected to the PWR-WiFi network and to this access point had no problems with data transmission. AP-44 provided throughout the day an average quality of data transmission at 97,14%. Such a high score shows that AP-44 had no serious problems with data transmission. Figure 5 shows the quality of data transmission for AP-44 throughout the day.

Figure 6 shows the quality of data transmission provided by all access point on the main campus on 13.03.2015. The average value of quality of data transmission was at 89,33% for that day.

Analyzing all collected data, the number of all measurements was obtained, which is 53 002 368. Based on average results for each day and individually for each access point, compartments have been developed that define whether the selected measurement has provided very good, good, medium, low or very low quality of data transmission.

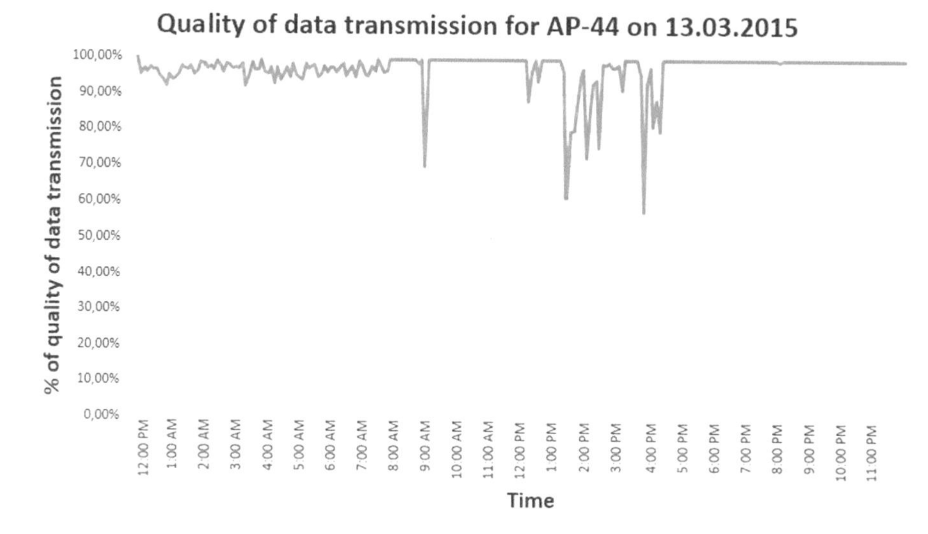

Fig. 5. Quality of data transmission for AP-44 access point on 13.03.2015

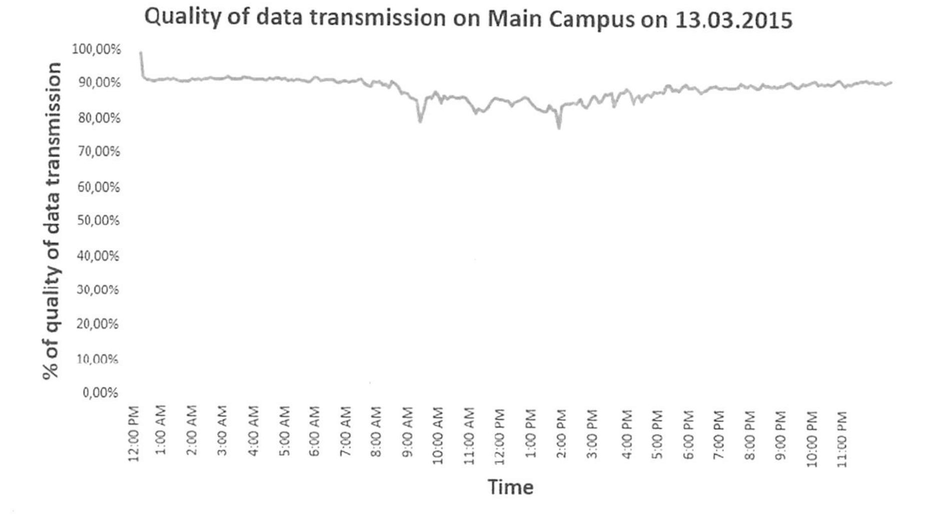

Fig. 6. Quality of data transmission on Main Campus on 13.03.2015

Proposed 5 different compartments:

- range of 100-90% - very good quality of data transmission,
- range of 89,99-80% - good quality of data transmission,
- range of 79,99-65% - average quality of data transmission,
- range of 64,99-45% - low quality of data transmission,
- range of 44,99-0% - very low quality of data transmission.

After analyzing the collected data for access points located on the main campus and calculating the average quality of data transmission, a result was obtained that the PWR-WiFi network provided users in the selected location with quality of data transmission at 87,14%.

Analyzing further, it can be shown that excluding the periods in which the network was overloaded with traffic, the quality of data transmission was provided at 98,01%. Similarly, when calculating only the transmission quality for the period in which network was overloaded, the value of 40,87% was obtained. For all collected data, including branches of the WUST and Geocentrum, the result was 84,78%, which indicates that the quality of transmission at this level was provided by the entire network.

5 Conclusions

Analyzing the quality of data transmission provided by PWR-WiFi network it could be claimed that it is on acceptable level. This paper proposes a method that calculates the quality of data transmission basing on the value of parameters collected at the access points. The difference between the proposed method and other methods mentioned in the paper and found in the literature is simplicity. The user does not need to analyze each parameter separately, gets result immediately, which defines the quality of data transmission.

In the case the user would like to know which factors affected the obtained result, could check the value of each factor. On this basis, it is possible to find out which parameters most affected the quality of data transmission in selected measurement, for a selected access point, for a selected time or period, for selected location or even for one access point in relation to all days in which it provided services to users.

After the research and analysis, it can be concluded that quality of data transmission in PWR-WiFi network was provided at a good, acceptable level. There were periods in which network was too overloaded, and its efficiency was substantially lower than expected, but overall result is high.

The average result of the quality of data transmission for the PWR-WiFi network at the main campus is 87,14%. This means that the wireless network has provided the quality of data transmission at that level. For the entire network the result is 84,78%.

The network behavior is quite predictable and simple to describe. The network shows practically zero activity at night. Same at holidays or other days off from classes. Definitely during the week, on weekdays, the number of users and network traffic is higher than on weekend or even during session and retake examinations. The statistical analysis conducted at the beginning also shows the predictability of the network during the day. Especially during the breaks between classes there is increased activity in the network.

The idea for further research and development of the presented method is the use of collected documentation about the location of access points and additional information for analysis of specific buildings, rooms and other areas.

The data can be presented in the spatial model to show areas with transmission problems. In the case of places where transmission quality was not satisfying, reason can be described and possible solution can be proposed. Rest on the collected data, it would be possible to create a 3D map that would show quality of data transmission for each area.

References

1. Mota, V.F.S., Macedo, D.F., Ghamri-Doudane, Y., Nogueira, J.M.S.: On the feasibility of WiFi offloading in urban areas: the Paris case study. In: 2013 IFIP Wireless Days, pp. 1–6 (2013)
2. Xing, X., Dang, J., Mishra, S., Liu, X.: A highly scalable bandwidth estimation of commercial hotspot access points. In: Proceedings - IEEE INFOCOM, vol. 2, pp. 1143–1151 (2011)
3. Gal, Z., Balla, T., Karsai, A.Sz.: On the WiFi interference analysis based on sensor network measurements. In: IEEE 11th International Symposium on Intelligent Systems and Informatics, pp. 215–220 (2013)
4. Kim, J., Song, N., Jung, B.H., Leem, H., Sung, D.K.: Placement of WiFi access points for efficient WiFi offloading in an overlay network. In: IEEE 24th Annual International Symposium on Personal, Indoor and Mobile Radio Communications, pp. 3066–3070 (2013)
5. Kaminska-Chuchmala, A.: Performance analysis of access points of university wireless network. Rynek Energii **1**, 120–124 (2016)
6. Mu, M., Pennarun, A.: Measuring Wireless Network connection quality. Technical Disclosure Commons, no. 127, pp. 1–11 (2016)
7. Senniappan, P.: Study of performance analysis in wired and wireless network. Am. J. Appl. Sci. **8**, 826–832 (2011)
8. Perez, E.M.: Frequency management in a campus-wide Wi-Fi deployment. Universitat Politecnica De Catalunya, vol. 5, pp. 37–48 (2013)
9. Yonis, A. Z.: Performance analysis of IEEE 802.11ac based WLAN in wireless communication systems, vol. 9, No. 2, pp. 1131–1136. Institute of Advanced Engineering and Science (2018)
10. Punithavanthani, D., Radley, S.: Performance analysis for wireless networks: an analytical approach by multifarious Sym Teredo. PubMed **11**(1), 1–8 (2014)

Information Systems Architecture and Technology Security Aspects Relating to the Usability Attributes and Evaluation Methods of Mobile Commerce Websites

Leila Goosen[1(✉)] and Sunday A. Ajibola[2]

[1] University of South Africa, Pretoria 0003, South Africa
GooseL@unisa.ac.za
[2] University of South Africa, Johannesburg 1710, South Africa

Abstract. It was demonstrated that many mobile commerce (m-commerce) websites failed to attract customers, due to not being user-friendly and a lack of security relating to user interfaces. Among research accessed, none conducted usability evaluations where security as attribute was evaluated on m-commerce applications. Therefore, a new usability model for m-commerce applications is proposed, including security as usability attribute having direct effects on the market behavior and efficiency of users' experiences with m-commerce applications. The research addressed the imbalance in literature by conducting a usability evaluation incorporating security, aiding validation of the proposed usability model. The study also addressed the lack of security attribute as instance of existing usability models and proposed a domain-specific evaluation method for usability evaluation of m-commerce websites. The outcomes showed that users consider security to be an important parameter when navigating m-commerce websites. Therefore, mobile application design and web design should look at optimizing the performance of security attributes for m-commerce websites, boosting customers' shopping online quality of service.

Keywords: Computer security systems · e-Business systems · Usability attributes · Evaluation methods · Mobile systems · M-Commerce

1 Introduction

Information Systems Architecture and Information and Communication Technologies (ICTs) are increasingly affecting "worldwide commercial and economic growth" [1].

There has been rapid growth of m-commerce applications "in recent years due to the development of mobile communication" technologies and e-commerce businesses [2, p. 149]: Statcounter [3] indicated that in September 2018, mobile devices represented approximately 52% of market share worldwide, 53% in China, 28% in Brazil, 76% in Nigeria, 67% in South Africa, 38% in the United Kingdom (UK) and 42% in the United States of America (USA). m-Commerce growth in 2017 contributed 34.5% of total e-commerce sales and it is estimated that it will account for 54% of total global e-commerce sales by 2021 [4]. Other statistics [5] showed that 76% of consumers state

L. Borzemski et al. (Eds.): ISAT 2019, AISC 1050, pp. 328–337, 2020.
https://doi.org/10.1007/978-3-030-30440-9_30

that it saves them time to shop on mobile devices. Furthermore, research [6] displayed that in the USA, $60.2 billion was spent in 2016, while it is estimated that such spending would grow to $93.5 billion and $175.4 billion in 2018 and 2022 respectively. m-Commence revenue shares of £36.18 billion for the UK and $23.97 billion for Germany in 2017 was demonstrated [7]. More research [8] disclosed that 18% of the South African and 14% of the Egyptian populations bought items through mobile devices in 2017.

Using m-commerce platforms is attractive, but many users can't effectively browse m-commerce portals, which is frustrating and dissatisfying [9], and "only about 30% of online users" conducting "online searching make online purchases" [2, p. 149].

This paper expands further on the earlier work of Ajibola and Goosen [10] and [11], in terms of describing trends in m-commerce usability in South Africa and Nigeria, as well as addressing missed opportunities in m-commerce usability.

2 Usability Attributes

The second author undertook this research to provide a comprehensive and complete list of quality factors and attributes, some of which identified were related to the scalability, security, and availability of e-commerce website software [12].

Security affects mobile communication and m-service, due to the unique features associated with mobile devices. Portable computers and mobile devices are more vulnerable to loss or damage, due to their physical features and the contexts in which these are used [13].

Kaur [14] suggested five factors that are fundamental when transferring websites from e-commerce to m-commerce. Security and privacy represent one of these essential factors that m-commerce applications should possess. Security measures in m-commerce applications are important, because of the vulnerability of applications on mobile devices. The world is a global village and security and privacy concerns have substantial effects on customers' perceptions of m-commerce applications. Thus, these concerns need to be addressed during the development of such applications [15].

The current study proposes the MObile Shoppers Application Development (MOSAD) usability model for m-commerce applications, which is user-centric and has security as one of its attributes. In the description of the factors of the MOSAD model, it is pointed out that due to the fact that m-commerce applications, which are a subset of mobile applications, are designed and developed to be small in size, physical desktop and laptop input methods, like the mouse and keyboard, are not applicable [16]. The description of the attributes of the MOSAD model also explained that the error rate refers to preventive errors, based on the ability of users to recover from such [17].

3 Usability Evaluation Methods

Using a focus group usually involves a number of participants from whom subjective data about the application can be collected during usability testing [18]. In terms of both qualitative and quantitative dimensions, the data collection instrument, like a log file or survey, gathers large quantities of data that can be numerically coded [19].

4 Heuristics for M-Commerce Applications

4.1 Heuristic Evaluation Method for M-Commerce Applications

m-Commerce application developers create m-commerce content with the defined purpose of enabling mobile shoppers to buy any goods they desire. "Shopping on mobile devices or smartphones has" certain "challenges, like trust, fear of security lapses, limited screen sizes," difficulty in "input mode and poor screen resolution, among others". Therefore, the application of "traditional usability heuristics in m-commerce applications is not appropriate and their usage would lead to" researchers missing "significant parts of the m-commerce applications" [20, p. 20].

4.2 M-Commerce Applications and Heuristics Used in Reviewed Studies

In terms of the usability evaluation methods used in m-commerce applications, the research source indicated for 'Cognitive Walkthrough and Nielsen's Heuristics' is [21].

4.3 The Proposed Usability Heuristics for M-Commerce Applications

In terms of a draft set of M-COMmerce (MCOM) heuristics, the last of these listed in [20, p. 24] deserves a mention in the current context: *"Ensure that users' privacy and security concerns are addressed"*.

5 Research Design and Methodology

It is worth noting that South Africa and Nigeria have the largest economies in Africa and their e-commerce sectors are relatively similar [22].

In order to obtain representative samples in terms of participants' demographic profiles for the asynchronous remote testing and heuristic evaluation methods, the criterion for sample selection was that all test participants must be Nigerian mobile shoppers, as they are the target users of the selected websites [23, 24].

In terms of the selection of test participants, according to StatCounter [25], 82.26% of the Nigerian population of 198 million are Facebook subscribers.

A mapping between the MCOM heuristics and those of Nielsen showed that one of the traditional heuristics is similar to the MCOM heuristics, while seven other heuristics were similar in content or definition, but not in the context of use. One of the new heuristics added was ensuring that user privacy and security concerns were addressed.

A post-evaluation questionnaire was completed by all test participants. The results helped to rate which among the four selected websites had the best features during test participants' interactions with these. The post-evaluation questionnaire contained seven open-ended questions and was additionally used to obtain the general usability details about the four selected websites, based on the test participants' ratings across six characteristics (*navigation, security and privacy,* purchasing process, architecture, internal search, and design) of the websites.

Earlier studies [26] classified the usability evaluation of commercial websites into seven problem areas, which are *navigation, content, security and privacy*, common look and feel, availability of tool, information content and compatibility. Based on the MCOM heuristics guidelines, this study was able to extend the number of identified usability problem areas categorized for the four selected m-commerce websites from seven to 10. This is due to the peculiarity in the features of the m-commerce websites uncovered through the course of the research. Therefore, the usability problem areas associated with m-commerce websites are *navigation, content, security and privacy*, purchasing process, architecture, internal search, design, accessibility and customer service, inconsistency and missing capability.

6 Discussion of Results

Table 1. The number of the identified false problems across the usability problem areas.

Usability problem area	Usability problems		
	Asynchronous testing	MCOM heuristics	False problems
Security & Privacy	0	4	4

Apart from the false problems across the usability problem area in Table 1, prior studies [27] used a metric of false problems to measure the effectiveness and comparison of the usability problems identified in user testing and heuristic evaluation methods.

Table 2. Usability evaluation methods and their respective problems.

Usability testing method	Usability problems (samples)	Sources
Features of usability problems discovered by the MCOM heuristic evaluation method	Connected to website interface quality and features	[26]
	Connected to website interface layout or appearance	[28]
	Inconsistency in webpage interface	[29]
	Problems with delay in response time when displaying results	[30]
	Compatibility related problem	[31]
	Privacy/security related problem	[32]

The usability testing method as shown in Table 2, like in other earlier studies, revealed additional usability problems, which were linked to the absence of facilities like help and feedback, navigation issues, the use of complicated terms, unsuitable choice of font size and consistency issues. Findings also confirmed usability problems, including aesthetics, inconsistency and design issues, and security and privacy problems.

Previous studies [28 29] displayed that the satisfaction of users with a particular website cannot be employed in judging the usability of that website. In addition, other studies [33] exposed that participants tended to be polite in their evaluations, by giving a website a higher rating, even when it cannot be used easily, due to a high number of usability problems. For example, a previous usability study disclosed that the results of users' evaluation ratings were positive, in spite of glaring poor user performance. These participants are usually regarded as 'Appeasers'. The study suggested that the reported irregularity may be associated with different users' cultural effects. This may be why several research participants acted extremely polite and refused to give negative ratings [34]. Another issue uncovered from data obtained from the satisfaction questionnaire is the failure of the method to determine particular usability problems. This is in terms of how inaccurate a website's internal search facility is, and privacy and security issues.

In terms of qualitative data obtained from the post-test questionnaire, the current study is in line with the results of previous research that applied open-ended questions in their satisfaction questionnaires [24] – these results are confirmed in Table 3.

Table 3. Identified usability problems from post-test questionnaire: qualitative data.

Usability problem areas	Associated usability problems
Accessibility and customer service	Only one language option is provided
Content	Wrong and inconsistent information resulting in missing products information
Design	Colors and font sizes of the content are not appealing
Inconsistency	Content/layout and design are inconsistent
Internal Search	Provided options are inadequate
Missing capabilities	The functions or information are missing
Navigation	Website links broke severally, resulting in orphaned webpages.
Purchasing process	Products ordering process is too extensive

Note: No usability problems were discovered in the architecture and privacy and security usability problem areas from the users' open-ended questionnaire. Therefore, only eight out of the ten identified usability problem areas are presented in Table 3.

With regard to the distribution of usability problems across usability problem areas, the analysis of results of the remote asynchronous testing and proposed domain-specific heuristic methods displayed that both methods were proficient in detecting usability problems relating to the nine problem areas, except security and privacy. The remote asynchronous (user) testing method completely failed to detect problems associated with the security and privacy problem area. The results are in line with previous research [26], in which user testing failed to detect problems in the security and privacy and compatibility attributes during the usability evaluation of tested websites.

Notably, three of the four identified usability problems with respect to security and privacy usability problem areas were major problems. This finding indicated that security and privacy were critical factors that need to be considered during the development and evaluation of the user interface for m-commerce websites in particular [35]. The general understanding is that mobile devices and m-commerce applications will be significantly sensitive to the impact of security, because these applications are used in different task settings and within various size limitations. Hence, it is important to discuss the significance of security as uncovered during the evaluation of the four selected m-commerce websites.

7 Framework: Usability Evaluation of M-Commerce Websites

Security as quality attribute was discovered during the application of MCOM heuristics in the evaluation of four selected m-commerce websites, as presented in Sect. 6. The significance of the security attribute in the evaluation of m-commerce websites was described in a sub-section of Sect. 6 of the current paper. Therefore, the following section will discuss the security attribute of the MOSAD model as important to be considered in the evaluation of m-commerce applications.

Security affects mobile communication, and this is due to the dimensions of mobile devices. Due to their physical features and the contexts in which these are used, portable computers are more vulnerable to loss or damage [13]. This aspect further affected the privacy concerns of users. For example, what happens when a user loses her/his mobile device that contains private information, such as messages, pictures, contacts, and other personal information? Although privacy problems occur in almost any interactive networked medium, it is more acute in the m-commerce application medium.

Furthermore, the capability of context-aware systems in revealing detailed information, which is private, has serious implications for users' privacy. Any sign of privacy violations could result in mobile users losing confidence in the application, resulting in them exiting the context-aware system. To this end, the transparency and controllability of mobile applications are fundamental. Research [36] revealed that in order to make the security mechanism of mobile applications acceptable to both users and corporations, certain steps need to be followed.

The SMS system of payment, though widely accessible and available, is vulnerable to security and congestion problems [37]. In the event of leakages of sensitive and private data arising from a security breach, the owner of the mobile application may be penalized by law; this could severely affect the organization's reputation [38].

While the mobile shopper interacts with the m-commerce application, the mobile context of use will have a significant impact on the security consciousness of the application. Therefore, it becomes imperative for usability professionals to consider security issues as important when developing and evaluating m-commerce applications. One approach to measure the security aspect of m-commerce applications is the use of questionnaires and inclusion in the heuristics to be used for the evaluation, as was done in MCOM heuristics [39].

Another approach is the use of the simple Goal-Question-Metric (GQM) technique [40]. GQM can be used as a subjective assessment tool to measure the security concerns of users while interacting with the system. In some cases, mathematical formulae (usually 3D, where x, y, and z represent the security factor, Human-Computer Interaction (HCI) and e/m-commerce requirements respectively) can be combined with GQM to obtain accurate evaluations [41].

8 Conclusions

The conclusions obtained from this research provide guidelines for website designers and developers, to construct m-commerce websites with good functionality and ease of use, to enhance users' experiences. A better user interface and strong security capability will motivate customers to shop online, thus, boosting m-commerce business.

In terms of accomplishing the research objective of determining the effectiveness of the proposed domain-specific evaluation method in the usability evaluation of m-commerce websites, the results of the structured comparison provided evidence that remote asynchronous testing with the Loop11 tool can provide faster, easier and cheaper indications of possible usability problem areas associated with the selected m-commerce websites. This can provide either an idea of these potential areas with respect to the overall website usability or help to identify specific webpages with usability problems. However, from the results, it was observed that the three identified problem areas relating to the absence of privacy and security, lack of capabilities and inconsistent design, could not be identified accurately.

From the results obtained, it was discovered that qualitative data obtained from MCOM heuristic evaluators was useful in highlighting specific usability problems, while the quantitative data from the heuristic checklist was unable to reveal specific usability problems. The results also showed that qualitative data obtained from the MCOM heuristic evaluators identified the majority of the minor usability problems (132) associated with all problematic areas. The crucial nature of privacy and security problems was highlighted by the MCOM heuristic evaluators. Essentially, the results demonstrated that the heuristic evaluators cannot play the role of actual users and to foretell actual problems that end users may face in real life, while evaluating the selected m-commerce websites.

One of the limitations of this research was that the snowball sampling approach used in the selection of participants might result in the researcher(s) having no idea regarding the actual population distribution in relation to the sample. It is, therefore, difficult to determine potential sampling errors and make statistical inferences (generalization) from the population sample. Hence, snowball sampling may not be considered as being a good representative of the population of the study undertaken [42].

References

1. Goosen, L., Van der Merwe, R.: e-learners, teachers and managers at e-schools in South Africa. In: Proceedings of the 10th International Conference on e-Learning (ICEL), Nassau (2015)
2. Yassierli, Y., Vinsensius, V., Mohamed, M.S.: The importance of usability aspect in M-commerce application for satisfaction and continuance intention. Makara J. Technol. **22**(3), 149–158 (2019)

3. Statcounter.: Desktop vs. Mobile vs. Tablet Market Share Worldwide - September 2018. StatCounter Global Stats Research (2018a). http://gs.statcounter.com/platform-market-share/desktop-mobile-tablet/worldwide. Accessed 25 Oct 2018
4. Mali, N.: Your M-Commerce Deep Dive: Data, Trends and What's Next in the Mobile Retail Revenue World. BigCommerce Research Publication (2018). https://www.bigcommerce.com/blog/mobile-commerce/#mobile-commerce-statistics. Accessed 25 Oct 2018
5. Pilewski, S.: 50 Research-backed Personalization Statistics. Dynamic Yield Research Publication (2018). https://www.dynamicyield.com/blog/50-most-important-dynamicyield-personalization-stats/. Accessed 15 Jul 2015
6. Lacy, L.: Mobile Shopping Is on the Rise. But Remains Split Between the Mobile Web and Apps. Adweek Research Publication (2018). https://www.adweek.com/digital/mobile-shopping-is-on-the-rise-but-remains-split-between-the-mobile-web-and-apps/. Accessed 26 Oct 2018
7. eMarketers.: m-Commerce in UK, Germany and France: €85 billion in 2018. eMarketers Research Publication (2018). https://ecommercenews.eu/mcommerce-uk-germany-france-e85-billion-2018/. Accessed 27 Oct 2018
8. Statista.: Share of population who bought something online via phone in the past month as of 3rd quarter 2017, by country. The Statistics Portal Research Publication (2018). https://www.statista.com/statistics/280134/online-smartphone-purchases-in-selected-countries/. Accessed 27 Oct 2018
9. Hult, G.T.M., Sharma, P.N., Morgeson, F.V., Zhang, Y.: Antecedents and consequences of customer satisfaction: do they differ across online and offline purchases? J. Retail. **13**(1), 1–14 (2018)
10. Ajibola, A.S., Goosen, L.: Trends of mobile E-commerce usability in South Africa. In: Proceedings of the International Conference on Advances in Engineering Sciences and Applied Mathematics (ICAESAM), Cape Town (2013)
11. Ajibola, A.S., Goosen, L.: Missed opportunities in mobile E-commerce usability. Int. J. Sci. Eng. Res. (IJSER) **5**(6), 954–956 (2014)
12. Rababah, O.M.A., Masoud, F.A.: Key factors for developing a successful e-commerce website. Commun. IBIMA **2010**, 1–9 (2010)
13. Alghamdi, A.S., Al-Badi, A.H., Alroobaea, R.S., Mayhew, P.J.: A comparative study of synchronous and asynchronous remote usability testing methods. Int. Rev. Basic Appl. Sci. **1**(3), 61–97 (2013)
14. Kaur, M.: A brief study of usability principles for mobile commerce. Int. J. Comput. Eng. Res. **4**(8), 20–25 (2014)
15. Gitau, L., Nzuki, D.: Analysis of determinants of M-commerce adoption by online consumers. Int. J. Bus. Humanit. Technol. **4**(3), 88–94 (2014)
16. Bicakci, K., Van Oorschot, P.C.: A multi-word password proposal (gridWord) and exploring questions about science in security research and usable security evaluation. In: Proceedings of the 2011 New security paradigms workshop, New York, NY, USA (2011)
17. Von Zezschwitz, E., Dunphy, P., De Luca, A.: Patterns in the wild: a field study of the usability of pattern and pin-based authentication on mobile devices. In: Mobile Human Computer Interaction - Security and Privacy, New York (2013)
18. Sieger, H., Möller, S.: Gender differences in the perception of security of mobile phones. In: Proceedings of the 14th International Conference on Human-Computer Interaction with Mobile Devices and Services Companion, New York, NY, USA (2012)
19. Jing, Y., Ahn, G.-J., Zhao, Z, Hu, H.: RiskMon: continuous and automated risk assessment of mobile applications. In: Conference on Data and Application Security and Privacy, New York, NY, USA (2014)

20. Ajibola, S., Goosen, L.: Development of heuristics for usability evaluation of M-commerce applications. In: Proceedings of the South African Institute of Computer Scientists and Information Technologists (SAICSIT), Thaba 'Nchu, South Africa (2017)
21. Yohandy, H.D., Setyohadi, D.: Usability evaluation using multi-method for improvement interaction in M-commerce. MATEC Web Conf. **218**(11), 1–7 (2018)
22. Amao, O.B., Okeke-Uzodike, U.: Nigeria, afro-centrism and conflict resolution: five decades after—how far, how well? Afr. Stud. Quart. J. **15**(4), 1–23 (2015)
23. Ozok, A.A., Wei, J.: An empirical comparison of consumer usability preferences in online shopping using stationary and mobile devices: results from a college student population. Electron. Commer. Res. **10**(2), 111–137 (2010)
24. Chin, E., Felt, A.P., Sekar, V., Wagner, D.: Measuring user confidence in smartphone security and privacy. In: Proceedings of the Eighth Symposium on Usable Privacy and Security, New York, NY, USA (2012)
25. Statcounter.: Social Media Stats Nigeria, StatCounter Global Stats, (2018b). http://gs.statcounter.com/social-media-stats/all/nigeria. Accessed 29 Oct 2018
26. Tan, W., Liu, D., Bishu, R.: Web evaluation: heuristic evaluation vs. user testing. Int. J. Ind. Ergon. **39**(4), 621–627 (2009)
27. Jaferian, P., Hawkey, K., Sotirakopoulos, A., Velez-Rojas, M., Beznosov, K.: Heuristics for evaluating IT security management Tools. In: Symposium on Usable Privacy and Security (SOUPS 2011), Pittsburgh, PA USA (2011)
28. Chen, S.Y., Macredie, R.D.: The assessment of usability of electronic shopping: a heuristic evaluation. Int. J. Inf. Manag. **25**(6), 516–532 (2005)
29. De Kock, E., Van Biljon, J. Pretorius, M.: Usability evaluation methods: mind the gaps. In: Proceedings of the Annual Research Conference of the South African Institute of Computer Scientists and Information Technologists, Vanderbijlpark, South Africa (2009)
30. Davids, M.R.: Development and usability evaluation of a multimedia e-learning resource for electrolyte and acid-base disorders, PhD Thesis. University of Stellenbosch, Stellenbosch, South Africa (2015)
31. Paz, F., Paz, F.A., Villanueva, D., Pow-Sang, J.A.: Heuristic evaluation as a complement to usability testing: a case study in web domain. In: 12th International Conference on Information Technology - New Generations, Lima, Peru (2015)
32. Singun, A.P.: The usability evaluation of a web-based test blueprint system. In: International Conference on Industrial Informatics and Computer Systems, Muscat, Oman (2016)
33. Riihiaho, S.: Experiences with usability testing: effects of thinking aloud and moderator presence, Ph.D. Thesis. Aalto University, Helsinki, Finland (2015)
34. Krishna, A., Prabhu, G., Bali, A., Madhvanath, S.: Indic scripts based online form filling - a usability exploration. In: 11th International Conference on Human-Computer Interaction (HCII 2005), Las Vegas, USA (2005)
35. Aziz, W., Hashmi, Y.: Usability principles for mobile commerce, M.Sc. Thesis. Luleå University of Technology, Luleå, Sweden (2009)
36. Bao, P., Pierce, J., Whittaker, S., Zhai, S.: Smart phone use by non-mobile business users. In: Proceedings of the 13th International Conference on Human Computer Interaction with Mobile Devices and Services, New York, NY (2011)
37. Oreku, G.S.: Mobile technology interaction to e-commerce in promising of u-commerce. Afr. J. Bus. Manag. **7**(2), 85–95 (2013)
38. Benou, P., Vassilakis, C., Vrechopoulos, A.: Context management for M-commerce applications: determinants, methodology and the role of marketing. Inf. Technol. Manag. **13**(2), 91–111 (2012)
39. Kainda, R., Flechais, I., Roscoe, A.W.: Security and usability: analysis and evaluation. In: International Conference on Availability, Reliability, and Security, Krakow, Poland (2010)

40. Yahya, F., Walters, R.J., Wills, G.B.: Using Goal-Question-Metric (GQM) approach to assess security in cloud storage. In: Enterprise Security, pp. 223–240. Springer, Heidelberg, Germany (2017)
41. Gonzalez, R., Martin, M., Munoz-Arteaga, J., Garcia-Ruiz, M., Álvarez-Rodriguez, F.: A measurement model for secure and usable e-commerce websites. In: Conference on Electrical and Computer Engineering, St. John's (2009)
42. Sharma, G.: Pros and cons of different sampling techniques. Int. J. Appl. Res. **3**(7), 749–752 (2017)

Author Index

L. Borzemski et al. (Eds.): ISAT 2019, AISC 1050, pp. 339–340, 2020.
https://doi.org/10.1007/978-3-030-30440-9

Printed in the United States
By Bookmasters